Microbial Processes in Reservoirs

Developments in Hydrobiology 27

Series editor
H.J. Dumont

Microbial Processes in Reservoirs

Edited by
Douglas Gunnison

1985 **DR W. JUNK PUBLISHERS**
a member of the KLUWER ACADEMIC PUBLISHERS GROUP
DORDRECHT / BOSTON / LANCASTER

Distributors

for the United States and Canada: Kluwer Academic Publishers, 190 Old Derby Street, Hingham, MA 02043, USA
for the UK and Ireland: Kluwer Academic Publishers, MTP Press Limited, Falcon House, Queen Square, Lancaster LA1 1RN, UK
for all other countries: Kluwer Academic Publishers Group, Distribution Center, P.O. Box 322, 3300 AH Dordrecht, The Netherlands

Library of Congress Cataloging in Publication Data

Library of Congress Cataloging in Publication Data
Main entry under title:

Microbial processes in reservoirs.

(Developments in hydrobiology)
1. Freshwater microbiology. 2. Reservoir
ecology. I. Gunnison, Douglas. II. Series.
QR105.5.M53 1985   576'.15   85-4307
ISBN 90-6193-525-3

ISBN 90-6193-525-3 (this volume)
ISBN 90-6193-751-5 (series)

Contents

Preface

The idea of producing a book on the activities of microorganisms in reservoirs had its origins in an article published by the editor in *ASM News* (December 1981, 47:527–531). Many individuals expressed an interest in having the article expanded into a book on this subject. Several people were contacted and asked if they would be willing to contribute chapters to the book. The interest displayed by many persons outside the area of reservoir microbiology was encouraging, as was the inspiration of the contributors themselves. We were subsequently approached by Dr. L. Harold Stevenson, Chairman of the Aquatic and Terrestrial Division of the American Society for Microbiology and asked to convene a Seminar on Microbial Processes in Reservoirs at the March 1984 national meeting of the Society. It became apparent that the book would have more appeal than we had previously believed, and we sought a publisher. Through discussion with several authorities in limnology and aquatic microbiology and after consultation with Mr. Wil Peters of Junk, we found that our book fit in well with the concepts embodied within the *Developments in Hydrobiology* series.

Reservoirs are indeed unique ecosystems. While the microbiology of reservoirs may not appear to differ extensively from that of natural lakes, the context of the reservoir ecosystem places certain constraints on microorganisms, the processes they carry out, and the fate and consequences of their activities. Thus, physical and hydrodynamic factors such as temperature, density, basin morphometry, the nature and placement of inflows, and the pre-

sence of bottom versus surface withdrawal all have a bearing on microbial processes. In addition, reservoirs are often constructed in areas where there are few, if any, natural lakes. In this regard, reservoirs are also often distinct from natural lakes, and to the extent that meteorologic, hydrologic, geologic, and edaphic factors make a difference, reservoir microbiology will also be different. Finally, the creation of a new reservoir offers the sediment microbiologist a unique opportunity to view the transformation of terrestrial environments into aquatic ecosystems. The inundation of previously dry or only periodically flooded terrestrial soils and vegetation and the accompanying successional processes have much to offer those interested in microbial ecology, decompositional processes, and the dynamics of nutrient regeneration and metal cycling.

In piecing together the various contributions to this book, the editor has attemted to dovetail the articles and to encourage the contributors to cross-reference their work. Through this process, the editor has endeavored to produce a book having continuity, rather than the shotgun appearance often characteristic of books and symposia constructed from a series of articles that all happen to be related to some central subject.

Particular thanks are given to the many contributors who helped make this book possible.

Douglas Gunnison
Editor

Introduction

DOUGLAS GUNNISON

Introduction

Reservoirs are important resources for modern society and may serve either single or multiple purposes ranging from flood protection, navigation, water supply, and fish and wildlife enhancement to hydropower, water quality management, and recreation. In many areas of the world, reservoirs are often the only large standing bodies of water present.

Microbial processes play a key role in the chemical and biological economies of reservoirs. The physical processes occurring in a reservoir are governed largely by hydrodynamics – the way water moves through the reservoir system. Microbial processes are, to a large extent, the chemical and biological counterpart to hydrodynamics. They determine how, when, and to what extent nutrients, metals, and many contaminants are released within the reservoir ecosystem, and through these processes, they mediate the availability of these substances to other components of the reservoir biota and to the waters that are eventually released from the reservoir.

We have organized this book so that the reader who is not familiar with reservoirs and their unique properties may rapidly become acquainted with these ecosystems. Chapter 1 is devoted exclusively to the environmental effects of reservoirs, both within the reservoir itself and on the once reservoirless stream. The second chapter deals with the reservoir as an ecosystem and with the physical, chemical, and biological processes occurring in the reservoir that form an environment for microorganisms. Although Chapters 1 and 2 have little to say about microbial processes, these sections set the stage for the rest of the book.

Reservoirs, being manmade, allow the scientist a first-hand opportunity to investigate the changes occurring when a standing body of water is suddenly created *de novo* from what was once a free-flowing stream. Ecologists have, for many years, observed, reported, and commented on the successional changes that accompany the biological, chemical, and physical processes occurring when a lake matures to form a pond which then becomes a swamp and, finally, a meadow. The forces of cultural eutrophication, through the disturbance of land that results in lake-filling erosion of soil and through the addition of plant-growth stimulating nutrients, often greatly accelerate this phenomenon. In the new reservoir, this process is reversed. What were once dry or only periodically flooded soils and bottomland and upland vegetation are all at once submerged. A flush of microbially-mediated decomposition occurs, utilizing much of the readily available organic matter in the flooded soils and vegetation and releasing large amounts of nutrients and metals. A period of eutrophy, often involving the first six to ten years of the life of the new reservoir, is subsequently followed by a decline in biological productivity and an approach to the mesotrophic state. These processes have collectively been termed the transition phase. Chapters 3 and 4 examine these processes in detail, with emphasis upon the activities of microorganisms associated with this phenomenon.

Once the transition phase has been completed, the reservoir settles down, and with few exceptions, decompositional processes no longer occupy the center stage from a microbial point of view. Instead, our attention is drawn to other areas, including the relationship between microorganisms and water quality (Chapter 5); interactions between the various microbial components of the reservoir ecosystem (Chapter 6); the effects of thermal additions upon microorganisms, particularly those having pathogenic importance (Chapter 8); and the effects of acidification upon the microbial ecology of reservoirs (Chapter 10). Microorganisms continue to play important roles in the nutrient and metal economies of the established reservoir. These roles are examined in detail from two different viewpoints – the dynamics of

mineral cycling (Chapter 7) and current attempts to model geomicrobial processes (Chapter 9).

This book is concerned with microbial processes in reservoirs. From this standpoint, emphasis is places upon the processes microorganisms carry out rather than upon the individual microorganisms responsible for them. In so doing, we do not intend to downplay the importance of the biology of any one microorganism or group of organisms. Rather, our efforts are focussed on the relationship of microbial processes to the functioning of the reservoir as an ecosystem.

Author's address:
Douglas Gunnison
Ecosystem Research and Simulation Division/EL
USAE Waterways Experiment Station
P.O. Box 631
Vicksburg, MS 39180
USA

CHAPTER 1

Environmental effects of reservoirs

R.M. BAXTER

Abstract. Dams and reservoirs contribute greatly to human prosperity and well-being but they have other effects which are deleterious to the environment and to human welfare. The initial flooding may bring about physical and chemical changes which influence the aquatic biota in various ways. The complex hydrodynamic characteristics of many reservoirs complicate matters further.

Severe effects may occur below a dam, sometimes for hundreds of kilometers. These are due to the action of the reservoir as a source or sink for heat, sediments and solutes, and to changes in the diurnal or annual flooding pattern. The movement of anadromous fish may be impeded or blocked.

Large impoundments may modify the climate in their vicinities or downstream. They may induce seismic activity under certain geophysical conditions which are not yet fully understood.

They may contribute to the spread of certain diseases, and reduce the incidence of others. They may cause grave dislocations in the social and economic life of people living in traditional societies, and sometimes necessitate the removal of large human populations.

Although much progress has been made in predicting the environmental impacts of dams and reservoirs, it is still not possible to predict with confidence all the effects, especially of very large projects.

Introduction

The art of controlling water by the construction of dams is one of the oldest branches of civil engineering, dating back to the third or even the fourth millenium B.C. (Biswas, 1975; Henderson-Sellers, 1979). It probably originated independently in several regions such as China, India, Mesopotamia and Egypt, and attained a level of development under the Roman Empire that was not reached again for several centuries.

The earliest dams were probably built to provide reliable supplies of water for domestic purposes and perhaps for irrigation, and this is still an important purpose of reservoirs throughout the world. A recent book on reservoirs (Henderson-Sellers, 1979) written from a British perspective deals very largely with water supply reservoirs. Other important purposes are the provision of reliable flows for navigation (or in earlier days in North America for the floating of logs), flood control, and the generation of energy. Energy generation was first carried out on a relatively small scale in grist mills and saw mills, but the development of hydroelectric power around the beginning of the present century gave a new impetus to dam-building, leading in the 1930's and 1940's to such achievements as the system of dams administered by the Tennessee Valley Authority, the Hoover Dam and the Grand Coulee Dam in the United States, and the great Volga reservoirs in the Soviet Union. After the Second World War even larger projects were undertaken, particularly

Gunnison, D. (ed.) Microbial Processes in Reservoirs.
© 1985, Dr W. Junk Publishers, Dordrecht, Boston, Lancaster. ISBN 90 6193 751 5.

the construction in Africa of several dams which impound quantities of water comparable to all but the largest natural lakes. In virtually all countries possessing rivers of sufficient flow and head, hydroelectric plants are now important sources of energy, and in such countries it is hydroelectric dams and reservoirs that have the greatest environmental impact. A publication on the effects of reservoirs in Canada (Baxter & Glaude, 1980) deals almost entirely with hydroelectric reservoirs.

Whatever the type of dam, it is almost always equipped with a spillway to allow excess water to pass over or around the dam, especially at times of flooding. There is also usually an outlet of some kind, to a water distribution system in a water supply reservoir or through a penstock to a turbine in a hydroelectric reservoir.

Not all dams are produced by human activities. Streams may be blocked temporarily or permanently by floating vegetation or landslides. Several substantial lakes throughout the world were formed by damming by laval flows (Hutchinson, 1957); an important example is Lake Tana in the Ethiopian highlands, the source of the Blue Nile (Cheesman, 1936). The environmental impacts of the dam-building activities of beavers are more extensive than is perhaps generally recognized (Naiman, 1983; New Scientist, 1983).

Dams may be built in various ways and from various materials depending on the purpose of the dam, the nature of the site, and the resources and materials available. One classification (Henderson Sellers, 1979) recognizes two fundamental types: earth- and rock-fill dams, and concrete and masonary dams. Earth- and rock-fill dams are essentially piles of carefully compacted earth or rock extending across the stream. Usually the core of the dam is composed of relatively impermeable material such as clay or occasionally concrete. The slope is fairly gentle both upstream and downstream. The upstream face is usually protected against erosion by rock, known as riprap, and the downstream face may also be covered with rock to increase the stability of the structure.

Concrete dams may be further sub-divided into three classes according to the manner in which they are held in place. Gravity dams, which are sup-

ported by their own weight, have a virtually vertical upstream face, while the downstream face slopes sufficiently to assure that the centre of gravity falls well within the base of this dam. Arch dams are convex in the upstream direction and are held in place by thrust against their abutments. Buttress dams are held in place by buttresses on the downstream face. Masonary dams, in the strict sence, are constructed of blocks of stone held together by mortar. Such dams were popular until the early part of the present century, but are seldom if ever built now (Henderson-Sellers, 1979).

The amount of water impounded differs enormously from one dam to another. The Owens Falls dam in Uganda is located at the outlet of Lake Victoria, which is technically the largest reservoir in the world (Mermel, 1983), with a volume of about $2 \times 10^{12} \, \text{m}^3$ (Serruya & Pollingher, 1983). However, since the lake was there before the dam was built and the construction of the dam has not had much effect on the hydrologic regime of the lake, the environmental impact of the dam was relatively small. Smaller lakes, however, can be seriously affected by the construction of dams at their outlets (Lindström), 1973). Dams built on rivers often impound vast amounts of water, as in the great African impoundments. This does not always happen however; in some instances it may be possible to operate a hydroelectric plant on the natural flow of the stream (a run-of-the-river plant), so engineering constructions are only necessary to provide an adequate head and direct the water into the penstock. Many of the more accessible rivers of the world are now regulated throughout most of their length by a series of dams so that virtually all the hydraulic head is exploited. Such an arrangement is known as a cascade. Sometimes the uppermost dam may store water, while those lower down operate on the run of the river.

Sometimes water is not impounded directly upstream of a dam but is rather diverted into an artificial or natural basin to form an off-channel reservoir.

Pumping water to a higher level is one of the most convenient ways of storing energy, and hydroelectric generators have the advantage over other types of generators that their output can be increased or decreased almost instanteneously. These considerations have led to the development of pumped storage reservoirs (American Fisheries Society, 1976), which provide a means of coping with variations in demand for electrical power. A coal-fired generator, for example, may be operated at a constant output, the excess power at periods of low demand being used to pump water from a lower reservoir to a higher. The pumps used can operate in reserve as turbines so that at times of high demand the water is allowed to run back and generate more power.

Still another type of dam, and one which is attracting increasing interest, is the tidal barrage, by which a bay of more commonly an estuary is cut off from the sea. The most dramatic purpose of such projects is the generation of electrical power (Gray & Gashus, 1972; Severn et al., 1979; International Symposium, 1978; Second International Symposium, 1981) but they may also be undertaken to provide storage reservoirs for water supplies or to protect the coastline (Corlett, 1979).

Ecological studies of reservoirs in North America began in the 1930's and 1940's (Richardson, 1934; Wiebe, 1939; Ellis, 1941; Krumholz, 1981). Similar investigations in the Soviet Union were begun about the same time (Zhadin & Gerd, 1963). Initially the emphasis was on maintaining and managing the fish stocks of the impounded waters. In the 1960's and 1970's with the construction of very large reservoirs in Africa and other parts of the world, the emphasis shifted to the possible deleterious effects of such projects, which were simetimes experienced thens or hundreds of kilometers away from the actual site of construction. The general topic of reservoir ecology has subsequently been the subject of several international symposia (Ackerman et al., 1973; Low-McConnell, 1966; Obeng, 1969; Driver and Wunderlich, 1979; Campbell et al., 1982) and others of more limited scope (American Fisheries Society, 1967; Efford, 1975; Oklahoma Geological Survey, 1976). Several reviews are also available dealing with the subject in general or with particular aspects (Baxter & Glaude, 1980; Lindström, 1973; Ellis, 1941; Geen, 1974–1975; McLachlan, 1974; Neel, 1966; Ridley & Steel, 1975; Baxter, 1977; Brocksen et al., 1982;

Goldman, 1976; Szekely, 1982; Rouse, 1980; Petr, 1978). In this chapter I shall attempt to provide a broad picture of the consequences that may ensue from dam construction and the impoundment of water, and so far as is possible relate these to general scientific principles.

An ecosystem is made up of a physical environment and the organisms living in it (Odum, 1971). When a stream is flooded, the first and most obvious event is the creation of a new physical environment. However, there is a change in the biological component of the ecosystem too. Many types of organisms, both plant and animal, which flourish in running water are poorly adapted to life in standing water (Hynes, 1969) and perish. In effect, a running water ecosystem is destroyed, and a standing water ecosystem is established. These two types of aquatic ecosystem differ in a number of ways. The most fundamental difference is perhaps in the manner in which solar energy, trapped by photosynthesis, is made available. A lake can, up to a point, be regarded as an isolated system, a microcosm (Forbes, 1887). Solar energy is trapped within the lake by photosynthesis by phytoplankton, which serve as food for zooplankton which are eaten in turn by larger animals, and so on. A stream cannot be viewed in this way; it cannot be considered in isolation but only as a part of a larger system which also includes its valley (Hynes, 1975). The population of phytoplankton in streams is usally small, and their communities are largely sustained by organic matter, such as leaves and other litter, entering it from the shore. This material is first acted upon by certain molds, known as Ingoldian fngi (Bärlocher, 1982). These partially decompose the material, making it more suitable as food for various invertebrates (Kaushik & Hynes, 1971). These then serve as food for larger organisms and so on. The conversion of a stream to a lake, therefore, ultimately involves the conversion of a heterotrophic stream ecosystem to an autotrophic lake ecosystem. Initially, however, there is a burst of increased heterotrophic activity, as the vegetation in and on the flooded ground decays. (See Gunnison et al., this volume, chapter 3; Godshalk & Barko, this volume, chapter 4).

The development of a new ecosystem involves a

fairly orderly process known as 'succession' whereby the character of the community changes as the species that appear in the earlier stages are more or less completely replaced by others. The general features of succession have been described by Odum (1969) and these can be observed in a newly impounded reservoir, although some of them may be delayed by the initial burst of heterotrophic activity resulting from flooding.

It is convenient to make a distinction between organisms that reproduce rapidly but are subject to high morality and those that breed more slowly but have a higher rate of survival. These are referred to as 'r-selected' and 'K-selected' species, respectively. 'r' is the intrinsic rate of increase and 'K' the asymptotic population size in the logistic equation: $N = K/(1 + e^{a-rt})$ where N is the population at time t. In new ecosystems the number of species present is likely to be few and competition not very intense, so that r-selected species are at an advantage. As the community becomes more diverse, however, and competition and predation more intense, the advantage shifts in favour of K-selected species. Besides this increase in diversity and shift in the reproductive pattern of the species present, there is a tendency for more and more of the inorganic nutrients to be present within organisms rather than free in solution, for the total organic material in the system to increase, and for a steady state to be approached in which the rate of production of organic material by photosynthetic organisms is about equal to its rate of degradation by herbivores and predators. There is also a tendency for smaller organisms to be replaced in part by larger.

The maturation of ecosystems can be prevented, and the system maintained in a fairly constant intermediate state, by periodic perturbation of the physical state of the system. This phenomenon, to which Odum (1969, 1971) has given the name of 'pulse stability' is very useful in interpreting many of the environmental effects of dam construction; the change in the annual flooding cycle below the dam allows previously pulse-stabilized systems to develop, and at the same time a new flooding regime above the dam may stabilize a new, immature, ecosystem there.

Off-chanel reservoirs may be completely artificial constructions, as regular in shape as swimming pools. Reservoirs formed by damming streams on the other hand, while their morphology is naturally dependent on the terrain in which they lie, may be very irregular. If the dammed stream had high steep banks, the new lake will have a long narrow form, whereas if the terrain is relatively flat, water will spread laterally and the new lake will be much wider than the previous stream. In any case, it will tend to take on a more or less dendritic form as water backs up into tributaries, and the shoreline development (i.e, the ratio of the length of the shoreline to the circumference of a circle of the same area as the lake) will usually be higher than for most natural lakes. This, and the straight or regularly curved downstream edge make reservoirs usually immediately recognizable as such from the air (Wood, 1975).

The shoreline will immediately become subject to modification under the influence of waves and currents, and ice if it is located in a cold region. The rate of change will depend on the energy available, the shape of the shoreline, and the nature of the material. The development and steepness of the shoreline will tend to decrease until an equilibrium is reached or approached.

Shoreline modification is particularly rapid and extensive in northern areas if the reservoir lies over permafrost since here the process involves not only erosion but melting and liquefaction of the frozen soil. Under these circumstances the establishment of a moderately stable new shoreline may take many decades (Newbury & McCullough, 1984; Newbury et al., 1978). A striking example of the effect of the nature of the shoreline material is provided by the Southern Indian Lake impoundment in northern Manitoba, Canada. This lies in part on permafrost and in part on bedrock. The new shoreline on bedrock differs little in appearance from a natural shoreline, whereas on permafrost it presents a chaotic picture of newly exposed varved clay, slumped banks and fallen trees (Baxter & Glaude, 1980).

Reservoirs also frequently differ from natural lakes in the shape of their longitudinal profiles. The deepest point is often at the downstream end, just above the dam; whereas natural lakes tend to be deepest somewhere near the middle. For this reason reservoirs have been referred to as 'half-lakes' (Ellis, 1941).

If a pre-existing lake is converted to a reservoir by damming its outlet, the morphological changes will range from negligible if the change in water level is slight to very extensive if the water level is substantially increased.

If a river were simply dammed and then left alone the new lake would probably eventually become little different from a natural lake as the shoreline eroded into a more regular form, and the longitudinal profile was modified by sedimentation. In practice this rarely, if ever, happens; the reservoir is managed to achieve the purpose for which it was built, such as flood control or the generation of electricity. This usually means that it is filled at times of high river flows (spring in temperate latitudes or the rainy season in the tropics) and partially emptied, or drawn down, during the following months. The biological consequences of this can be considered in relation to the concept of the ecotone (Odum, 1971; Margalef, 1968). An ecotone is a boundary or transition between two habitats or communities, such as forest and field or land and water. It is a matter of common observation that such boundaries often support a richer variety of living things than either of the adjoining zones, a phenomenon sometimes referred to as the 'edge effect'. The practice of draw-down can have a devastating effect on the shoreline ecotone of reservoirs, particularly in high latitudes (Ryder, 1978); Swedish limnologists refer to the draw-down zone as the *Aridal* (Lindström, 1973), thus calling attention to the barren character of the region.

Another consequence of managing reservoirs is that water is often drawn from a point near the bottom of the dam to provide the greatest possible head for the generation of electricity. This has an effect on the physical limnology of the reservoir and also on the environment downstream.

Development and stabilization of a new reservoir ecosystem

Let us consider the most drastic case: a large river is

dammed and a large quantity of water impounded to produce a large artificial lake.

The first and most obvious effect is the flooding of a considerable area of land. The consequences of this will depend on a variety of factors, including the character of the terrain, the geochemical nature of the flooded material, the type of vegetation, whether or not the site has been cleared, and the prevailing climate. (See Gunnison et al., this volume, chapter 3). The impact is not necessarily proportinal to the area flooded; the linear extent of the submerged shoreline must be taken into consideration because of the extremely disruptive effect on the shoreline ecotone.

Most of the trees in the permanently flooded area are likely to be killed, although some flood tolerant species such as willow (*Salix*) and various shrubs and bushes may survive (Harms et al., 1980; Harris, 1975; Connor et al., 1981). Even in the areas that are only flooded for part of the year many of the trees and other vegetation are likely to be killed. (See Godshalk & Barko, this volume, chapter 4). Even in areas that are not flooded, the vegetation may be greatly affected by the undercutting of the banks by erosion (Grelsson, 1981).

It is usual to carry out a certain amount of pre-clearing for esthetic reasons, to allow access to the water for fishing, to facilitate navigation, to provide suitable areas where fish may spawn or for other reasons. It is sometime possible to practice selective pre-clearing, removing the trees likely to be killed by flooding and leaving the flood-tolerant species (Allen & Aggus, 1983). In high latitudes, clearing may be carried out after the reservoir has been filled. If the water level is lowered after a substantial ice cover has formed, sufficient force may be exerted to break large trees (Bollulo, 1980). The resulting debris may be collected in the spring and burned or disposed of in other ways.

Non-woody plants and the softer parts of trees submerged in reservoirs quickly decay. In some instances, such as in the tropical artifical Lake Brocopondo in Surinam, South America, this lead to the production of large quantities of hydrogen sulfide which gave offense to residents many miles downwind, rendered the water below the dam unsuitable for drinking, and corroded metallic parts of the dam (Szekely, 1982; Canfield, 1983). The woody parts of trees may last much longer. In high latitudes they may persist with little change for decades (Richardson, 1934; Van Coillie *et al.*, 1983). In warmer climates they are probably destroyed more quickly, partly by the action of insects (Petr, 1970). If the flooded area includes sphagnum bogs, large patches of material may rise to the surface as floating islands that may persist for some time (Zhadin & Gerd, 1963; Nault, 1980).

Meanwhile, erosion of the banks will be proceeding, with transport of material from the shoreline to deeper water. The slowing of the flow of incoming water will also lead to the deposition of any suspended material that may be present.

The restoration of a physically stable and esthetically attractive new shoreline is made more difficult by the practice of drawdown. This exposes different areas to erosive forces at different times, thus probably causing erosion to be more severe, and to continue for a longer time, than it otherwise would. At the same time the vegetation of the shoreline is subjected to an essentially unnatural hydrological regime. In unregulated rivers, the banks are usually flooded for a fairly short period during the spring floods or rainy season, depending on the latitude. Reservoirs, however, are usually filled at these times, and drawn down only slowly during the subsequent months. This is of less significance in the tropics, where the climate permits growth on the draw-down zone as it is laid bare, than in higher latitudes where the shoreline may be laid bare too late in the season for growth to occur. The selection of plants capable of growing on the draw-down zones of reservoirs has received, and continues to receive, a good deal of study (Conner *et al.*, 1981; Gill, 1977; Perrault, 1980). Not only terrestrial shoreline plants, but also aquatic vascular plants, may be adversely affected by draw-down (Ryder, 1978; Kogan & Kemzhayev, 182). If floating peat islands are present, these may show a richer aquatic vegetation than he shore, because they rise and fall with the water and so are largely unaffected by draw-down (Vogt, 1978).

In small hydroelectric reservoirs, and particularly in pumped-storage reservoirs (American Fisheries Society, 1976) the water level may fluctu-ate over a period of several hours or a day, with consequent very severe effects on the shoreline ecotones.

Another possible consequence of flooding is the forcing of water into aquifers. Under certain circumstances this may be regarded as a benefit; in arid regions, or in regions where the demand for water has led to depletion of ground-water, water may be made available through wells a considerable distance away (Lagler, 1969; Eisenhauer *et al.*, 1982). In other areas, deleterious effects may be observed such as destruction of trees by waterlogging of the soil (Avakyan, 1975), soil salinization, flooding of basements, and damage to the foundations of buildings (Razumov & Medovar, 1980). If the impounded water is of poor quality, the water in the aquifer may be impaired (Hoffman & Meland, 1973).

The physical limnology of natural lakes is reasonably well understood, at least in broad outlines (Hutchinson, 1957). In most lakes in the temperate regions it is dominated by the thermal stratification cycle; during the summer there is a well marked discontinuity between the warmer upper layer (epilimnion) and the cooler lower layer, (hypolimnion). As the epilimnion cools in the fall the two layers gradually mix until the whole body of water reaches $4°C$, the temperature of maximum density. Then a reserve stratification occurs, with colder water on top, until the spring warming causes mixing again and the establishment of the summer stratification. Very shallow lakes do not become stratified, whereas lakes that are deep in proportion to their width, or in which the bottom water contains a greater concentration of solutes than the upper water, may maintain a permanent stratification (meromixis).

Tropical lakes may also show a seasonal pattern of stratification and mixing, although this is less clear-cut and regular than the pattern in temperate lakes (Wood *et al.*, 1976; Beadle, 1981).

The behaviour of reservoirs on the other hand is sometimes such as to defy analysis (Neel, 1966). There are probably a number of reasons for this. Reservoirs may be built in areas where there are no natural lakes because these could not long exist on account of rapid evaporation or high load of sediment in the streams. The morphology of reservoirs, particularly their 'half-lake' longitudinal profile (Ellis, 1941) also contributes by making possible the accumulation of a stagnant pool of bottom water against the inner face of the dam (Fiala, 1966).

The fact that water is often drawn from a point well below the surface is another complicating factor. Moreover, the retention time of water in many reservoirs is fairly short, so that what are called near-field effects, associated with inflow and outflow, are likely to be of more significance relative to the far-field effects of thermal circulation and wind-generated currents than they are in most natural lakes. In pumped storage reservoirs, pumpback may cause mixing of the water and prevent or very much decrease the development of thermal stratification (Potter et al., 1982).

When the water entering a lake or reservoir is different in density from the water already present because of differences in temperature or a heavy load of dissolved or suspended material, the inflowing water may retain its identity for some time and flow through the standing water as a density current, called an underflow, an interflow or an overflow, depending on its position (Neel, 1966). The first density currents in a reservoir were observed in Norris Reservoir in Tennesee (Wiebe, 1939) and they have subsequently been found to play an important role in the physical limnology of many reservoirs (Gloss et al., 1980). The situation may be rendered still more complex by the action of internal waves, which may transport water from dense underflows to the surface (Fischer & Smith, 1983).

The difficulty of predicting the hydrodynamic behaviour of reservoirs is well exemplified by two reservoirs in Tasmania, Lake Barrington and Lake Gordon. The older of these, Lake Barrington, promptly ecame meromictic. This was attributed to its morphometry and to increased solutes in the lower waters as a result of decomposition of flooded vegetation (Tyler & Buckney, 1974). It was expected that Lake Gordon would show similar behaviour. However, it was found that for six months of the year a cold density current displaced older water upward, producing a complex stratification pattern and preventing permanent meromixis (Steane & Tyler, 1982).

It will be apparent from the foregoing discussion that impoundment may have a profound effect on water chemistry. One important effect, already referred to, is the depletion of oxygen and the formation of reduced substances, such as sulfide, in the lower layers. This is caused by the decomposition of flooded trees and other vegetation; organic matter in the flooded soil makes a smaller contribution (Sérodes, 1982). Inorganic substances may also be leached from the soil, causing further deterioration of water quality. In small water supply reservoirs, it is sometimes worthwhile to remove the vegetation and topsoil completely before flooding, in order to obtain water of good quality as soon as possible (Campbell et al., 1975; 1976). Activities associated with the construction of the dam, such as forest clearing and the construction of access roads, may lead to increased erosion and soil leaching within the watershed, contributing further to the deterioration of water quality (Kelly et al., 1980).

If an stream is impounded to form a lake, a population of plankton is likely to develop (McLachlan, 1974). If a natural lake is converted to a reservoir, such a population is likely to be already present. In either case, the growth of phytoplankton is favored by the release of nutrients such as phosphate from the flooded plants and soil. This often leads to a burst of increased primary productivity, which has been referred to as the trophic upsurge (Baranov, 1961; Ostrofsky & Duthie, 1980; Grimard & Jones, 1982). This usually lasts only a year or two, often to be followed by a much longer period of lower productivity or trophic depression, and eventually a smaller increase (Baranov, 1961).

This applies to temperate reservoirs; in tropical reservoirs the whole process is more rapid and the period of depression less pronounced (McLachlan, 1974). Even in the temperate zone, not all reservoirs follow this sequence. If impoundment also leads to an increase in turbidity (Kelly et al., 1980) which will decrease the light available for photosynthesis, or of dissolved humic material (Vogt, 1978; Ostrofsky & Duthie, 1975; Jackson & Hecky, 1981) which will decrease the light available and also bind nutrients and make them unavailable, the trophic upsurge may not occur or may be only of brief duration.

Nutrient dynamics in reservoirs, like many other aspects of their limnology, are likely to be influenced by the way in which they are operated. In natural lakes, nutrients are taken up by planktonic algae in the epilimnion. The algae gradually settle into the hypolimnion where they die and decompose, releasing nutrients into the water, to be brought again to the surface when the lake mixes in the fall. Nutrients entering the lake in the inflowing water enter this cycle, so the total concentration of nutrients in the lake increases with time. In other words, there is a tendency for lakes to become more eutrophic with age.

It was pointed out by Wright (1967) that reservoirs with low-level discharge would be expected to behave differently. If water is withdrawn from the hypolimnion then nutrients released there will no longer be retained to be recycled, but will be carried out of the reservoir into the stream below the dam. A number of studies (e.g., Martin & Arneson, 1978; Priscou et al., 1982) have shown hat this actually occurs. This may be in part responsible for the decrease in primary productivity (trophic depression) that sometimes occurs as reservoirs mature (Baranov, 1961).

The phenomenon of trophic depression, however, seems to be of less common occurrence than that of trophic upsurge (Bayne et al., 1983). If the reservoir receives a sufficient loading of nutrients from external sources, the net effect is likely to be increasing eutrophication (Soltero et al., 1974).

Other substances can be mobilized from the flooded soil. It is now widely recognized that the concentration of mercury in fish in new reservoirs is likely to be higher than in rivers or natural lakes (Abernathy & Cumbie, 1977; Bodaly et al., 1984). This is probably not due to pollution by human activities since it has been observed in areas where such pollution is extremely unlikely; rather, it seems due to microbial action on mercury in the flooded terrain. Mercury apparently enters the food chain directly without passing into solution in the water, since concentrations in the reservoir water do not become elevated (Cox et al., 1979; Meister et al., 1979; Waite et al., 1980). The concentration in fish can become sufficiently high to cause a significant elevation in the concentration of mercury in the hair of perople who habitually eat them (Lodenius & Seppänen, 1982; Lodenius et al., 1983).

Impoundment in a mature reservoir, where any flooded vegetation has largely decomposed and most of the soluble material has been leached from the soil, often improves water quality (Purcell, 1939). Suspended material can settle out, the bacterial count can decrease, dissolved organic material can decompose, and the dissolved oxygen concentration can increase. In a prairie reservoir where the total dissolved solids concentration of the inflowing water was high, impoundment improved overall water quality by diluting the inflow with better quality water from the spring runoff (Allan, 1978). A similar improvement of water quality by averaging over time has been achieved in a river contaminated by acid drainage from abandoned coal mines (Sheer & Harris, 1982).

Large populations of floating macrophytes may develop in reservoirs, particularly in warm regions. These appear to have an advantage over rooted plants because, like the floating sphagnum island previously mentioned, they are not susceptible to the effects of changes in the water level (McLachlan, 1974).

The zooplankton of reservoirs may show certain differences from that of natural lakes. The number of rotifers is often high relative to crustaceans in new reservoirs (Potter & Meyer, 1982; Pinel-Alloul et al., 1982). Rotifers are highly prolific (r-selected) (Potter & Meyer, 1982) and thus well adapted to colonize new ecosystems. They are probably also favoured by the growth of bacteria on which they may feed, in the course of the decomposition of flooded vegetation (Pinel-Alloul et al., 1982). Even mature reservoirs may show unusual features that are difficult to explain. Many Spanish reservoirs contain a species of copepod which appears to be absent or rare in natural waters (Margalef, 1973). The zooplankton of a Central American reservoir was found to include substantial numbers of water mites (Gliwicz & Biesiadeka, 1975).

The development of the benthos of new reservoirs has received much attention. Insects, particularly the larvae of certain chironomids (midges) are very often dominant in the early stages. Insects are generally highly prolific, and since they generally have a winged adult stage they can easily reach a new habitat to colonize it. Some species of chironomids have the further advantage of being able to tolerate the low oxygen tensions often found in the waters of new impoundments. Certain species seem to be characteristic of reservoirs in northern regions (Rosenberg *et al.*, 1984).

The chironomid populations of some of the large reservoirs of the USSR were extraordinarily high in the first few years (Morduchai-Boltovskoi, 1961) to the degree that the vast swarms of midges that emerged caused alarm among the populatin and reduced visibility for vessels (Zhadin & Gerd, 1963). Considerable numbers of chironomids also appeared in two tropical African reservoirs, Lake Kariba and Lake Volta (McLachlan, 1974). In temperate reservoirs the very high chironomid populations do not persist, but give way to smaller populations of other types of organisms, such as oligochaete worms and molluscs (Nursell, 1952, 1969; Krzyzanek, 1970). In tropical reservoirs, however, moderately large chironomid populations may persist for some time, and oligochaete populations do not become very high (McLachlan, 1974). Submerged or partially submerged trees were found to be an extremely important habitat in the African impoundments. In Lake Volta in particular, these made possible the development of an extremely high population of nymphs of a species of mayfly that burrows in wood (Pet, 1970). At the same time the growth of periphytic algae ('aufuchs') represents a substantial part of the primary production of the lake (Petr, 1975). The various types of organisms living in this habitat interact in a complex and fasciating way (McLachlan, 1974; Beadle, 1981).

The development of a fishery in a reservoir, although it is probably always of secondary importance to such purposes as the generation of electricity and flood control, is still a very valuable resource in certain areas. In North America the emphasis has been largely but not exclusively on sport fishing, but in other parts of the world commercial fisheries in reservoirs have contributed significantly to the economy and the food supply.

It has been known for a long time that the population of fish in new reservoirs may be very high (Ellis, 1941). There is a number of reasons for this. Certainly the high initial primary productivity plays a part by providing ample food for planktivorous species, as does the high population of benthic invertebrates for bottom feeders. In Lake Volta in Africa the presence of large numbers of standing trees appears to have contributed because the abundant 'aufwuchs', and invertebrate animals feeding on it, contributed substantially to the supply of fish food (Petr, 1975). Flooded trees and lower vegetation probably also provide shelter for young fish and reduce the rate of predation.

The nature of the fish population in an impoundment is usually different from that in the river. Species that were rare in the river may develop large populations in the reservoir with surprising speed (Mahon & Ferguson, 1981). The circumstances that may give one species a competitive advantage over another may be extremely subtle. In some reservoirs in the southern USA, the brook silverside was almost completely replaced by the very similar inland silverside over a period of about two years, perhaps because the latter species is better adapted than the former for capturing copepods (McComas & Drenner, 1982).

Fish species well adapted to life in reservoirs are not always present in the stream before impoundment. In this case, the fish yields in the new impoundment will be relatively low. This has been observed in many tropical impoundments other than those in Africa (Fernando & Holcik, 1982). In such cases it may be desirable to stock the reservoir with suitable species. In the tropics various species of cichlids, such as tilapia, could be used. The exceptionally good sport fisheries, particularly for centrarchids (bass and sunfish), in many reservoirs in the midwestern and southwestern US are based on introduced species, sometimes supplemented by periodic stocking, and often feeding on introduced species (Martin *et al.*, 1982; Rinne *et al.*, 1981).

The effects of impoundment on fish populations are not entirely positive, even for those species well adapted to such environments. The greatest problem us that of draw-down. If fish spawn near the shore their eggs or young may be stranded and perish when the water level drops. This can be counteracted by the construction of floating spawning platforms (Zhadin & Gerd, 1963) or of small sub-impoundments that remain filled when the water level falls (Gagnon, 1980) but the best procedure is to follow a draw-down regime which maintains a high water level at the critical breeding time, to the extent that this is compatible with the purposes for which the reservoir was constructed (Allen & Aggus, 1983).

A serious and unexpected effect of the Southern Indian Lake impoundment in northern Manitoba, Canada, has been the collapse of the Lake Whitefish Fishery in the late (Bodaly et al., 1984). This reservoir is not drawn down, and the collapse of the fishery seems to have been due to emigration of fish from the lake and blockage by the dam of their migration between the lake and the lower Churchill River.

The very large and rapid fluctuations that occur in pumped storage plants probably have a somewhat adverse effect on fish populations. However, some species at least spawn at greater depths in such environments than they do in lakes with stable water levels, and standing crops comparable to those in nearby unregulated waters are maintained (Bennet et al., 1979). Rather surprisingly, perhaps, the growth of bass in one pumped storage system was accelerated, perhaps because the very large draw-down concentrated both the bass and their prey, making predation easier (Heisey et a.., 1980).

There is also a danger of increased infection by parasites in reservoirs, for a number of reasons. Increased fish populations may mae it easier for parasites to spread from one fish to another, and impoundment favours the development of copepods, which are the intermediate hosts for many fish parasites (Hoffman & Bower, 1971; Mackie et al., 1982). In projects involving diversion from one watershed to another there is a distinct danger that certain parasites may be introduced into an area from which they have previously been absent. At least one project has been suspended partly on the basis of this consideration (Arai & Mudry, 1983).

The effects of dams on anadromous and catadromous fish will be discussed in the following section.

Impoundments can also affect other vertebrates.

The initial flooding may destroy the habitat of large numbers of animals and birds and drown many of them. Sometimes large-scale rescue operations have been undertaken, but it is doubtful if these are really very effective (Monosowski, 1983). At the time of the filling of the LG2 reservoir in northern Quebec, Canada, local trappers were encouraged to trap as many beaver as possible in the area to be flooded because it was believed that they would all die in any case (Tessier, 1980).

Birds may nest in the draw-down zone, only to have their nests flooded when the water rises (Allen & Aggus, 1983).

On the other hand reservoirs, if suitably managed, can provide food and habitat for waterfowl (Allen & Aggus, 1983). In regions where the climate permits growth of grass and other plants on the draw-down zone, these may provide grazing for large animals (McLachlan, 1974; Fowler & Whelan, 1980).

In certain reservoirs in Quebec, Canada, the fish community associated with submerged trees supports large numbers of mergansers, ospreys, and otters and floating peat islands provide a habitat for ducks (Verdon & Demers, 1982).

The development of a new ecosystem in estuarine impoundment depends on the purpose for which it is intended (Corlett, 1979). If the purpose is to provide a basin for the storage of fresh water, it will obviously be desirable to replace the impounded sea water with fresh water as quickly as possible. Marine organisms that cannot escape will soon die, and their decay may cause some depletion of oxygen. Eventually a freshwater flora and fauna will be established, and the impoundment will come to resemble any other freshwater reservoir. The wider impact of the project will depend on its location. The moviment of migratory fish, if there are any using the estuary, will be impeded so that it may be desirable to provide some kind of fish pass. If wading birds havve been using the shore of the impounded estuary as a feeding area they will be displaced. The overall effect on the bird population will depend on the availability of other feeding grounds.

The tidal regime on the seaward side of the dam will be affected, which will in turn bring about morphological changes as the erosion and sedimentation patterns change (Van Howeninge & De Graauw, 1982).

The construction of dams for coastal protection and land reclamation has been carried out on a particularly wide scale in the Netherlands. In such impoundments the replacement of sea water by fresh water occurs much more slowly, over a period of years. The mass destruction of marine organisms will therefore not likely occur. The same general considerations will, however, apply as to storage impoundments.

The power of the tides must have impressed people from the earliest times and indeed in earlier days some use was made of it to run mills in Europe and New England (Charlier, 1982). In modern times the possibility of using tidal energy to generate electricity has been considered in may parts of the world. In North America the very high tides in the Bay of Fundy, in eastern Canada, and in Passamaquoddy Bay, between the province of New Brunswick and the state of Maine have looked particularly attractive (Atlantic Tidal Power Programming Board, 1969; Daborn, 1977; International Joint Commission, 1961). However, at the present time there exist only one commercial and three experimental tidal power generating plants (Banal, 1982). The commercial plant has been operating on the Rance River in Brittany, France, since 1966 (Mauboussin, 1972; Cotillon, 1979). The environmental impacts do not seem to have been severe; some changes have been observed in the fish species in the vicinity, some morphological changes have occurred, and the tidal range has been decreased by reducing the maximum and increasing the minimum (Charlier, 1982). In impoundments of this kind, unlike the previous two, the water will remain saline so a massive change in the flora and fauna will not occur (Corlett, 1979).

The experimental plants are in the USSR, China, and Canada, at Annapolis in Nova Scotia on the Bay of Fundy. Several other possible sites are being considered throughout the world (Banal, 1982). The reason that tidal power is not yet used on a wider scale is probaly the great cost of construction combined with the fact that they produce power according to a lunar cycle, so that other construction, such as pumped storage facilities, is required to provide power according to the solar cycle followed by most human activities.

A good deal of research has been carried out on the possible environmental impacts of a large-scale tidal power project on the Bay of Fundy (Daborn, 1977; Yeo & Risk, 1979; Surette, 1983). Drastic changes in the sediment regime would be expected (Yeo & Risk, 1979; Amos, 1977, 1979). These might well have an adverse effect on the productivity of the intertidal community, leading to a decrease in the supply of food for birds and fish (Yeo & Risk, 1979). Considerable numbers of fish might be killed in the turbines; this is a matter of particular concern because it is believed that a substantial part of the shad population of eastern North America spends part of their lives in the Bay of Fundy (Surette, 1983). The natural oscillation period of the Bay of Fundy and Gulf of Maine would be altered, with consequent increases in the amplitude of the tides, perhaps as far down the eastern seaboard as Boston (Greenberg, 1977). The dynamics of ice movement during the winter would be changed (Gordon & Desplanque, 1983).

Many of these considerations are of concern for tidal power projects in general and have been raised for example in connection with a proposed scheme for the Severn Estuary in the United Kingdom (Mettam, 1978).

Downstream effects of impoundments

The downstream effects of dams may be drastic and far-reaching. They are sometimes unexpected, and the net effects may be the result of opposing causes which make them difficult to predict. Many of then can be understood as the results of two phenomena:
1. The action of the reservoir as a source or a sink.
2. Alterations in the flow regime.
Reservoirs often act as sinks for suspended sediment, as the result of the decrease of the rate of flow and consequently of the capacity of the water to maintain material in suspension. The clear water discharged from the dam may then be extremely erosive, and cause rapid degradation of the bed and

banks of the stream below, as its increased velocity enables it to pick up a new load of suspended material (Taylor, 1978; Petts, 1980b). Finer bed material will be carried away until the bed is covered with coarser material which is less easily moved, and at the same time the roughness of the surface may increase so the water velocity at the interface is decreased (Hammad, 1972). Bank erosion may then occur with consequent changes in the morphology of the stream (Kerr, 1973). Considerable downstream erosion occurred in the Nile after the closure of the Aswan Dam (Hammad, 1972). The Nile Delta has also been affected as the effects of coastal erosion are no longer balanced by silt from the river (Petr, 1978; Kashef, 1981). The damming of the Volta River in Ghana has led to an increase in erosion on the Ghanaian coast (Ly, 1980).

Reservoirs can also serve as sources of sediments. This may happen abruptly and on a massive scale if reservoirs are flushed out to maintain their storage capacity, and can be very destructive to aquatic organisms downstream (Gray & Ward, 1982; Hesse & Newcomb, 1982). If flood water carries very fine suspended material into a reservoir and this turbid water remains at the level of the outlet, the reservoir may act as a source of sediment for a long time (Larson, 1982).

The production of reducing conditions by the decomposition of flooded vegetation has already been mentioned. Although the massive production of hydrogen sulphide is probably chiefly a problem in the tropics, the depletion of oxygen can occur even in high latitudes (Lemire, 1982), sometimes completely deoxygenating streams many kilometers below dams (Vogt, 1978). The problem is not limited to new reservoirs; established reservoirs, if they develop a thermal stratification and discharge from the hypolimnion, very often decrease the downstream oxygen concentration (Brocksen *et al.*, 5ot e5e3; Cada *et al.*, 1983). The removal of nutrients from a reservoir by hypolimnion discharge (Wright, 1967) can lead to increased primary productivity below the dam. A dramatic demonstration of this is provided by Lake Mohave, which receives nutrient-rich hypolimnion water from Lake Mead and is extraordinarily productive (Priscou *et al.*, 1982).

On the other hand reservoirs may serve as nutrient sinks and decrease the primary productivity downstream. Lake Nasser, the reservoir of the High Aswan Dam, seems to have this effect on the Nile (Petr, 1978; Balba, 1979; Talling, 1980). The supply of nutrients to Kootenay Lake, in British Columbia, Canada, has been diminished by the construction of dams on its inflows (Daley *et al.*, 1981).

Construction of a tidal barrage may decrease the flushing rate of the estuary below it, delaying the removal of pollutants and leading to a reduction in the dissolved oxygen concentration and other changes in the chemistry of the waters (Connell *et al.*, 1981).

Stratified reservoirs with hypolimnion discharge can have a considerable influence on the water temperature below the dam. During the summer most of the solar energy falling on the reservoirs is retained as heat in the epilimnion. Consequently, the water leaving the reservoir is cooler than it would be from a natural lake discharging from the surface (Wright, 1967). Under certain circumstances this may be desirable, by making it possible to maintain a fishery for cold-water fish, such as trout, below the dam (Pfitzer, 1967). In other situations the water may be too cold, requiring special constructions to permit the discharge of warmer water (Peters, 1979).

Some of the heat trapped in summer serves to warm the discharge in winter, since the hypolimnion will then be warmer than the epilimnion. In high latitudes this may prevent the stream from freezing many kilometers below the dam (Keenham *et al.*, 1982).

Changes in the flow regime below a dam may occur over two very different time scales. There may be an increase in the amplitude of diurnal variations in water level, as the amount of water passed through the turbines is varied in response to diurnal changes in the demand for electricity, and there may be a damping in the amplitude of annual variations in water level as the water entering the reservoir in the spring or the rainy season is retained and gradually released throughout the year. These 'first-order impacts' (Petts, 1980a) bring about 'second-order impacts' such as channel changes or changes in invertibrate populations which in turn lead to 'third-order impacts' such as changes in first habitat.

The diurnal variations can be very destructive to aquatic organisms (Fisher & LaVoy, 1972; Trotzky & Gregory, 1974; Ward, 1976) and cause changes in the stream morphology, as well as interfering with human activities below the dam. However, it is the change in the annual flood pattern that has greatest impact. This may interact with the clear water erosion previously discussed to bring about large-scale morphological changes in the stream channel, as material brought into suspension in some places in deposited further downstream (Petts, 1980a, 1980b). Sediment deposited in the main stream by tributaries may accumulate in the absence of an annual freshet to sweep it away (Petts, 1980b) and become stabilized by the growth of vegetation on it (Kerr, 1973; Kellerhalls & Gill, 1973). The overall effect is extremely complex and difficult to predict.

Among the most dramatic effects of abolition of an annual flood may be the tendency toward maturation of ecosystems previously maintained in an immature state by pulse stabilization. A well documented example of this is the impact of a dam on the Peace River in British Columbia, Canada, on the Peace-Athabasca Delta in Alberta, several hundred kilometers downstream. This is not a marine delta, but a large area of marshland interspersed with small lakes and ponds at the western end of Lake Athabasca between the Peace and Athabasca Rivers (Peace-Athabasca Delta Project Group, 1972; Hughes & Cordes, 1981), maintained in this condition by spring floods. After closure of the dam it was observed that the delta appeared to be drying up and the marshland changing into meadow. Because the inhabitants of the area subsisted to a large extent on fishing and trapping in the marsh, and because of the esthetic value of the delta in its original state, a large-scale and intensive study was undertaken leading to the installation of submerged weirs (Peace-Athabasca Delta Project Group, 1973; Alberta Environment Conservation Authority, 1974). These seem to have been successful in restoring the delta to something approaching its original state, although they appear to impede the movements of fish to some degree (Kristensen & Summers, 1978).

Similar instances could be cited from many parts of the world. The damming of the Zambezi River in Africa to form Lake Karina prevented the annual flooding of the floodplain below and led to a change in the vegetation towards species less suitable for the large numbers of animals living there (Attwell, 1970). Here the loss was perhaps partly compensated for by the development of a new pulse-stabilized system in the drawdown zone of the reservoir (McLachlan, 1974). The loss of the annual refertilization of the floodplain of the Nile as a result of the removal of sediment and prevention of annual flooding by the Aswan Dam (Kashef, 1981) might be considered another example.

Estuaries are ectones between fresh water and salt water. These, together with the lower reaches of rivers entering them and an indefinite area of the sea adjacent to them, are potentially very sensitive to flow changes. During the filling of a reservoir, if there is no significant inflow of water below the dam from tributaries, the downstream flow may be greatly reduced or virtually abolished. Sea water can then move up the river from its mouth to the probable detriment of freshwater organisms. When the LG2 reservoir on La Grande River in northern Quebec, Canada was filled, this was done in the winter in the expectation that the ice would provide a partial barrier against this saline encroachment. Perhaps for this reason, the effect on fish seems to have been small (Caron & Roy, 1980; Roy, 1982). When the reservoir is filled and flow is restored, the effect may be to reduce the extent of saline encroachment which previously occurred at periods of low flow. Lakes in the Nile Delta have become less saline (Banoub, 1979) and more productive of fish (Shaheen & Yosef 1979) since the closure of the Aswan High Dam. Both the initial increased encroachment and its later decrease were observed on the Volta River in Ghana below the Akosombo Dam (People & Ragoyska, 1969; Petr, 1978; Beadle, 1981), and gave cause for concern because of their effect on the breeding of a commercially important mollusc, *Egeria radiata*, which requires rather specific conditions of salinity to spawn. During the filling of the reservoir their spawning grounds moved about 50 km up-river, and when filling was complete they moved down again to about 10 km from the mouth. If hydrological conditions had been only slightly different, this organism could have been eliminated altogether, with consequent economic hardship for those engaged in the fishery.

When fresh water enters the sea a complex system of currents known as haline circulation is set up (Neu, 1982a, 1982b). The seaward flow of the lighter surface layer of fresh water is balanced by a deeper landward flow of denser sea water. Thus deep ocean water, rich in nutrients, is constantly transported onto the continental shelf. Consequently, any reduction in the flow of fresh water into the sea gives rise to the possibility of a reduction in productivity in the sea off the river mouth and for some distance away from it. It has been suggested that the regulation of rivers flowing into the Gulf of St. Lawrence has reduced the catches of some species there (Sutcliffe, 1973). Phytoplankton blooms in the Mediterranean Sea near the mouth of the Nile have been much diminished since the closing of the Aswan High Dam, and perhaps as a consequence sardine catches have greatly decreased (Aleem, 1972; Pandian, 1980). Regulation of the Dnieper and the Don has been blamed for serious decline in fish catches in the Sea of Azov and the Black Sea (Neu, 1982; Tolmazin, 1979).

It will be evident from the foregoing that the biological impacts of a dam on the stream below are exceedingly complex. There is often a decrease in diversity (Ward, 1976). Many invertebrates require temperature changes at various stages in their metamorphoses to trigger the transformation to the next stage, so they may be unable to develop if the thermal regime is changed (Ward, 1975; Gore, 1977). If these are important as fish food, fish populations may be diminished, but usually the species that disappear are replaced by others better adapted to the new temperature regime and other altered conditions (Walburg et al., 1981; Spence and Hynes, 1971a). The total biomass may be increased (Walburg et al., 1981) or decreased (Lehmkuhl, 1972). The decreased turbidity that often occurs may lead to an increase in primary productivity and hence the supply of fish food, but it may expose young fish to a greater risk of predation (Geen, 1974). It appears to favor certain species of fish, such as trout, over others (Walburg et al., 1981). However, the diversity of fishes, like that of invertebrates, is often decreased below dams (Spence and Hynes, 1971b; Edwards, 1978). The release of deoxygenated hpolimnion water can be very destructive, not only on account of the low oxygen concentration but also on account of the presence of reduced substances such as sulphide, ferrous, and manganous ions and ammonia, many of which are toxic (Walburg et al., 1981; Oglesby et al., 1978).

Still another peril for fish below a dam is gas bubble disease (American Fisheries Society, 1980). Water plunging over a spillway may entrain air bubbles and carry them to a considerable depth where the hydrostatic pressure is sufficient to dissolve the atmospheric gases to concentrations significantly higher than the equilibrium concentration at atmospheric pressure. If fish ingest such water the gases may come out of solution as bubbles in the tissues which may cause injury or death (Harvey, 1975; Bouck, 1980). Suitable spillway design can reduce the extent of gas supersaturation to some extent (H.A. Smith, 1974; P.M. Smith, 1976) but the problem is still a cause for concern (Bouck et al., 1980).

In spite of the various hazards which they cause, dams often have a net favourable effect on fish and fisheries in the stream below (Wiebe, 1960; Crisp et al., 1983).

Dams and reservoirs pose special problems for migratory fish, especially anadromous species, i.e., those that live in the sea and spawn in fresh waters, such as salmon and shad.

Dams may delay the passage of adult fish upstream (Haynes & Gray, 1980). Fish ladders may make them less formidable as mechanical barriers, but other effects are less easily avoided. Fish return to the streams in which they were hatched following subtle tactile and olfactory clues (Hasler et al., 1978). Regulation of stream flows may obscure these clues, causing the fish to become disoriented (Haynes & Gray, 1980). Once the dam has been passed, the fish may seek cool deep water in the reservoir and have difficulty finding their way out again. Delays for whatever reason are particularly dangerous to such species as Pacific Salmon which do not feed on the way to the spawning grounds and must rely on the limited energy reserves within their bodies.

If a stream is even temporarily blocked, as during dam construction, for a period equal to or greater than the period of sojourn of the fish in the sea, the stock in the stream may be effectively destroyed.

If spawning has been successful, the young fish must return to the sea. These are exposed to many perils on dammed rivers (Ruggles, 1980; Raymond, 1979; Gloss & Wahl, 1983). Like adults, they may seek cold water in the depths of reservoir and be delayed in their passage or even spend their lives there. Some may be killed in passing over spillways, and more in passing through turbines. In some species, such as Atlantic Salmon, the adults do not die after spawning and so must also negotiate these obstacles on the way back to the sea (Ruggles, 1980).

In some species, the young fish may spend part of their lives in the estuary feeding on crustaceans which feed on detritus (Healey, 1979; Sibert, 1979; Naiman & Sibert, 1979). If a dam on the river diminished the supply of detritus to the estuary, that might be expected to have an adverse effect on the fish population.

Catadromous fish, those that live in fresh water and spawn in the sea, appear to be less susceptible to the effects of dams than are anadromous fish. The most important catadromous fish are eels. Adult eels may be killed in passing through turbines, but even if all the eels in a particular stream were prevented from returning to the sea, the stream would be repopulated annually as long as a sufficient breeding stock existed in another stream. Young eels can negotiate obstacles that would be impassable to adult salmon. When the Moses-Saunders dam was constructed on the St. Lawrence River it did not completely block the migration of young eels, because the annual eel catch on Lake Ontario did not decrease significantly, but there is some evidence that their movement was impeded to a certain extent (Kolenosky & Hendry, 1982). An eel ladder has now been installed (Eckersley, 1982), which has apparently reduced or abolished the blockage (Liew, 1982). In general, eels seem to

pose more of a problem for hydroelectric installations than the installations do for eels, because they tend to swim into turbine housings in large numbers when the flow is reduced and cause clogging. Since eels are repelled by light, one approach to this problem is the use of strobe lights (Patrick *et al.*, 1982).

Other effects of dams and impoundments

Natural lakes modify the climate in their vicinities (Richards, 1969; Phillips & McCulloch, 1972; Phillips & Irbe, 1978) and reservoirs can be expected to do so too, according to their size and geographical location. The effects may include the induction of local winds as land breezes and lake breezes, more frequent occurrences of fog in the vicinity and a general damping of temperature variations, possibly leading to an extension of the frost-free season. Many of these effects are related to changes which a reservoir may induce in the absorption of solar radiant energy in the flooded area. Such changes depend on changes in the albedo, which is the fraction of the incident energy that is reflected from the surface (Vowinckel & Orvig, 1980). The albedo of water is very low, so flooding is likely to increase the amount of energy absorbed. However, the albedo of snow or ice is high, so in cold regions the result of flooding may be a net decrease of energy absorbed over the whole year. The most dramatic possibilities for climatic change as a result of impoundments are those associated with the damming of large northward flowing rivers. In the Mackenzie River in Northern Canada, and probably in others in similar regions, the spring breakup of ice in the lower reaches is hastened by the spring floods (Gill, 1971). These by their hydrostatic pressure break the ice, and also wet it, increasing the absorption of radiant energy because wet ice has a lower albedo than has ice covered with snow.

The possibility of regulating large northward-flowing rivers has been considered in the USSR for many years. The main purpose of such schemes would be to make more water available in the south for various purposes, including the restoration of the water level of the Caspian Sea which has fallen as a result of the dams along the Volga River which flows into it (Water Power & Dam Construction, 1983). The climatic consequences of such a project could be extremely serious. By decreasing the amount of water in northern soils and thereby lowering its heat reserve, it might make the climate more severe. Still more disturbing, however, are the possible consequences of decreasing the amount of fresh water entering the Arctic Ocean. These are a matter of debate, but it is thought possible that the circulation and stratification patterns might be so disturbed as to lead to significant melting of the polar ice cap and possible far-reaching changes in the climate of the northern hemisphere (Water Power and Dam Construction, 1983; Aagard & Coachman, 1975; Micklin, 1977; Lamb, 1971; Gribbin, 1979).

In 1939 seismic activity was observed in the vicinity of Lake Mead, the reservoir of the Hoover Dam, and attributed to the presence of the reservoir, which had been filled three years before. For many years this was considered to have been a unique occurrence, but during the 1960's severe earthquakes were observed at reservoirs in China, Greece, India, and Rhodesia (now Zimbabwe). By 1976 more than 30 cases of seismic activity induced by impoundments had been reported throughout the world (Simpson, 1976); many more have been reported since then (Buchbinder, 1977; Buchbinder et al., 1981; Simpson et al., 1981; Adams, 1983) and the latest count exceeds 100 (Adams, 1983). At least two have led to loss of life (Koyna, India and Kremasta, Greece). In 1963 a large landslide into the reservoir at Vajont Dam in Italy caused a large flood below the dam which killed about 2,000 people. It is possible that the landslide was the result of induced seismic activity. In some instances induced seismicity has caused damage to the dam (Simpson, 1976). The phenomenon is more likely to occur in areas where spontaneous seismic activity has been observed in the past, so it is often difficult to be sure whether an earthquake occurring after the filling of a dam has, in fact, been induced by the impoundment or would have occurred in any event (Stein et al., 1982). It is unlikely that the strains caused by an impoundment would in themselves be sufficient to cause movement of the rocks below; more likely the additional strain triggers movement where strains already exist, by the vertical pressure exerted by the mass of water and an increase in the pore pressure in the rocks as water is gradually forced into them (Smith, 1982; Castle et al., 1980; Gough and Gough, 1970, Leith et al., 1981). Water may also serve to lubricate the rocks and in the longer term bring about chemical changes that weaken them (Smith, 1982; Leith et al., 1981). Earthquakes may occur as the reservoir is being filled (Buchbinder et al., 1981; Simpson et al., 1981; Leith et al., 1981) or not until several years after filling (Adams, 1983). Although most induced earthquakes have been associated with reservoirs at least 100 m deep, one seems to have been induced by Lake Nasser behind the Aswan Dam, which is only 72 m deep (Adams, 1983). The particular pattern of induced seismicity at any particular site depends very much on the local geological conditions (Smith, 1982; Leith et al., 1981; Keith et al., 1982).

Many of the most serious diseases that affect mankind depend in one way or another on water for their transmission. The incidence of those transmitted by drinking water (e.g., cholera, typhoid, dysentery) is not likely to be much affected by reservoir construction except incidentally; the crowding together of construction workers under conditions of poor sanitation while the dam is being built can lead to an increase in these diseases, as well as others (Goldman, 1976), whereas increased prosperity as a result of the construction of a dam might lead to improved sanitation and a decrease in the incidence of diseases.

A number of other diseases, however, are related to water in a more complicated way, and the possible effects on these of dam construction have caused considerable concern. Three of the most important of these are onchocerciasis, malaria, and schistosomiasis.

Onchoceriasis, also known as river blindness, is caused by a small worm which is transmitted by the bite of certain species of blackfly (*Simulium*). The worms may lodge in various parts of the body, causing debilitation, and if they lodge in the eye they may cause blindness. In certain parts of West Africa the incidence of the disease approaches 100 percent (Worthington, 1978).

Blackflies do not breed in standing waters, but in fastrunning, well-oxygenated waters. The construction of a dam is likely to decrease, rather than increase, the number of suitable breeding sites by flooding rapids upstream. A few new sites may be provided by spillways, but this can be minimized by suitable design and operation (Ripert *et al.*, 1979). It might be expected that construction of a reservoir might diminish the incidence of river-blindness, and this was observed when the Akosombo Dam was built on the Volta River in Ghana and when the Kainji Dam was built on the Niger in Nigeria (Worthington, 1978).

Malaria, on the other hand, is likely to increase as a result of impoundment. The mosquitoes which transmit malaria breed in standing waters, so the potential breeding area is likely to increase. The presence of submerged terrestrial vegetation, or the growth of rooted or floating aquatic macrophytes, further favour the breeding of mosquitoes by providing shelter.

The possibility of an increase in malaria was a matter of concern to the Tennessee Valley Authority when the dams of the Tennessee Valley were being constructed. The problem was solved by carefully clearing the edges of reservoirs, and periodically altering the water level to strand mosquito larvae (Worthington, 1978; Egouniwe, 1976). A similar approach would probably be effective in many other places, although a fluctuating water level might be favourable to some species of mosquitoes (Waddy, 1975).

Schistosomiasis, also known as Bilharziasis, is transmitted in a different way. This disease, or rather group of diseases, is caused by parasitic worms. Infected individuals excrete eggs of the parasite in their feces or urine. If these enter the water, larvae emerge which infect certain species of aquatic snails. After a period of development in the snail, the larvae emerge in a different form which is capable of entering the human body through the skin, thus establishing a new infection. This disease is little known to people in temperate regions, but in tropical and subtropical regions it is exceedingly wide spread. It is not generally fatal but is very debilitating and it is difficult to cure, so that it causes great economic loss through loss of

productivity in regions where it is common. It is probably for this reason that is has aroused more concern than any other disease associated with reservoirs.

The incidence of schistosomiasis was considerably increased by the construction of the Akosombo Dam in Ghana (Worthington, 1978; Waddy, 1975). It was feared that it might also increase after the construction of the Aswam High Dam in Egypt, but this did not happen (Miller *et al.*, 1978; Walton, 1981).

The control of schistosomiasis is a complex task, requiring treatment where possible of the people infected, improved sanitary conditions and knowledge of hygiene among the population at risk, and control of the snail vectors. In the particular case of reservoirs the same procedures used to control mosquitoes are effective against snails.

The aim of controlling snails by varying the water level in the reservoir may conflict with the aim of controlling blackflies by avoiding as far as possible the fast flow of water over the spillway. It has been suggested (Diamant, 1980) that this conflict can be resolved by the installation of enclosed automatic spillway siphons so the flow occurs in an enclosed space inaccessible to flies.

The modern era of dam construction on a large scale can perhaps be considered to have begun in the 1930's with the Tennessee Valley Authority in the USA and the construction of large dams on the Volga River in the USSR. These projects undoubtedly caused great disruption in the lives of the people in the regions involved. On the other hand, they brought great benefits to these same people, such as new opportunities for employment, electricity for domestic use, and so on, and they are looked upon with justifiable pride in their respective countries.

The situation has been very different with many subsequent projects. The people whose lives have been most disrupted have often not stood to gain very much from the projects, which have often been most useful for people living at some distance from the site. For example, the construction of certain reservoirs in Finland uprooted several hundred people who had been leading self-sufficient lives herding reindeer, and had great difficulty in adjusting to the change (Vogt, 1978). The four great African impoundments, Aswan in Egypt, Volta in Ghana, Kainji in Nigeria and Kariba in Zimbabwe and Zambia, together caused the evacuation of more than a quarter of a million people (Deudney, 1981). The Tucurui Dam in Brazil will require the resettling of 15,000 people, many of whom have already suffered a great deal as a result of contact with the outside world (Canfield, 1983).

It is difficult for North Americans and Europeans to appreciate the distress that such evacuations may cause to people who live in tightly-knit tribal societies and who may never have travelled more than a few kilometers from the communities where their families have lived for generations, even if the evacuation is carried out smoothly and efficiently and the people receive adequate compensation for their economic loss. Not only do they suffer grief at the loss of their homes and other places such as the burial grounds of their ancestors to which they feel emotionally attached, and anxiety as to their future both economically and culturally, but they may lose confidence in their local leaders and develop a sense of dependency on the authorities responsible for their transfer (Futa, 1983).

Fortunately, the lessons to be learned from these earlier experiences have not been lost on those responsible or more recent projects, and the possible social impacts have been considered from the start, and measures to mitigate them have been included in the plans. The Jams Bay project in northern Quebec, Canada, has had a profound effect on the lives of the people living there who are mostly Cree and Inuit (Eskimos) who have traditionally maintained themselves by hunting, trapping, and fishing. At the beginning of the project, very detailed negotiations were carried out involving the Federal Government, the Quebec Government, representatives of the Crees and Inuit, and the Crown corporations responsible for the undertaking. These led to an agreement (James Bay & Northern Quebec Agreement, 1976) which it was hoped would protect, as far as possible, the traditional way of life of the people and assure that they shared in the benefits of the project. This seems to have been reasonably successful, although there is

some evidence that the impacts have been more severe than was anticipated (Berkes, 1981, 1982).

A new hydroelectric project is being undertaken in Ghana, downstream from the Akosombo Dam. This will involve the resettlement of 6,000 to 7,000 people, much fewer than were affected by the earlier dam, but still a substantial number. In an attempt to avoid as much as possible the kind of distress that was caused by the earlier evacuation, those responsible have budgetted for the cost of resettlement in negotiating the loans to finance the project, and have required the consultants responsible for the hydroelectric project to assist also in the resettlement project (Futa, 1983).

Dam and reservoir construction can have a very significant influence on international affairs (Deudney, 1981). Many of the large rivers of the world constitute or cross international boundaries, so they can only be exploited by agreement of the countries involved. The St. Lawrence Seaway with its dams and power plants provides an example. The desire to benefit from a shared resource has provided an impetus for cooperation between countries that have been traditionally much less friendly than Canada and the United States, in South America (Cano, 1976), Europe (Kovacs & David, 1977) and elsewhere.

Dams have been shown to be vulnerable to attack in wartime (Collins, 1982), and their destruction or damage can cause terrible devastation in the valley below. Thus to construct a large dam is in a sense to give hostages to potential enemies, and may well provide an additional motive for a country to seek a peaceful solution to its disputes (Deudney, 1981).

Conclusions

In principle, reservoirs are constructed because it is believed that they will, overall, lead to an increase in wealth and in human well-being. Those involved in such a project, however, may have a variety of motives not necessarily very closely related to this overall goal.

The workmen who do the actual physical work are concerned about earning a living, and no doubt are also motivated by a sense of craftsmanship and inspired by the sense of participating in something big. The engineers who plan and direct the construction are certainly motivated to a considerable degree by the 'existential pleasures' (Florman, 1976) of engaging in what has been described as 'one of the most thrilling tasks an engineer can undertake' (Mary, 1965). In many parts of the world, the decision to build a particular dam is a political one. In the eyes of many, a dam is an almost unique symbol of progress (Dickinson, 1980), so that a critical examination of the harsh economic realities of a project may not be very appealing.

The outcome of all this is:

1 A project may not even achieve its primary goal of being profitable; Hart's (1980) study of the Volta River project provides a well-documented example of the economic complexities involved in the construction of a large dam in a developing country.
2 A project may appear to provide a net gain, but this may be limited to one segment of the population while another segment suffers a net loss.
3 Nowhere in the process, as traditionally carried out, is there any provision for an examination of the ecological impact of a project except for the most obvious considerations, such as flooding.

Circumstances like these have led to a widespread reaction against the whole idea of building dams, which have come to be looked upon by some as a kind of man-made disaster. Much of the criticism of dam-building is no doubt justified, but much also is ill-informed and irrational. In any case it has unquestionably served a useful purpose. It is to be hoped that future projects will be given a more searching economic scrutiny before being undertaken than some have had in the past, and as we have seen, the social problems associated with the displacement of populations by flooding have become a matter of prime concern.

The assessment of environmental impacts of reservoirs has become a well-developed art (Rees, 1980; Canter, 1983), but is still by no means an exact science. Hecky *et al.* (1984), in a summary of a decade's study of the effect of the southern Indian Lake impoundment and diversion, listed eight correct predictions, two incorrect predictions, and seven unpredicted effects. The most important of these, from the practical point of view, were increased mercury concentrations in fish and the collapse of the Lake Whitefish Fishery.

To the ecologist, an impoundment should provide both an opportunity and a challenge; an opportunity to use what is in effect a large-scale ecological experiment to advance human knowledge of natural processes, and a challenge to use existing knowledge to anticipate and mitigate the possible undesirable consequences of the project.

References

Aagaard, H. & Coachman, L.K., 1975. Toward and ice-free Arctic Ocean. Eos 56:484–86.

Abernathy, A.R. & Cumbie, P.M., 1977. Mercury accumulation by largemouth bass (*Micropterus salmoides*) in recently impounded reservoirs. Bull. envir. Contam. Toxicol. 17:595–602.

Ackerman, W.C., White, G.F., Worthington, E.B. & Ivens, J.L., (eds.), 1973. Man-made lakes: Their problems and environmental effects. American Geophysical Union, Washington, DC: 847 p.

Adams, R.D., 1983. Incident at the Aswan Dam. Nature 301–14.

Alberta Environment Conservation Authority, 1974. The restoration of water levels in the Peace-Athabasca Delta. Report and recommendations. Edmonton, Alta.: 136 p.

Aleem, A.A., 1972. Effect of river outflow management on marine life. Mar. Biol. 15:200–8.

Allan, R.J., 1978. Natural controls on dissolved solids in Boundary Reservoir. Saskatchewan. Can. Water Resour. J. 3(3):78–96.

Allen, H.H. & Aggus, L.R. (eds.), 1983. Effects of fluctuating reservoir water levels on fisheries, wildlife, and vegetation; Summary of a workshop, 24–26 February 1981. Miscellaneous Paper E–83–2, US Army Engineer Waterways Experiment Station, CE, Vicksburg, Miss.

American Fisheries Society, 1967. Reservoir Fishery Resources Symposium. Athens, Georgia: Univ. Georgia: 569 p.

American Fisheries Society, 1976. Biological considerations of pumped storage development. Trans. am. Fish. Soc. 105:155–80.

American Fisheries Society, 1980. Special section: Gas bubble disease. Trans. am. Fish. Soc. 109:657–771.

Amos, C.L., 1977. Effects of tidal power structures on sediment transport and loading in the Bay of Fundy-Gulf of Maine system. In G.R. Daborn (ed), Fundy tidal power and the environment. Proc. Workshop on the Environmental Implications of Fundy Tidal Power, Wolfville, N.S., Nov. 4, 5 1976. Acadia Univ. Inst., Wolfville, N.S.: 233–253.

Amos, C.L., 1979. Sedimentation resulting from Fundy tidal

power. In Severn, R.T., Dineley, D.L., & Hawker, L.E., (eds.), Tidal Power and Estuary Management, Proceedings of the Thirtieth Symposium of the Colston Research Society. Scientechnica, Bristol, UK: 296 p, p. 173–179.

Arai, H.P. & Mudry, D.R., 1983. Protozoan and metazoan parasites of fishes from the headwaters of the Parsnip and McGregor rivers. British Columbia: a study of possible parasite transfaunations. Can. J. Fish. aquat. Sci. 40:1676–1684.

Atlantic Tidal Power Programming Board. Feasibility of tidal power development in the Bay of Fundy. October 1969.

Attwell, R.I.G., 1970. Some effects of Lake Kariba on the ecology of a floodplain of the mid-Zambezi valley of Rhodesia. Biological Conservation 2:189–196.

Avakyan, A.B., 1975. Problems of creating and operating reservoirs. Sov. Hydrol. 1975(3):194–199.

Balba, A.M., 1979. Evaluation of changes in the Nile water composition resulting from the Aswan High Dam. J. envir. Qual. 8:153–156.

Banal, M., 1982. L'énergie marémotrice en 1982. La Houille Blanche No. 5/6–1982, pp. 433–439.

Banoub, M.W., 1979. The salt regime of Lake Edku (Egypt) before and after the construction of Aswan's High Dam. Arch. Hydrobiol. 85:392–399.

Baranov, I.V., 1961. Biohydrochemical classification of the reservoirs in the European USSR. In Tyurin, P.V. (ed.), The Storage Lakes of the USSR and Their Importance for Fishery. Israel Program for Scientific Translations, Jerusalem: 139–183.

Bärlocher, F., 1982. On the ecology of Ingoldian Fungi. Bioscience 32:581–586.

Baxter, R.M., 1977. Environmental effects of dams and impoundments. Annu. Rev. Ecol. Syst. 8:255–283.

Baxter, R.M. & Glaude, P., 1980. Environmental effects of dams and impoundments in Canada: experience and prospects. Can. Bull. Fish. Aquat. Sci. 205:34 p.

Bayne, D.R., Lawrence, J.M. & McGuire, J.A., 1983. Primary productivity studies during early years of West Point Reservoir, Alabama-Georgia. Freshw. Biol. 13:477–489.

Beadle, L.C., 1981. The inland waters of tropical Africa, Second edition. Longmans, London: 475 p.

Bennet, D.H., Raleigh, R.F. & Maughan, O.E., 1979. Effects of pumped storage project operations on the spawning success of centrarchid fish in Leesville Lake, Virginia. In Driver, E.E. & Wunderlich, W.O. (eds.). Environmental Effects of Hydraulic Engineering Works. Proceedings of an International Symposium held at Knoxville, Tennessee, USA. Sept. 12–14, 1978. pp. 125–134.

Berkes, F., 1981. Some environmental and social impacts of the James Bay hydroelectric project, Canada. J. environ. Management 12:157–172.

Berkes, F., 1982. Preliminary impacts of the James Bay hydroelectric project, Quebec, on estuarine fish and fisheries. Arctic 35:524–530.

Biswas, A.K., 1975. A short history of hydrology, p. 57–79. In Biswas, A.K., (ed.), Selected Works in Water Resources. International Water Resources Association, Champaign, Ill.: 382 p.

Bodaly, R.A., Hecky, R.E. & Fudge, R.J.P., 1984. Increases in fish mercury levels in lakes flooded by the Churchill River diversion, northern Manitoba. Can. J. Fish. Aquat. Sci. 41:692–700.

Bodaly, R.A., Johnson, T.W.D., Fudge, R.J.P. & Clayton, J.W., 1984. Collapse of the Lake Whitefish (*Coregonus clupeaformis*) Fishery in Southern Indian Lake, Manitoba, following lake impoundment and river diversion. Can. J. Fish. Aquat. Sci. 41:692–700.

Bollulo, D.T., 1980. Le dèboisement écologique et l'élimination des débris ligneux flottants au reservoir de LG 2. Eau du Québec, 13(1):47–53.

Bouck, G.B., D'Aoust, B., Ebel, W.J. & Rulifson, R., 1980. Atmospheric gas supersaturation: educational and research needs. Trans. am. Fish. Soc. 109:769–771.

Bouck, G.R., 1980. Etiology of gas bubble disease. Trans. am. Fish. Soc. 109:703–707.

Brocksen, R.W., Fraser, M., Muraka, I. & Hildebrand, S.G., 1982. The effects of selected hydraulic structures on fisheries and limnology. CRC Crit. Rev. Environmental Control, 12:69–89.

Buchbinder, G., 1977. Earthquakes in a Québec hydro development. Geos, Fall 1977. pp. 6–8.

Buchbinder, G.G.R., Anglin, F.M. & McNicoll, R., 1981. La séismicité provoquée au réservoir LG-2. Canadian J. Earth Sci. 18:693–98.

Cada, G.F., Kumar, K.D., Solomon, J.A. & Hildebrand, S.G., 1983. An analysis of dissolved oxygen concentrations in tail waters of hydroelectric dams and the implications for small-scale hydropower development. Water Resources Res. 19:1043–1048.

Campbell, P.G., Bobee, B., Caillé, A., Demalsy, M.J., Demalsy, P., Sasseville, J.L. & Visser, S.A., 1975. Pre-impoundment site preparation: A study of the effects of topsoil stripping on reservoir water quality. Verh. int. Ver. Limnol. 19:1768–1777.

Campbell, P.G., Bobée, B., Caillé, A., Demalsy, M.J., Demalsy, P., Sasseville, J.L., Visser, S.A., Couture, P., Lachance, M., Lapointe, R. & Talbot, L., 1976. Effects du décapage de la cuvette d'un réservoir sur la qualité de l'eau emmagasinée: élaboration d'une méthode d'étude et application au réservoir de Victoriaville (riviére Bulstrode, Québec). INRS-Eau, Rapp. Sci. 37:304 p. 3 append. (For the ministére des Richesses naturelles, Québec).

Campbell, P.G., Perrier, R., & Cantin, M., (eds.), 1982. International Symposium on reservoir ecology and management, Laval University, Quebec, Canada, June 1981. Canadian Water ResourcesJ. 7(1 and 2).

Cano, G.J., 176. Argentina, Brazil, and the de la Plata River Basin: A summary review of their legal relationship. Natural Resources J. 16:863–882.

Canter, L.W., 1983. Impact studies for dams and reservoirs. Water Power and Dam Construction, July 1983, pp. 18–23.

Caron, O & Roy, D., 1980. Coupure de la Grande Rivière: période critique pour la faune aquatique en aval du barrage de LG2. Eau du Québec 13(1):23–28.

Castle, R.O., Clark, M.M. Grantz, A. & Savage, J.C., 1980. Tectonic state: its significance and characterization in the assessment of seismic effects associated with reservoir impounding. Engineering Geol. 15:53–99.

Caufield, C., 1983. Dam the Amazon, full steam ahead. Natural History 7/83:60–67.

Charlier, R.H., 1982. Oceans and electrical power (Part II). Intern. J. Environmental Studies 19: 7–16.

Cheesman, R.E., 1936. Lake Tana and the Blue Nile. Reprinted 1968. Cass, London: 400 p.

Collins, A.R., 1982. The origins and design of the attack on the German dams. Proc. Instn. Civ. Engrs. Part 2 73: 383–405.

Connell, D.W., Bycroft, B.M., Miller, G.J. & Lather, P., 1981. Effects of a barrage on flushing and water quality in the Fitzroy River Estuary, Queensland. Aust. J. mar. Freshwat. Res. 32:57–63.

Conner, W.H., Gosselink, J.G. & Parrondo, R.I., 1981. Comparison of the vegetation of three Louisiana swamp sites with different flooding regimes. Amer. J. Bot. 68:320–331.

Corlett, J., 1979. The likely consequences of barrages on estuarine biology. In Severn, R.T., Dineley, D.L. & Hawker, L.E. (eds.), Tidal Power and Estuary Management. Proceedings of the Thirtieth Symposium of the Colston Research Society. Scientechnica, Bristol, U.K. 1979: 296 p, p. 228–234.

Cotillon, J., 1979. La Rance tidal power station review and comments. In Severn, R.T., Dineley, D.L. & Hawker, L.E. (eds.), Tidal Power and Estuary Management. Proceedings of the Thirtieth Symposium of the Colston Research Society. Scientechnica, Bristol, UK: 296 p, p. 49–66.

Cox, J.A., Carnahan, J., DiNunzio, J., McCoy, J. & Meister, J., 1979. Source of mercury in fish in new impoundments. Bull. envir. Contam. Toxicol. 23:779–783.

Crisp. D.T., Mann, R.H.K. & Cubbym P.R., 1983. Effects of regulation of the River Tees upon fish populations below Cow Green Reservoir. J. Appl. Ecol. 20:371–386.

Daborn, G.R. (ed.), 1977. Fundy tidal power and environment. Proc. workshop on the environmental implications of fundy tidal power, Wolfville, N.S. Nov. 4–5, 1976. Acadia Univ. Inst., Wolfville, N.S.: 304 p.

Daley, R.J., Carmack, E.C., Gray, C.B.J., Pharo, C.H., Jasper, S. & Wiegans, R.C., 1981. The effects of upstream impoundments on the limnology of Kootenay Lake, B.C. NWRI Scientific Series No. 117, 98 p, National Water Research Institute, Vancouver, B.C., Canada.

Deudney, D., 1981. An old technology for a new era. Environment 23(7):17–20, 37–45.

Diamant, B.Z., 1980. Environmental repercussion of irrigation development in hot climates. Environmental Conservation 7:53–58.

Dickinson, H., 1980. Foreword to Hart, 1980.

Driver, E.E. & Wunderlich, W.O., (eds.), 1979. Environmental effects of hydraulic engineering works. Proceedings of an International Symposium held at Knoxville, Tennessee, USA September 2–14, 1978. Tennessee Valley Authority, Knoxville, Tennessee.

Eckersley, M.J., 1982. Operation of the eel ladder at the Moses-Saunders Generating Station, Cornwall 1974–79. In Ontario Ministry of Natural Resources. Proceedings of the 1980 North American Eel Conference. Ont. Fish. Tech. Rep. Ser. No. 4: vi and 97, pp.4–7.

Edwards, R.J., 1978. The effect of hypolimnion reservoir releases on fish distribution and species diversity. Trans. am. Fish. Soc. 107:71–77.

Efford, I.E., (ed.), 1975. Environmental impact assessment and hydroelectric projects: hindsight and foresight in Canada. J. Fish. Res. Board Can. 32:97–209.

Egouniwe, N., 1976. Public health aspect of tropical water resources development. Water Resour. Bull. 12:393–98.

Eisenhauer, D.E., Manbeck, D.M. & Stork, T.H., 1982. Potential for groundwater recharge seepage from flood-retarding reservoirs in south central Nebraska. J. Soil Water Conservation 37:57–60.

Ellis, M.M., 1941. Freshwater impoundments. Trans. am. Fish. Soc.. 71:80–93.

Fernando, C.H. & Holcik, J., 1982. The nature of fish communities: a factor influencing the fishery potential and yields of tropical lakes and reservoirs. Hydrobiologia 97:127–140.

Fiala, L., 1966. Akinetic spaces in water supply reservoirs. Verh. int. Ver. Limnol. 16:685–92.

Fischer, H.B. & Smith, R.D., 1983. Observations of transport to surface waters from a plunging inflow to Lake Mead. Limnol. Oceanogr.28:258–272.

Fischer, S.G. & LaVoy, A., 1972. Differences in littoral fauna due to fluctuating water levels below a hydroelectric dam. J. Fish. Res. Board Can. 29:1472–1476.

Florman, S.C., 1976. The existential pleasures of engineering. St. Martin's Press, New York. 160 p.

Forbes, S.T., 1887. The lake as a microcosm. Bull. Peoria, Ill., Scientific Association 1887:77–87.

Fowler, D.K. & Whelan, J.B., 1980. Importance of inundation-zone vegetation to white-tailed deer. J. Soil Water Conservation 35:30–33.

Futa, A.B., 1983. Water resource development – organization of a resettlement programme (a case study of the Kpong resettlement programme in Ghana). Water International 8:98–108.

Gagnon, R., 1980. Une digue à des fins écologiques. Eau du Québec 13:62–64.

Geen, G.H., 1974. Effects of hydroelectric development in western Canada on aquatic ecosystems. J. Fish. Res. Bd. Can. 31:913–927.

Geen, G.H., 1975. Ecological consequences of the proposed Moran Dam on the Fraser River. J. Fish Res. Board Can. 32:126–135.

Gill, C.J., 1977. Some aspects of the design and management of reservoir margins for multiple use. Appl. Biol. 2:129–182.

Gill, D., 1971. Damming the Mackenzie: a theoretical assessment of the long-term influence of river impoundment on the ecology of the Mackenzie River Delta. In Proc. Peace-Athabasca Delta Symp. Jan. 14–15, 1971. Univ. Alberta, Edmonton, Alta. 359 p., 204–222.

Gliwicz, A.M. & Biesiadka, E., 1975. Pelagic water mites (Hydracarina) and their effect on the plankton community in a neotropical manmade lake. Arch. Hydrobiol. 76:65–88.

Gloss, S.P. & Wahl, J.R., 1983. Mortality of juvenile salmonids passing through Ossberger crossflow turbines at small-scale hydroelectric sites. Trans. am. Fish. Soc. 112:194–200.

Gloss, S.P., Mayer, L.M. & Kidd, D.E., 1980. Advective control of nutrient dynamics in the epilimnion of a large reservoir. Limnol. Oceanogr. 25:219–228.

Goldman, C.R., 1976. Ecological aspects of water impoundment in the tropics. Revista de Biologia Tropical 24 (supl.1), 87–112.

Gordon, D.C. Jr. & Desplanque, C., 1983. Dynamics and environmental effects of ice in the Cumberland Basin of the Bay of Fundy. Can. J. Fish. aquat. Sci. 40:1331–1342.

Gore, J.A., 1977. Reservoir manipulations and benthic macroinvertebrates in a prairie river. Hydrobiologia 55:113–123.

Gough, D.I. & Gough, W.I., 1970. Load-induced earthquakes at Lake Kariba-II. Geophys. J. Roy. Astron. Soc 21:79–101.

Gray, L.J. & Ward, J.V., 1982. Effects of sediment release from a reservoir on stream macroinvertebrates. Hydrobiologia, 96:177–184.

Gray, T.J. & Gashus, O.K., (ed.), 1972. Tidal Power. Proceedings of an International Conference on the Utilization of Tidal Power, May 24–29, 1970, Halifax, N.S. Plenum Press, New York and London: 630 p.

Greenberg, D., 1977. Effects of tidal power development on the physical oceanography of the Bay of Fundy and Gulf of Maine. In Daborn, G.R., (ed.), Fundy Tidal Power and the Environment. Proc. Workshop on the Environmental Implications of Fundy Tidal Power, Wolfville, N.S., Nov. 4–5, 1976. Acadia Univ. Inst., Wolfville, N.S.: 200–232.

Grelsson, G., 1981. Patterns of bank vegetation and erosion along a North Swedish river barrage reservoir. Wahlenbergia 7:81–88.

Gribbin, J., 1979. Climatic impact of Soviet river diversions. New Scientist, December 6, 1979. pp. 762–765.

Grimard, Y. Jones, H.G., 1982. Trophic upsurge in new reservoirs: a model for total phosphorus concentrations. Can. J. Fish. aquat. Sci. 39:1473–1483.

Hammad, H.Y., 1972. River bed degradation after closure of dams. J. Hydraul. Div. Am. Soc. Civ. Engin. April 1972: pp. 591–605.

Harms, W.R., Schreuder, H.T., Hook, D.D., Brown, C.L. & Shropshire, F.W., 1980. The effect of flooding on the swamp forest in Lake Ocklawaha, Florida. Ecology 6:1412–1421.

Harris, M.D., 1975. Effects of initial flooding on forest vegetation at two Oklahoma lakes. J. Soil Water Conservation 30:294–295.

Hart, D., 1980. The Volta River project. A case study in politics and technology. The University Press, Edinburgh: 131 p.

Harvey, H.H., 1975. Gas disease in fishes – a review. In Adams, W.A., Greer, G., Desnoyers, J.E., Atkinson, G., Kell, G.S., Oldham, K.B. & Walkley, J., (eds.), Chemistry and Physics of Aqueous Gas Solutions. Electrochemical Society, Princeton, NJ: 450–485.

Hasler, A.D., Scholz, A.T. & Horrall, R.M., 1978. Olfactory imprinting and homing in salmon. American Scientist 66: 347–355.

Haynes, J.M. & Gray, R.H., 1980. Influence of Little Goose Dam on upstream movements of adult Chinook Salmon, Oncorhynchus tshawytascha. Fish. Bull. USA 78:185–190.

Healey, M.C., 1979. Detritus and juvenile salmon production in the Nanaimo Estuary: I. Production and feeding rates of juvenile chum salmon (Oncorhynchus keta). J. Fish. Res. Bd. Can. 36:488–496.

Heckey, R.E., Newbury, R.W., Bodaly, R.A., Patalas, K. & Rosenberg, D.M., 1984. Environmental impact prediction and assessment: the Southern Indian Lake experience. Can. J. Fish. Aquat. Sci. 41:720–732.

Heisey, P.G., Mathur, D. & Magnusson, N.C., 1980. Accelerated growth of Smallmouth Bass in a pumped storage system. Trans. am. Fish. Soc. 109:371–77.

Henderson-Sellers, B., 1979. Reservoirs. MacMillan Press Ltd., London and Basingstoke, UK: 128 p.

Hesse, L.W. & Newcomb, B.A., 1982. Effects of flushing Spencer Hydro on water quality, fish, and insect fauna in the Niobrana River, Nebraska. North American J. Fisheries Management 2:45–52.

Hoffman, G.L. & Bower, O.N., 1971. Fish parasitology in water reservoirs: a review. Reservoir Fisheries and Limnology. Am. Fish Soc. Spec. Publ. No. 8, pp. 495–511.

Hoffman, J.I. & Meland, N., 1973. The effect of an artificial lake development complex on the groundwater system. In Groundwater pollution. Underwater Research Institute, St. Louis, Mo.: 1–12.

Hughes, F.M. & Cordes, L.D., 1981. Peace-Athabasca Delta-Wetland in transition. Geographical Magazine 53:890–896.

Hutchinson, G.E., 1957. A treatise on limnology. Vol. 1. John Wiley & Sons, NY: 1015 p.

Hynes, H.B.N., 1969. Life in freshwater communities. In Obeng, L.E., (ed.) Man-made lakes; the Accra Symposium. Ghana Universities Press, Accra, Ghana: 25–31.

Hynes, H.B.N., 1975. The stream and its valley. Verh. int. Ver. Limnol. 19:1–15.

International Joint Commission, 1961. Investigation of the International Passamoquoddy Tidal Power Project. Report of the International Joint Commission Docket 72. Investigations of the International Passamoquoddy Engineering and Fisheries Boards, April 1961, 271 p.

International Symposium on Wave and Tidal Energy, Canterbury, England. Organized by BHRA Fluid Engineering, Cranfield, Bedford, England, September 1978.

Jackson, T.A. & Hecky, R.E., 1981. Depression of primary productivity by humic matter in lake and reservoir waters of the boreal forest zone. Can. J. Fish. aquat. Sci. 37:2300–2317.

James Bay and Northern Quebec Agreement, 1976. Éditeur officiel du Québec, Quebec City, Que. 455 p.

Kashef, A.-A.I., 1981. Technical and ecological impacts of the High Aswan dam. J. Hydrology 53:73–84.

Kaushik, N.K. & Hynes, H.B.N., 1971. The fate of dead leaves that fall into streams. Arch. Hydrobiol. 68:65–515.

Keenham, T., Panu, U.S. & Kartha, V.C., 1982. Analysis of freeze-up ice jams on the Peace River near Taylor, British Columbia. Can. J. Civ. Eng. 9:176–188.

Keith, C.M., Simpson, D.W. & Soboleva, O.V., 1982. Induced seismicity and style of deformation at Nurek Reservoir, Tadjik. SSR. J. Geophysical Research 87(B6): 4605–4624.

Kellerhalls, R. & Gill, D., 1973. Observed and potential downstream effects of large storage projects in northern Canada. In Proc. eleventh meeting of the Commission Internationale des grands barrages, Madrid, Spain: 731–754.

Kelly, D.M., Underwood, J.K. & Thirumurthi, D., 1980. Impact of construction of a hydroelectric project on the water quality of five lakes in Nova Scotia. Can. J. Civ. Eng. 7:173–184.

Kerr, J.A., 1973. Physical consequences of human interference with rivers, pp. 664–696. In Fluvial processes and sedimentation: Proc. hydrology symp., Edmonton, Alta., May 8–9, 1973. Prepared for the Subcommittee on Hydrology by Inland Waters Dir., Environ. Can. 759 p.

Kogan, Sh.I. & Kemzhayev, M.A., 1982. The effect of hydrological factors on overgrowth of the Khauz-Khan reservoir. Hydrobiological J. 18(5):34–38.

Kolenosky, D.P. & Hendry, M.J., 1982. The Canadian Lake Ontario Fishery for American Eel (*Anguilla rostrata*). In Ontario Ministry of Natural Resources. Proceedings of the 1980 North American Eel Conference. Ont. Fish. Techn. Rep. Ser. No. 4: vi and 97 pp. 8–16.

Kovacs, I. & David, L., 1977. Joint use of International Water Resources. Ambio 6:87–90.

Kristensen, J. & Summers, S.A., 1978. Fish populations in the Peace-Athabasca Delta and the effects of water control structures on fish movements. Fisheries and Marine Service Manuscript Report No. 1465. Department of Fisheries and the Environment, Winnipeg, Manitoba, Canada.

Krumholz, L.A., 1981. Abraham Herbert Wiebe (Obituary) Fisheries 6:28–29.

Krzyzanek, E., 1970. Formation of bottom fauna in the Goczalkowice dam reservoir. Acta Hydrobiol. 12:399–421.

Lagler, K.F., 1969. Man-made lakes. Planning and development. FAO, Rome: 71 p.

Lamb, H.H., 1971. Climate-engineering schemes to meet a climatic emergency. Earth-Science Reviews 7:87–95.

Larson, D.W., 1982. Comparison of reservoirs with dissimilar selective withdrawal capabilities: effects on reservoir limnology and release water quality. Canadian Water Resources J. 7(2):90–110.

Lehmkuhl, D.M., 1972. Change in thermal regime as a cause of reduction of benthic fauna downstream of a reservoir. J. Fish. Res. Bd. Can. 29:1329–1332.

Leith, W., Simpson, D.W. & Alvarez, W., 1981.Structure and permeability: geological controls on induced seismicity at Nurek reservoir, Tadjikistan, USSR, Geology 9:440–444.

Lemire, R., 1982. Etude physico-chimique des eaux du Réservoir Desaulnier (Territoire de la Baie James). (Abstract) Canadian Water Resources J. 7(1):452–453.

Liew, P.K.L., 1982. Impact of the eel ladder on the upstream migrating eel (*Anguilla rostrata*) population in the St. Lawrence River at Cornwall: 1974–1978. In Ontario Ministry of Natural Resources. Proceedings of the 1980 North American Eel Conference. Ont. Fish. Tech. Rep. Ser. No. 4 vi and 97. p. 17–22.

Lindström, T., 1973. Life in a lake reservoir. Ambio 2:145–153.

Lodenius, M. & Seppänen, A., 1982. Hair mercury contents and fish eating habits of people living near a Finnish man-made lake. Chemosphere 11:755–759.

Lodenius, M., Seppänen, A. & Herranen, M., 1983. Accumulation of mercury in fish and man from reservoirs in northern Finland. Wat. Air Soil Pollut. 29:237–246.

Low-McConnell, R.H., (ed.), 1966. Manmade lakes. Academic, London: 218 p.

Ly, C.K., 1980. The role of the Akosombo Dam on the Volta River in causing coastal erosion in central and eastern Ghana (West Africa). Marine Geology 37:323–332.

Mackie, G.L., Morton, W.B. & Ferguson, M.S., 1983. Fish parasitism in a new impoundment and differences upstream and downstream. Hydrobiologia 99:197–205.

Mahon, R. & Ferguson, M., 1981. Invasion of a new reservoir by fishes: species composition, growth, and condition. Canadian Field Naturalist 95:272–75.

Margalef, R., 1968. Perspectives in Ecological Theory. Univ. Chicago Press, Chicago: 111 p.

Margalef, R., 1973. Plankton production and water quality in Spanish reservoirs. First report on a research project. Paper prepared for XI Congress, International Commission on Large Dams, Madrid.

Martin, D.B. & Arneson, R.D., 1978. Comparative limnology of a deep-discharge reservoir and a surface-discharge lake on the Madison River, Montana. Freshwat. Biol. 8:33–42.

Martin, W.E., Bollman, F.H. & Gum, R.L., 1982. Economic value of Lake Mead fishery. Fisheries 7(6):20–24.

Mary, M., 1965. Les barrages. Presses Universitaires de France, Paris. 126 p.

Mauboussin, G., 1972. L'usine marémotrice de La Rance. In Gray, T.J. & Gashus, O.K., (ed.). Tidal Power. Proceedings of an International Conference on the Utilization of Tidal Power held May 24–29, 1970, Halifax, Nova Scotia. Plenum Press, New York and London: 630 p, pp. 189–214.

McComas, S.R. & Drenner, R.W., 1982. Species replacement in a reservoir fish community: silverside feeding mechanics and competition. Can. J. Fish. aquat. Sci. 39:815–821.

McLachlan, A.J., 1974. Development of some lake ecosystems in tropical Africa, with special reference to the invertebrates. Biol. Rev. Cambridge Philos. Soc. 49:365–97.

Meister, J.F., DiNunzio, J. & Cox, J.A., 1979. Source and level of mercury in a new impoundment. J. Am. Water Works Assoc. 71:574–76.

Mermel, T.W., 1983. Major dams of the world, 1983. Water Power & Dam Construction. August 1983, pp. 43–49.

Mettam, C., 1978. Environmental effects of tidal power generating schemes. Hydrobiol. Bull. 12:307–321.

Micklin, P.P., 1977. International environmental implications of Soviet development on the Volga River. Human Ecology 5:113–135.

Miller, F.D., Hussein, M., Mancy, K.H. & Hilbert, M.S., 1978. Aspects of environmental health impacts of the Aswan high dam on rural population in Egypt. Prog. Wat. Techn. 11:173–180.

Monosowski, E., 1983. The Tucurui experience. Water Power & Dam Construction. July, 1983, pp. 11–14.

Morduchai- Boltovskoi, F.D., 1961. Die Entwicklung der Bodenfauna in den Stauseen der Wolga. Verh. int. Ver. Limnol. 14:647–51.

Naiman, R.J., 1983. The structure of biotic communities in large rivers: a watershed approach (abstract). 22nd Congress of the International Associations of Limnology: Abstracts p. 252.

Naiman, R.J. & Sibert, J.R., 1979. Detritus and juvenile salmon production in the Nanaimo Estuary: III. Importance of detrital carbon to the estuarine ecosystem. J. Fish. Res. Bd. Can. 36:504–520.

Nault, R., 1980. Surveillance écologique intensive du réservoir de LG 2. Eau du Québec 13(1):33–43.

Neel, J.K., 1966. Impact of reservoirs. Limnology in North America. Frey, D.G., (ed.), pp. 575–93. Univ. Wisconsin Press, Madison: 734 p.

Neu, H.J.A., 1982a. Manmade storage of water resources – a liability to the ocean environment? Part I. Marine Pollution Bulletin 13:7–12.

Neu, H.J.A., 1982b. Manmade storage of water resources – a liability to the ocean environment? Part II. Marine Pollution Bulletin 13:44–47.

New Scientist, 1983. Beavers destroyed European forests. September 22, 1983, p. 849.

Newbury, R.W. & McCullough, G.K., 1984. Shoreline erosion and restabilization in the Southern Indian Lake reservoir. Can. J. Fish. Aquat. Sci. 41:558–565.

Newbury, R.W., Beaty, K.G. & McCullough, G.K., 1978. Initial shoreline erosion in a permafrost affected reservoir, Southern Indian Lake, Canada. In Proc. 3rd Int. Conf. on Permafrost, Edmonton, Alta., July 10–13, 1978. Vol. 1. National Research Council, Ottawa, Ontario:833–839.

Nursall, J.R., 1952. The early development of a bottom fauna in a new power reservoir in the Rocky Mountains of Alberta. Can. J. Zool. 30:387–409.

Nursall, J.R., 1969. Faunal changes in oligotrophic man-made lakes: experience on the Kananaskis River system. In L.E. Obeng (ed), Man-made lakes: the Accra Symposium Ghana Universities Press, Accra, Ghana: 163–175.

Obeng, L.E., (ed.), 1969. Man-made lakes: the Accra Symposium. Ghana Universities Press. Accra, Ghana: 398 p.

Odum, E.P., 1969. The strategy of ecosystem development. Science. 164:262–70.

Odum, E.P., 1971. Fundamentals of ecology. 3rd ed. W.B. Saunders, Philadelphia, Pa.: 574 p.

Oglesby, G.B., Noell, W.C., Delo, H.O., Dyer, W.A., England, R.H., Fatora, J.R., Grizzle, J.M. & Deutsch, S.J., 1978. Toxic substances in discharges of hypolimnetic waters from a seasonally stratified impoundment. Environ. Conserv. 5:287–293.

Oklahoma Geological Survey. 1976. Oklahoma Reservoir Resources. Okla. Geol. Surv., Oklahoma Acad. Sci. Publ. No. 5. Norman, Okla: 151 p.

Ostrofsky, M.L. & Duthie, H.C., 1975. Primary productivity, phytoplankton, and limiting nutrient factors in Labrador lakes. Int. Revue ges. Hydrobiol. Hydrogr. 60:145–58.

Ostrofsky, M.L. & Duthie, H.C., 1980. Trophic upsurge and the relationships between phytoplankton biomasss and productivity in Smallwood Reservoir, Canada. Can. J. Bot. 58:1174–1180.

Pandian, T.J., 1980. Impact of dam-building on marine life. Helgoländer Meeresuntersuchungen 33:415–421.

Patrick, P.H., Sheehan, R.W. & Sim, B., 1982. Effectiveness of a strobe light eel exclusion scheme. Hydrobiologia 94:269–277.

Peace-Athabasca Delta Project Group. 1972. The Peace-Athabasca Delta, a Canadian resource. Prepared for the Canada, Alberta, and Saskatchewan environment ministries. 144 p.

Peace-Athabasca Delta Project Group, 1973. The Peace-Athabasca Delta Project. Technical report. Prepared for the Canada, Alberta, and Saskatchewan environment ministries. 176 p.

People, W. & Rogoyska, M.S., 1969. The effect of the Volta River hydroelectric project on the salinity of the Lower Volta River. Ghana J. Sci. 9:9–20.

Perreault, R., 1980. Eménagement des habitats riverains du réservoir du LG 2. Eau du Québec 13(1):54–60.

Peters, J.C., 1979. Modification of intakes at Flaming Gorge Dam, Utah, to improve water temperature in the Green River. In Proc. Int. Symp. on Environmental Effects of Hydraulic Engineering Works, Sept. 12–14, 1978, Knoxville, Tenn: 295–304.

Petr. T., 1970. Macroinvertebrates of flooded trees in the man-made Volta lake (Ghana) with special reference to the burrowing Mayfly Povilla adusta Navas. Hydrobiologia 36:373–98.

Petr. T., 1975. On some factors associated with the initial high fish catches in new African man-made lakes. Arch. Hydrobiol. 75:32–49.

Petr., T., 1978. Tropical man-made lakes – their ecological impact. Arch. Hydrobiol., 81:368–385.

Petts, G.E., 1980a. Long-term consequences of upstream impoundment. Environmental Conservation 7:325–332.

Petts, G.E., 1980b. Implications of the fluvial process – channel morphology interactions below British reservoirs for stream habitats. Science of the Total Environment 16:149–163.

Pfitzer, D.W., 1967. Evaluation of tailwater fishery resources resulting from high dams. In Reservoir Fishery Resources Symposium. American Fisheries Society, Washington, DC. 569 p, p. 477–488.

Phillips, D.W. & McCulloch, J.A.W., 1972. The climate of the Great Lakes basin. Atmos. Environ. Serv. Climatol. Stud. 20:40 p.

Phillips, D.W. & Irbe, J.G., 1978. Lake to land comparison of wind, temperature and humidity on Lake Ontario during the International Field Year for the Great Lakes. Atmos. Environ. Serv. CLI–2–77:51 p.

Pinel-Alloul, B., Magnin, E. & Codin-Blumer, G., 1982. Effets de la mise en eau du réservoir Desaulniers (Territoire de la Baie de James) sur la zooplancton d'une rivière et d'une tourbière reticulée. Hydrobiologia 86:271–296.

Potter, D.U. & Meyer, J.L., 1982. Zooplankton communities of a new pumped storage reservoir. Water Resources Bull. 18:635–642.

Potter, D.U., Stevens, M.P. & Meyer, J.L., 1982. Changes in physical and chemical variables in a new reservoir due to pumped storage operations. Water Resources Bull. 18:627–633.

Priscou, J.C., Verduin, J. & Deacon, J.E., 1982. Primary productivity and nutrient balance in a lower Colorado River reservoir. Arch. Hydrobiol. 94:1–23.

Purcell, L.T., 1939. The aging of reservoir waters. J. Am. Water Works Assoc. 31:1775–1806.

Raymond, H.L., 1979. Effects of dams and impoundments on migrations of juvenile Chinook salmon and Steelhead from the Snake River, 1966–1975. Trans am. Fish. Soc. 108:505–529.

Razumov, G.A. & Medovar, Y.A., 1980. Formation of an artificial aquifer in the shore area of a reservoir. Water Quality Bulletin 5(4):84–95, 105.

Rees, C.P., 1980. Guidelines for environmental impact assessment of dam and reservoir projects. Prog. Wat. Tech. 13:57–71.

Richards, T.L., 1969. The Great Lakes and the weather. In Anderson, D.V. (ed.), The Great Lakes as an Environment. Great Lakes Inst. Rep. PR 39:189 + cvii p.

Richardson, L.R., 1934. Observations on the effects of dams on lakes and streams. Trans. am. Fish. Soc., 64:457–58.

Ridley, J.E. & Steel, J.A., 1975. Ecological aspects of river impounments. In Whitton, B.A., (ed.), River Ecology. Univ. California Press, Berkeley: 565–587.

Rinne, J.N., Minkley, W.L. & Bersall, P.O., 1981. Factors influencing fish distribution in two desert reservoirs, Central Arizona. Hydrobiologia 80:31–34.

Ripert, C., Same-Ekobo, A., Enyong, P. & Palmer, D., 1979. Évaluation des répercussions sur les endémies parasitaires (malaria, bilharziose, onchercercose, dracunculose) de la construction de 57 barrages dans les monts Mandara (Nod-Cameroun). Bulletin de la Societé de Pathologie Exotique 72:324–339.

Rosenberg, D.M., Bilyj, B. & Wiens, A.P., 1984. Chironomidae (Diptera) emerging from the littoral zone of reservoirs, with special reference to Southern Indian Lake, Manitoba. Can. J. Fish. Aquat. Sci. 41:672–681.

Rouse, H., 1980. Hydraulic structures and/or the environment? CRC Critical Reviews in Environmental Control 10:271–77.

Roy, D., 1982. Répercussions de la coupure de la Grande Rivière à l'aval de LG2. Nat. can. 109:883–891.

Ruggles, C.P., 1980. A review of the downstream migration of Atlantic Salmon. Can. Techn. Rep. Fish. Aquat. Sci. No. 952 IX + 30 p. Dept. of Fisheries and Oceans, Halifax, Nova Scotia, Canada.

Ryder,R.A., 1978. Ecological heterogeneity between north-temperate reservoirs and glacial lake systems due to differing succession rates and ultural uses. Verh. int. Ver. Limnol. 20:1568–1574.

Second International Symposium on Wave & Tidal Energy, Cambridge, England. Organized by HBRA Fluid Engineering, Cranfield, Bedford, England, September 1981.

Sérodes, J.B., 1982. Demande en oxygène des sols et arbres noyés du réservoir La Grande 2, baie James. Nat. can. 109:857–867.

Serruya, C. & Pollingher, U., 1983. Lakes of the warm belt. Cambridge University Press.

Severn, R.T., Dineley, D.L. & Hawker, L.E., (eds.), 1979. Tidal Power and Estuary Management. Bristol, Scientechnica. 296 p.

Shaheen, A.H. & Yosef, S.F., 1979. The effect of the cessation of Nile flood on the fishery of Lake Manzala, Egypt. Arch. Hydrobiol 85:166–191.

Sheer, D.P. & Harris, D.C., 1982. Acidity control in the North Branch Potomac. J. Wat. Pollut. Cont. Fed. 54:1441–146.

Sibert, J.R., 1979. Detritus and juvenile salmon production in the Nanaimo Estuary: II. Meiofauna available as food to juvenile chum salmon (*Oncorhynchus keta.*) J. Fish. Res. Bd. Can. 36:496–503.

Simpson, D.W., 1976. Seismicity changes associated with reservoir loading. Eng. Geol. 10:123–150.

Simpson, D.W., Hamburger, M.W., Pavlov, V.D. & Nersesov, I.L., 1981. Tectonics and seismicity of the Toktogul reservoir region, Kirgizia, USSR. J. Geophysical Research 86B:345–358.

Smith, H.A., 1974. Spillway redesign abates gas super-saturation in Colombia River. Civ. Eng. (NY) Sept. 1974, pp. 70–73.

Smith, P.J., 1982. Reservoirs and the triggering of earthquakes. Nature 295:9.

Smith, P.M., 1976. Spillway modification to reduce gas supersaturation. In Symp. on Inland Waterways for Navigation, Flood Control, and Water Diversions. Vol. 1. American Society of Civil Engineers, New York, NY: 667–671.

Soltero, R.A., Gasperino, A.F. & Graham, W.G., 1974. Chemical and physical characteristics of a eutrophic reservoir and its tributaries: Long Lake, Washington. Water Res. 8:419–431.

Spence, J.A. & Hynes, H.B.N., 1971a. Differences in benthos upstream and downstream of an impoundment. J. Fish. Res. Bd. Can. 28:35–43.

Spence, J.A. & Hynes, H.B.N., 1971b. Differences in fish populations upstream and downstream of a mainstream impoundment. J. Fish. Res. Bd. Can. 28:45–46.

Steane, M.S. & Tyler, P.A., 1982. Anomalous stratification behaviour of Lake Gordon, headwater reservoir of the Lower Gordon River, Tasmania. Aust. J. mar. Freshwat. Res. 33:739–760.

Stein, S., Wiens, D.A. & Fujita, K., 1982. The 1966 Kremasta reservoir earthquake sequence. Earth & Planetary Science Letters 59:49–60.

Surette, R., 1983. Bay of Fundy full of surprises. Canadian Geographic, Oct./Nov. 1983, pp.70–77.

Sutcliffe, W.H., Jr., 1973. Correlations between seasonal river discharge and local landings of American lobster (*Homarus americanus*) and Atlantic halibut (*Hippoglossus hippoglossus)* in the Gulf of St. Lawrence. J. Fish. Res. Bd. Can. 30:856–59.

Szekely, F., 1982. Environmental impact of large hydroelectric projects on tropical countries. Water Supply & Management 6:233–242.

Talling, J.F., 1980. Some problems of aquatic environments in Egypt from a general viewpoint of Nile ecology. Water Supply & Management 4:13–20.

Taylor, K.V., 1978. Erosion downstream of dams. In Environmental Effects of Large Dams. Report by the Committee on Environmental Effects of the United States Committee on Large Dams. American Society of Civil Engineers, New York: 225 p.

Tessier, A., 1980. La SOTRAC et la mise en eau du réservoir de LG2. Eau du Québec 13(1):30–32.

Tolmazin, D., 1979. Black Sea – dead sea? New Scientist, December 6, 1979. pp. 766–769.

Trotzky, H.M. & Gregory, R.W., 1974. The effects of water flow manipulation below a hydro-electric power dam on the bottom fauna of the upper Kennebec River, Maine. Trans. am. Fish. Soc. 103:318–324.

Tyler, P.A. & Buckney, R.T., 1974. Stratification and biogenic meromixis in Tasmanian reservoirs. Aust. J. mar. Freshwat. Res. 25:299–313.

Van Coillie, R., Visser, S.A., Campbell, P.G.C. & Jones, H.G., 1983. Evaluation de la dégradation du bois de conifères immergés durant plus d'un demi-siècle dans un réservoir. Annls. Limnol. 19:129–134.

Van Houweninge, G. & De Graauw, A., 1982. The closure of tidal basins. Coastal Engineering 6:331–360.

Verdon, R. & Demers, C., 1982. Les répercussions environmentales associées aux réservoirs de la Côte-nord de l'estuaire du Saint-Laurent. Eau du Québec: 15:221–227.

Vogt, H., 1978. An ecological and environmental survey of the humic man-made lakes in Finland. Aqua Fenn. 8:12–24.

Vowinckel, E. & Orvig, S., 1980. Meteorological consequences of natural or deliberate changes in the surface environment (with particular reference to the James Bay region). Polar Geography and Geology 4(2):87–110.

Waddy, B.B., 1975. Research into the health problems of man-made lakes, with special reference to Africa. Trans. R. Soc. Trop. Med. Hyg. 69:39–50.

Waite, D.T., Dunn, G.W. & Stedwill, R.J., 1980. Mercury in Cookson Reservoir (East Poplar River). Saskatchewan Environment, Water Pollution Control Branch WPC-23.

Walburg, C.H., Novotny, J.F., Jacobs, K.E., Swink, W.D.,

Campbell, T.M., Nestler, J. & Saul, G.E., 1981. 'Effects of Reservoir Releases on Tailwater Ecology: A Literature Review,' Technical Report E–81–12, prepared by US Department of the Interior, Fish and Wildlife Service, National Reservoir Research Program, East Central Reservoir Investigations, and Environmental Laboratory, US Army Engineer Waterways Experiment Station, for the US Army Engineer Waterways Experiment Station, CE, Vicksburg, Miss.

Waton, S., 1981. Egypt after the Aswan Dam. Environment 34(4):30–36.

Ward, J.V., 1976. Comparative limnology of differentially regulated sections of a Colorado mountain river. Arch. Hydrobiol. 78:319–342.

Ward, J.V., 1976. Effects of flow patterns below large dams on stream benthos: a review. In Orsborn, J.F. & Allman, C.H., (eds.), Instream Flow Needs Symp. Am. Soc.: 235–253.

Water Power and Dam Construction, 1983. Reversing the flow of Soviet rivers. (35(5), May, 1983, pp. 53–57.

Wiebe, A.H., 1939. Density currents in Norris Reservoir. Ecology, 20:446–450.

Wiebe, A.H., 1960. The effects of impoundments upon the biota of the Tennessee River system. Proc. 7 th Tech. Meet. of the International Union for the Conservation of Nature and Natural Resources. Athens, Sept. 11–19, 1958. IV:101–117.

Wood, E.A., 1975. Science from your airplane window. Dover Publications, New York: 227 p.

Wood, R.B., Prosser, M.V. & Baxter, R.M., 1976. The seasonal pattern of thermal characteristics of four of the Bishoftu crater lakes, Ethiopia. Freshwat. Biol. 6:519–530.

Worthington, E.B., 1978. Some ecological problems concerning engineering and tropical diseases. Prog. Wat. Techn. 11:5–11.

Wright, J.C., 1967. Effect of impoundments on productivity, water chemistry, and heat budgets of rivers. In Reservoir Fisheries Resources Symposium. University of Georgia, Athens, Ga.; 188–199.

Yeo, R.K. & Risk, M.M., 1979. Fundy tidal power: environmental sedimentology. Geoscience Canada 6:115–121.

Zhadin, V.I. & Gerd, S.V., 1963. Fauna and flora of rivers, lakes and reservoirs of the USSR. Jerusalem: Israeli Program Sci. Transl. 626 p.

Author's address:
R.M. Baxter
National Water Research Institute
P.O. Box 5050
Burlington, Ontario
Canada L7R 4A6

CHAPTER 2

Characterization of the reservoir ecosystem

ROBERT H. KENNEDY, KENT W. THORNTON and DENNIS E. FORD

Abstract. Reservoirs are valuable environmental resources, the management of which will require a sound understanding of processes influencing their water quality. While these aquatic ecosystems are formally defined as lakes, the great importance of riverine flows, and a larger basin and watershed size set them apart from the small natural lakes upon which much of our understanding of lake water quality processes is based.

Reservoirs are greatly influenced by tributary inflows and their water quality conditions reflect geographic, climatic, and watershed characteristics. The occurrence of density currents, sedimentation and mixing, and the fact that reservoirs often exhibit a long, narrow morphology frequently lead to the establishment of distinct longitudinal gradients in water quality characteristics.

Described in this chapter are processes and physical characteristics of importance in the establishment of longitudinal gradients in reservoirs. Only through an understanding of these processes and characteristics can sound management strategies for reservoirs be developed and successfully implemented.

Introduction

Reservoirs provide an economic, recreational, and environmental resource of ever increasing value. Although once constructed primarily for flood control, power generation, navigation, and water supply, reservoirs now meet a number of other needs including flow augmentation, wildlife and fish habitat, recreation, and water quality management. Their increased value and public appeal is linked to an increased demand for water-based recreational sites and a progressive deterioration in the quality of natural lakes readily accessible to population centers. This has prompted a heightened awareness of reservoir water quality problems and fostered interest in the development of management strategies for reservoirs and reservoir-systems. However, if sound management strategies are to be formulated and successfully implemented, our un-

derstanding of these important aquatic ecosystems must be improved.

Constructed by man to store, divert or otherwise control the flow of streams and rivers, reservoirs are geologically younger than natural lakes. Reservoirs are typically constructed along narrow and steeply-sloping river basins and, therefore, exhibit markedly different hydrologic and morphologic characteristics than do lakes of natural origin and longer geologic history (Thornton *et al.*, 1981). A more complex shape and more developed shoreline for reservoirs reflect the topography of the pre-impoundment river basin.

Pre-impoundment features also influence the hydrologic character of reservoirs. Unlike natural lakes, which often receive inflows from relatively diffuse sources, reservoirs commonly receive a majority of their water and material loads via a single, large tributary entering at a point distant from the

Gunnison, D. (ed.) Microbial Processes in Reservoirs.
© 1985, Dr W. Junk Publishers, Dordrecht, Boston, Lancaster. ISBN 90 6193 751 5.

impounding dam. Together these characteristics influence the distribution of influent materials and provide a physical setting for the establishment of longitudinal patterns in limnological conditions (Thornton *et al.*, 1981).

Reservoirs also differ with respect to the manner in which water is discharged. While natural lakes discharge surface water by uncontrolled overflow, the withdrawal of water from reservoirs commonly occurs by the controlled use of submerged gates located at various depths in the water column. The limnological implications of this are great. For example, surface withdrawal from stratified lakes results in the storage of material and the dissipation of heat, while bottom withdrawal from reservoirs results in the dissipation of material and the storage of heat (Martin & Arneson, 1978; Wright, 1967). Since the storage of heat reduces stability, the potential for wind-generated mixing is increases. This, in turn, may lead to increased nutrient recycling (Gaugush, 1984).

Although ultimately governed by identical physical, chemical, and biological processes, geologic history and setting, morphology, and hydrology may exert differing degrees of influence over the quality of waters impounded in reservoirs and natural lakes. This chapter explores these differing influences and their potential impact on the limnological character of reservoirs.

Geographic and climatic considerations

Distinct patterns in the world-wide distribution of lakes are apparent (Schuiling, 1976). Maxima in the percentage of land surface covered by lakes occur at 40–55 degrees latitude in both the southern and northern hemispheres; a third maxima is observed near the equator. Similar patterns are observed for natural lakes of the United States, but not for reservoirs.

A comparison of natural lakes and US Army Corps of Engineers (CE) reservoirs sampled during the National Eutrophication Survey (NES) indicates a bimodel distribution for natural lakes and a nearly unimodel distribution for CE reservoirs (Walker, 1981) (Fig. 1). While natural lakes pre-

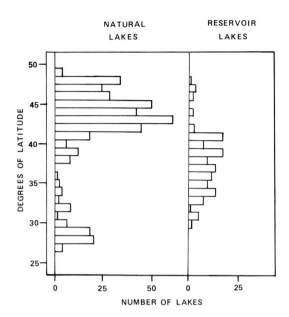

Fig. 1. Distribution of natural lakes and US Army Corps of Engineer reservoirs. Based on data reported in Walker (1981).

dominate in the northern, glaciated portion of the United States and in the extreme south (i.e. the Florida solution lakes), reservoirs are most abundant in the middle latitudes of the United States. The small secondary mode in the reservoir distribution (at approximately 45 degrees N latitude) represents reservoirs on the Missouri and Columbia River systems and several small New England flood control reservoirs. Thus, the maximum in the distribution of CE reservoirs coincides with the minimum in the distribution of natural lakes in the United States.

These geographic differences in the distribution of reservoirs and natural lakes are related to patterns in geology, climatic conditions, and water requirements. In general, reservoirs are located in regions where (1) water excesses result in flooding or (2) water shortages, either seasonal or longterm, require the storage of water. Natural lakes, on the other hand, are most prevelant in regions where adequate and relatively stable water supplies exist. In the eastern United States, for example, precipitation generally exceeds evaporation, water supply is plentiful, and lakes are abundant. In the western United States, evaporation exceeds precipitation, water is generally scarce, and reservoirs are prevalent.

Geographic location plays an important role in determining limnological conditions in lakes and reservoirs. Climatic and geologic patterns lead to differences in the rates at which chemical and mechanical weathering occur, and thus, the rates at which materials are transported to lakes and reservoirs. Geologic patterns also determine the quality of these materials. In turn, the quantity and quality of materials transported from watershed to lake or reservoir dictate general water qualty conditions (Gibbs, 1970). In regions in which evaporation exceeds precipitation, evaporative losses from impounded surface waters lead to increases in salinity and potential changes in plant and animal species composition. Regional patterns in the occurrence of extreme precipitation events influence regional patterns in hydrology. Since they are constructed in areas where extreme flows must be impounded or otherwise controlled,reservoirs may experience higher and more variable flushing rates than do natural lakes.

A large number of reservoirs have been constructed in warm, wind-swept areas of the midwestern and southwestern United States where topographic relief is moderate to low. These conditions foster weak thermal gradients and the frequent occurrence of mixing during stratified periods. Because of this, and high rates of erosion in the surrounding watershed, many of these reservoirs are periodically or perpetually turbid.

Thus, geographic location and climatic condition play important roles in determining the limnological characteristics of lakes and reservoirs. Therefore, it is noteworthy that reservoirs and natural lakes exhibit differing geographic and climatic distributions. These differences may help to explain the differing importance of various processes influencing water quality in reservoirs.

Reservoir-watershed interactions

Lakes and reservoirs are dependent upon their surrounding watershed for the supply of water and material, and it is through this linkage that potential water quality characteristics are established. As previously mentioned, lakes and reservoirs often differ in the number and spatial distribution of tributary streams. This reflects differing watershed geometries and dissimilarities in the origin of lakes and reservoirs. Natural lakes occupy depressions in the local topography and thus, are centrally located in relatively symmetrical drainage basins. Tributary streams enter at several points around the periphery of the lake and a large percentage of the drainage area is contiguous with the lake. Reservoirs, on the other hand, are designed to control flow and thus, are located at the downstream boundary of major drainage basins. Because of this, reservoir drainage basins are narrow and elongated, and only a small percentage is contiguous with the reservoir. In most instances, a single, large tributary supplies a majority of the annual water and material inputs.

Reservoir watersheds are larger than those for natural lakes. A comparison of 309 natural lakes and 107 reservoirs (Thornton et al., 1981) indicates that the average size of a reservoir watershed is approximately an order of magnitude larger than that of a natural lake. Since watershed area is related to annual discharge, reservoirs have considerably greater water loads than lakes. For example, a consideration of the exponential relationship between drainage area and flow (Chow, 1964) explains the three-fold higher areal water load for reservoirs than for natural lakes.

Contrasts in watershed sizes and geometries, and differences in stream order may also lead to differences in the quantity and quality of material loads to lakes and reservoirs. Higher order streams are fed by a network of lower order streams, so loads transported by a reservoir's major tributary are potentially exposed to a high degree of chemical, physical, and biological processing. In addition to natural inputs of nutrients and organic matter, larger streams and rivers are more likely to receive direct inputs from industrial and municipal point-sources. Streams tributary to natural lakes are generally of lower stream order and the processing of materials in these streams would be limited by stream length. A higher percentage of contiguous watershed area would imply that direct inputs would also have a greater impact on the material budgets of lakes than on those of reservoirs.

Considerations of the river continuum concept (Vannote *et al.*, 1980) would imply differences in the contribution of various forms of organic carbon supplied to lakes and reservoirs. While there is a continuing discussion of the importance of heterotrophy vs autotrophy in stream systems (Cummins, 1974; Fisher & Likens, 1973; Minshall, 1978; Naiman & Sedell, 1980; Vannote *et al.*, 1980), there is general agreement that autotrophic production increases with stream order, at least up through ninth order streams (Naiman & Sedell, 1981). A positive correlation between watershed area or wetted upstream bottom area and the export of benthic algae was found for several Iowa streams (Swanson & Bachmann, 1976). Minshall (1978) also indicated that autotrophic production was an important contributor to the organic carbon pool in the larger streams of forested areas and in streams of all sizes in arid and semiarid climates. The location of these latter streams generally corresponds with the distribution of many reservoirs in the western and southwestern United States.

Sediment, particulate organic matter and absorbed constituents are transported primarily during storm events or elevated flows (Bilby & Likens, 1979; Johnson *et al.*, 1976; Kennedy *et al.*, 1981; Sharpley & Seyers, 1979; Verhoff & Melfi, 1978). For large streams and rivers, this transport may occur through a series of storm events with periods of deposition and material processing occurring between events (Verhoff & Melfi, 1978); Verhoff *et al.*, 1979). Since fine particulate organic matter (FPOM) is more susceptible to transport than is coarse particulate organic matter (POM) (Bilby & Likens, 1979), and since stream processing increases the concentration of FPOM, reservoirs may receive relatively higher proportions of FPOM and dissolved organic matter than to natural lakes. Terrestrial contributions to the FPOM pool decrease with increases in stream order (Naiman & Sedell, 1979), so a large percentage of the FPOM loaded to reservoirs may be of autotrophic origin.

As would be expected from a comparison of watershed areas, nutrient and sediment loads are higher for reservoirs than for natural lakes. Nutrient loading estimates for lakes and reservoirs sampled during the NES indicate average phosphorus and nitrogen loading rates for reservoirs which are 95 and 55 percent higher, respectively, than for natural lakes surveyed (Thornton *et al.*, 1981). Reservoirs also experience high loading rates for sediment and organic matter (Groeger & Kimmel, 1984). Since sediment carrying capacity is high for large streams and rivers, many reservoirs located on major drainages receive heavy sediment loads, which when deposited lead to losses in storage volume and the creation of extensive deltas in the upstream portions of the reservoir.

Inflow mixing processes

The characteristics of the interface between tributary stream and receiving lake or reservoir play a critical role in determining the ultimate impacts of water and material loads. As discussed above, natural lakes often receive inputs from several small and diffuse sources, while reservoirs commonly receive these inputs from a single, large tributary. For small natural lakes, this means that influent water and materials enter the lake through shallow littoral areas where chemical exchanges with sediments, biological uptake, and/or sedimentation potentially modify the quantity and quality of material loads. In areas where macrophytic plants are abundant, flows may be greatly reduced. Reservoir tributaries, on the other hand, are larger, exhibit higher flows, and tend to retain their riverine characteristics for considerable distances into the reservoir.

Hydraulic interactions between tributary and reservoir water masses result in the establishment of currents which often play an important role in the initial transport and distribution of influent materials. The characteristics of these currents are dependent upon basin morphology, tributary flow, and the relative densities of the interacting water masses.

Density differences between tributary and reservoir water, which commonly arise from differences in temperature and/or dissolved and suspended solid concentrations, dictate the vertical placement of inflowing tributary waters. As a tributary inflow enters a reservoir, it displaces reservoir water ahead of it until increases in water depth and basin cross-sectional area result in decreased flow velocities. If there is no density difference between tributary and reservoir waters, flows progress through upstream portions of the reservoir as a plug flow. However, as in open channel flow, the point of maximum velocity remains near the water surface. Mixing resulting from this interaction occurs throughout the water column.

In most instances, however, tributary and reservoir waters differ in density. Depending on the magnitude of this difference, density currents can enter the epilimnion, metalimnion, of hypolimnion (Fig. 2). When the inflow density is less than the surface water density, the inflow will float on the water surface as an overflow. In an overflow, excess hydrostatic pressure in the density current causes the current to flow in all directions not obstructed by boundaries. Because of their near-surface location, overflows are susceptible to mixing from wind-induced mechanisms and diel heating and cooling processes. In many cases, vertical mixing resulting from either wind stress or heat loss, can weaken the density difference. This may result in the mixing of overflowing waters throughout the water column and thus a weakening of the density current.

If the density of the inflow is greater than that of the reservoir's surface waters, reservoir waters will be displaced until the buoyancy forces cause inflowing waters to plunge beneath the less dense surface waters. The location of the point at which this occurs, termed the plunge point, is determined by a balance between stream momentum, the pressure gradient across the interface separating tributary and reservoir waters, and resisting shear forces. Changes in the bed slope, bed friction, and cross-sectional area may also affect its location.

The potential for mixing at the plunge point is significant because of the large eddies formed by flow reversals and the pooling of the inflowing water (Akiyama & Stefan, 1981). Ford & Johnson (1983) estimated this mixing to be on the order of 25 percent of the inflow volume. Knapp (1942) noticed that the flow in the vicinity of the plunge-point occurred at the bottom of this pooled mixing zone. Ford *et al.* (1980) and Kennedy *et al.* (1983)

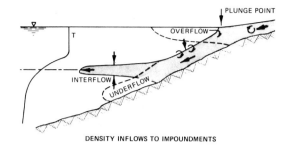

Fig. 2. Density inflows to reservoirs.

substantiated this pooling phenomenon during dye studies on DeGray Lake, Arkansas, and West Point Reservoir, Georgia, respectively, when the dye clouds appeared to have stalled at the plunge point.

After the inflow plunges, it follows the old river channel as an underflow. The speed and thickness of the underflow is determined by a flow balance between the shear forces and the acceleration due to gravity. An underflow will entrain overlying reservoir water due to shear and turbulance generated by bottom roughness. When densities throughout the water column are lower than that of the underflowing water, the density current may continue to propagate downslope until it reaches the impounding dam.

An interflow or intrusion occurs when a density current leaves the river bottom and propagates horizontally into a stratified body of water. Intrusions differ from overflows and underflows because an intrusion moves through a reservoir at an elevation where the intrusion and reservoir densities are similar. Intrusions require a continuous inflow and/or outflow for movement, or they stall and collapse (i.e., dissipate). Turbulence is usually quickly dissipated in an intrusion since the metalimnetic density gradient creates strong buoyancy forces which inhibit mixing. Mixing still occurs, however, because of the flow gradient.

Material transport

The transport and initial distribution of influent nutrients, organic matter, and suspended sediments, both of which are greatly influenced by hydraulic interactions between tributary and reservoir, have important implications for the establishment of spatial patterns in material deposition and in heterotrophic and autotrophic activity. While it is often assumed that material inputs are instantaneously and completely mixed within the receiving lake, the observation of riverine conditions in headwater areas of reservoirs and the establishment of well-defined density currents suggest that such views are erroneously over-simplistic.

Patterns in the distribution of influent materials can be predicted based solely on considerations of flow and basin morphology. Since carrying capacity is related to flow velocity and turbelence, suspended particulate concentrations are potentially highest in the vicinity of the tributary inflow. This is also true for nutrients, metals, organic matter and contaminants because of their association with organic and inorganic particulates. However, as water column depths and basin widths increase, decreases in flow velocity and turbulance reduce carrying capacity and particulate concentrations are reduced due to the effects of sedimentation. For narrow, shallow basins strongly influenced by riverine currents, this often occurs at locations well within the reservoir. In such basins, particulate concentrations remain relatively high in the riverine portion of the reservoir. For broad basins in which water column depth increases rapidly downstream from the tributary inflow, concentrations decline more sharply and resultant gradients in particulate concentrations are most pronounced in extreme upstream portions of the reservoir.

The presence and eventual loss of particulate material from the water column have several direct and indirect effects on the physical and chemical characteristics of the reservoir and on biological activity. High concentrations of inorganic suspended material reduce water clarity and limit light availability for algal growth. The accumulation of deposited material leads to losses in reservoir volume and the creation of submerged deltas. The deposition of oxygen-demanding organic materials reduces oxygen stores in upstream portions of the reservoir hypolimnion, and nutrients, metals, and contaminants are potentially exchanged with the overlying water column. In areas protected from the direct influence of flow, extensive growths of macrophytic plants may develop. This portion of the reservoir is, therefore, most impacted by tributary loads. It is also processes occuring in this reach of the reservoir which act to modify both the quality and quantity of these loads.

Density currents have several impacts on limnological conditions within the receiving reservoir. During periods of overflow, material inputs enter the upper, well-lighted portion of the water column and are thus potentially available for biological uptake and utilization. However, during periods of interflow or underflow, upper, more productive strata are deprived of external sources of materials. Under such circumstances, algal populations must rely on recycling for their supply of required nutrients.

Underflows and interflows may contribute to or alleviate the development of anoxia. Underflows containing high organic matter concentrations may, through microbial decomposition, deplete hypolimnetic oxygen supplies, particularly in the upstream portion of the reservoir where hypolimnetic volume is small. Underflows may also displace anoxic water further into the reservoir, transport resolubilized constituents such as phosphorus, iron, manganese, and lower molecular weight organic compounds, and promote additional hypolimnetic anoxia further into the reservoir. However, oxygenated underflows may reduce anaerobic conditions and lead to the oxidation and precipitation of reduced iron and manganese. Interflows may contribute to metalimnetic oxygen minima by introducing organic matter to the metalimnion. High nutrient interflows may be entrained into the mixed layer during hydrometeorological events and initiate plankton blooms.

Density currents can also produce flow reversals in the reservoir that cause constituents to be transported upstream. When the inflow plunges beneath the surface and enters the reservoir as an interflow or as an underflow, some of the reservoir water is entrained and a counteracting circulation pattern is set up in the overlying surface waters. This reverse flow pattern transports floating debris upstream to the plunge point where it remains at the surface.

Reservoir mixing

In both lakes and reservoirs, mixing results from the cumulative effects of differential surface heat exchange, absorption of solar radiation with depth, wind magnitude and direction, inflow magnitude, density and location, outflow magnitude and location, and changes in project operation (i.e., withdrawal depth, pool level changes etc.). Mixing is therefore dynamic since its effectiveness varies in response to these dynamic forcing events.

Wind is the major source of energy for many physical phenomena which either directly or indirectly cause mixing. When the wind blows across a water surface, it creates a shear stress which transmits energy to the water body. The fraction of the wind shear that is used to generate currents is unknown. It is, however, usually assumed that the surface current speed is 1 to 3 percent of the wind speed. Bengtsson (1978) and others have found that: (1) the percentage is not a constant even for a specific case; (2) the percentage decreases with increasing wind speed; and (3) the percentage increases with increasing lake dimensions.

In reservoirs, the existence of well defined inflow density currents (Ford & Johnson, 1983), withdrawal zones and possibly pumpback jets (Roberts, 1981) probably significantly perturb classical wind-generated circulation patterns but little is known of these interactions. In addition, reservoir operating procedures that result in pool level fluctuations also generate currents that interact with the wind-generated current field. These forces (i.e., inflow, outflow, pumpback) may result in the formation of internal waves.

All stably stratified fluids possess the ability to sustain internal wave motion. The waves can be generated locally by turbulence or externally by an outside pertubation such as hydropower operation. Internal waves transport momentum and can exist without breaking and forming turbulence. It is only after they break that turbulence is generated and mixing is possible.

Internal waves impact the mixing regime because they radiate energy away from the place where they were generated. They can reduce entrainment into the mixed layer by creating a local energy loss from the mixed layer (Kantha et al., 1977). Internal waves can also increase mixing by transporting energy into a region where it would not otherwise be available.

Seiche activity is another mechanism that may result in reservoir mixing. When a steady wind blows across a water surface, water is transported to the windward side of the lake and the water surface is tilted by the wind force. When the wind force is removed, the potential energy associated with the tilting of the water surface and the epi-hypolimnetic interface is converted into kinetic energy and flow. The result is an oscillating motion called seiching which decays with time. A seiche is one indirect mechanism for energy from the wind to become available for mixing in the hypolimnion of a lake.

These mixing mechanisms can significantly influence water quality and microbial processes in reservoirs. The seiche and associated upwelling represent one mechanism where metalimnetic and hypolimnetic waters are temporarily exposed to the surface waters or epilimnion. Addition of this nutrient rich water to the euphotic zone may stimulate plankton production and succession. The correlation between upwelling and phytoplankton blooms is well established for coastal zones (Rao, 1977) and natural lakes (Coulter, 1968).

Turbulent mixing and the entrainment of metalimnetic water into the epilimnion has been observated following the passage of storms or frontal systems (Carmack & Gray, 1982). Turbulent mixing in the hypolimnion is patchy and may result from the breaking of internal waves on the reservoir bottom (Imberger, 1979). Many reservoirs are not completely cleared of timber before innundation. These trees may serve to generate additional turbulence and create additional mixing through the breaking of surface and internal waves on the trees. This turbulence can act to redistribute water and materials accumulating in the hypolimnion. This is of potential significance since high constituent concentrations and anoxia often occur at the sediment-water interface. Upward transport of nutrients and reduced materials may increase metalimnetic oxygen consumption and increase nutrient transfers to surface waters.

Outflows generate withdrawal currents and turbulence which contribute to the inlake mixing regime. Withdrawal currents are generated because the potential energy associated with the water level (i.e., the head) is converted into kinetic energy and causes water to flow through an outlet. The characteristics of the withdrawal currents will be dependent on the withdrawal rate, ambient inlake stratification, outlet location, and lake geometry. In cases where the outlet is located in the hypolimnion of a strongly stratified lake, the withdrawal zone and associated mixing may be limited to the hypolimnion. Withdrawal zones can also be limited to the epilimnion or metalimnion of a lake depending on the outlet location, flow rate, and ambient stratification.

Hypolimnetic withdrawal can, as previously mentioned, reduce reservoir nutrient concentrations by discharging anoxic, high-nutrient water downstream. Nutrients available for redistribution throughout the reservoir at fall turnover, therefore, can be significantly reduced through hypolimnetic withdrawal.

Reservoir water quality patterns

Reservoirs exhibit limnological conditions reflective of their climatic and geologic setting, their unique hydrologic and morphometric characteristics, and the relative importance of an advective flow regime. As previously discussed, geographic location and the fact that reservoirs are commonly constructed at the downstream end of long, narrow watersheds, are factors greatly influencing the quantity, quality and seasonality of external material load, and thus the potential concentrations of various water quality constituents. However, it is processes influencing their distribution, utilization, deposition, and redistribution that govern actual concentrations. These processes include advective transport, sedimentation, diffusion, biological uptake, absorption and desorption, oxidation and reduction, and losses due to discharge.

While these processes are common to both natural lakes and reservoirs, differences in their relative importance in determining water quality in reservoirs are apparent. In natural lakes, for instance, material loads are commonly received from relatively diffuse sources and wind-generated circulation is the major process influencing their intitial distribution. In reservoirs, however, advective currents play an important role in the transport and distribution of biotic and abiotic materials, and are directly or indirectly responsible for frequently observed water quality patterns in reservoirs. Longitudinal patterns are most pronounced in reservoirs since these types of lakes are commonly long, narrow, and fed by a single large tributary located a considerable distance from the dam. In most cases, the input of materials from the surrounding watershed also occurs via this single tributary. Thus, allochthonous materials, whether dissolved or particulate, are potentially transported throughout the reservoir by flows occurring primarily along the major axis of the reservoir.

As previously discussed, however, reservoir morphology and the occurrence of thermal stratification act to modify riverine flows. No longer confined to river or stream channels, inflowing waters spread laterally allowing energy dissipation and a marked reduction in velocity. Since suspended solid carrying capacity is related to velocity, dramatic decreases in suspended particulate concentrations occur near the location of the tributary inflow. It is in these areas that a large percentage of the reservoir's material load is lost due to sedimentation. Thus, it is interactions between morphology, advective forces and sedimentation which lead to the establishment of physical conditions conducive to the establishment of chemical and biological gradients.

Recognition of the importance of advective transport, sedimentation, and morphology in the establishment of commonly observed water quality patterns in reservoirs led Thornton et al. (1981) to propose a heuristic model which divides a reservoir into a riverine, transition, and lacustrine zone. The definition of these zones is dependent upon the manner in which inflows enter and move through a reservoir. In the riverine zone, current velocities are decreasing but the advective forces are still sufficient to maintain a well-mixed environment. In the transition zone, the buoyancy forces begin to dominate and the inflow plunges. The upstream and downstream boundaries of this zone may correspond with the location of the plunge point under low-flow and high-flow conditions, respectively. This zone is also the zone of sedimentation. In the lacustrine zone, buoyancy forces dominate and inflows move through the reservoir in well-defined horizontal layers as interflows and underflows.

Chemical and biological conditions in the riverine zone, the zone most strongly influenced by tributary flow, reflect conditions of the inflowing river and the influence of a relatively short water residence time. Nutrient concentrations in surface waters are highest in this portion of the reservoir and are often highly variable due to seasonal and/or event-related changes in loading.

The transition zone is potentially the most dynamic, most diverse reach of the reservoir. It is in this portion of the reservoir that dramatic changes in the physical environment occur. Riverine flow velocities decline here due to the influences of a changing basin morphology and, because of this, sedimentation of allochthonous material is often high. Under stratified conditions, under- and interflowing density currents are established and material loads are transported downward. This, and the fact that reduced carrying capacity results in losses in particulate material leads to improved water clarity in surface strata. Since nutrient concentrations here continue to be relatively high, this increase in water clarity often promotes increases in algal productivity.

The deposition in the transition zone of large quantities of particulate material, much of which is organic detritus of terrestrial origin, has several consequences of potential importance to reservoir nutrient dynamics, and to autotrophic and heterotrophic activity. Since hypolimnetic thickness and volume in this reach of the reservoir are minimal, inputs of large quantities of organic matter often lead to dramatic declines in dissolved oxygen stores. In many cases, this portion of the reservoir may become anoxic, even when more-downstream locations do not (Hannan & Cole, 1985). Under such conditions the release from sediments of nutrients, metals and contaminants may play an important role in the recycling of these materials.

Kennedy et al. (1983) report such occurrences in DeGray Lake, a large, hydroelectric reservoir in southcentral Arkansas. Despite its mesotrophic characteristics throughout much of the pool, conditions are clearly eutrophic to hypereutrophic in upstream areas. Oxygen declines in early summer are rapid and complete, and accompanied by significant releases of phosphorus, iron and manganese. The accumulation of these materials immediately below the thermocline allows for their vertical entrainment to the euphotic zone. Such occurrences are apparently common since algal biomass in this reach of the reservoir exceeds that which would be predicted from considerations of nutrient loading from external sources.

Similar observations have been documented for West Point Lake, a large reservoir located on the Chattahoochee River in western Georgia (Kennedy et al., 1982). In this reservoir, however, the transition zone is more extensive and located at a greater distance within the pool owing to the higher flow of this nutrient- and sediment-laden river. Algal biomass, while minimal at extreme upstream and downstream locations due to high levels of turbidity and reduced nutrient concentrations, respectively, is highest at a midpool location coincident with the transition zone. Anoxic conditions, which develop during summer months, occur in deep areas of the reservoir downstream from the plunge point and transition zone.

Conditions in the lacustrine zone are most lake-like. Thermal stratification is well established and the effects of flow from the tributary stream are minimal. Processes occurring upstream in the riverine and transition zone have impacted both the quantity and quality of material loads to the reservoir. As a result, material recycling plays an important role relative to the supply of external material and nutrients, and autotrophic production may exceed the supply of allochthonous organic matter. However, the deposition and retention of nutrients in upstream areas, and the transport of influent nutrients to sub-surface strata by density currents frequently leads to nutrient impoverishment. This, in turn, imposes limits on algal production.

Patterns in limnological conditions along the

length of reservoirs are often reflected in sediment characteristics. These patterns are of significance in understanding benthic community structure and sediment-water interactions. Hakanson (1977) describes the existence of three energy environments which affect patterns in sediment deposition and quality. In high energy environments, exemplified by areas constantly impacted by riverine flows, sediment deposition is discouraged by turbulence and flow. In moderate energy environments, sediments are deposited and subsequently eroded by periodic flows or wave action. In areas uneffected by flow or wave action (i.e. low energy environments), conditions favor sediment accumulation.

In a reservoir setting in which flows decrease along the longitudinal axis, a range of sedimentary environments are observed (Gunkel *et al.*, 1984). Sediments in the riverine portion of the reservoir exhibit low moisture content and large particle size, and generally have low concentrations of nutrients, metals, contaminants and other materials associated with fine sediments. In reservoirs in which morphology and hydrology favor the establishment of a lacustrine zone, fine sediments, which are frequently of autochthonous origin, accumulate. These sediments commonly have high organic, nutrient and metal concentrations, and small particle sizes. Sediments accumulating in the transition zone reflect the combined influences of riverine transport, autochthonous production, and a moderate energy environment. These sediments are potentially rich in nutrients and organic material deposited by tributary inflows. The fact that algal production is high in this portion of the reservoir, also explains the high organic and nutrient concentrations of these sediments.

While it is clear that patterns in the distribution of sediment are reflective of physical, chemical and biological occurrences in the water column, the quality of sediments may influence water column conditions through sediment/water interactions. The deposition of influent materials in upstream areas leads to the accumulation of organic matter, nutrients and metals. Microbial activity, supported by a continuous influx of organic carbon, can lead to the release of labile material and the establishment of anoxic conditions and attendant releases of nutrients and metals.

The above discussion of water and sediment quality patterns in reservoirs has been aided by the conceptual model of Thornton *et al.* (1981); however, boundaries between each of these zones are highly dynamic and difficult to deliniate. Thus, while the definition of distinct zones within a reservoir provides a nomenclature for such discussions, it must be noted that longitudinal patterns are more appropriately viewed as a continuum of changing conditions.

Summary

Reservoirs are dynamic ecosystems exhibiting a diversity of aquatic habitats. Key to the development of sound management practices for these complex systems is our understanding of interactions between watershed and reservoir, and of processes occurring within the reservoir. Reservoirs are greatly influenced by tributary flows and the material loads they transport. This is due, in part, to the fact that reservoir tributaries drain proportionately larger watersheds. Riverine influences in upstream areas are pronounced in reservoirs having a long, narrow morphology and a single, large tributary. Because of this, external loads are transported great distances within the reservoir and longitudinal gradients in material concentrations, sediment deposition, and biological activity are frequently observed. (Fig. 3).

While it is commonly assumed that external material loads to lakes are instantaneously and completely mixed with receiving waters, the occurrence of gradients and density currents suggest that this assumption is invalid for most reservoirs. The quantity and quality of external loads to reservoirs are often greatly modified by processes occurring in upstream areas. During periods of stratification, inflowing waters and the loads they transport are potentially carried to strata well below the photic zone. Under such circumstances, reservoir productivity and the supply of material from the watershed are not directly coupled.

Comparisons made here between natural lakes and reservoirs are not intended to suggest that these two types of aquatic systems are inherently

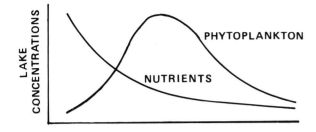

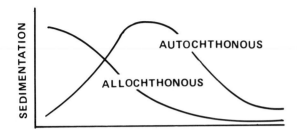

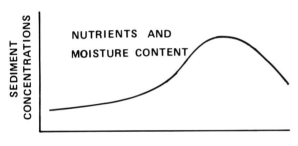

Fig. 3. Relative changes in water quality conditions, sediment deposition, and sediment characteristics along the length of a typical reservoir.

different. Rather, it is intended as a means by which the understanding of lakes can be reshaped to address environmental concerns in reservoirs. Only through an improved understanding of processes of importance in defining limnological conditions in reservoirs can sound management practices be developed and successfully implemented.

References

Akiyama, J. & Stefan, H.G., 1981. Theory of plunging flow into a reservoir. St. Anthony Falls Hydraulic Laboratory Internal Memorandum I–97. Minneapolis, Minn.

Bengtsson, L., 1984. Wind reduced circulation in lakes. Nordic Hydrol. 9:75–94.

Bilby, R.E. & Likens, G.E., 1979. Effect of hydrologic fluctuations on the transport of fine particulate organic carbon in a small stream. Limnol. Oceanogr. 24:69–75.

Carmack, E.C. & Gray, C.B.J., 1982. Patterns of circulation and nutrient supply in a medium residence-time reservoir Kootenay Lake, British Columbia. Can. Wat. Res. J. 7:51–70.

Chow, V.T., (ed.), 1964. Handbook of Applied Hydrology. McGraw-Hill Book Co. New York.

Coulter, G.W., 1968. Hydrological processes and primary production in Lake Tangayika. Proc. 11th Conf. Great Lakes Res. International Association Great Lakes Res. pp. 609–626.

Cummins, K.W., 1974. Structure and function of stream ecosystems. Bioscience 64:631–641.

Fisher, S.G. & Likens,G.E., Energy flow in Bear Brook, New Hampshire: an intergrative approach to stream ecosystem metabolism. Ecol. Monogr. 43:421–439.

Ford, D.E., Johnson M.C., & Monismith, S.G., 1980. Density inflows to DeGray Lake, Arkansas. Proc. 2nd. Int. Symp. on Stratified Flows. Int. Assoc. Hydraulic Res. 2:977–987.

Ford, D.E. & Johnson, M.C., 1983. An assessment of reservoir density currents and inflow processes. Technical Report E–83–7. US Army Engineer Waterways Experiment Station, Vicksburg, Miss.

Gaugush, R.G., 1984. Mixing events in Eau Galle Lake. Proc. Int. Symp. on Lake and Reservoir Management. EPA 440/5/84–001. US Environmental Protection Agency, Washington, DC.

Gibbs, R.J., 1970. Mechanisms controlling world water chemistry. Science 170:1088–1090.

Groger, A.W. & Kimmel, B.L., 1984. Organic matter supply and processing in lakes and reservoirs. Proc. Int. Symp. on Lake and Reservoir Management. EPA 440/5/84–001. US Environmental Protection Agency, Washington, DC.

Gunkel, R.C., Gaugush, R.F., Kennedy, R.H., Saul, G.E., Carroll, J.H. & Gauthey, J.E., 1984. A comparative study of sediment quality in four reservoirs. Technical Report E–84–2. US Army Engineer Waterways Experiment Station, Vicksburg, Miss.

Hakanson, L., 1977. The influence of wind, fetch, and water depth on the distribution of sediments in Lake Vanern, Sweden. Can. Earth Sci. 14.

Hannan, H.H. & Cole, T., 1985. Dissolved oxygen dynamics in reservoirs. In Thornton, K., (ed.) Perspectives in Reservoir Limnology. John Wiley, New York, NY.

Imberger, J., 1979. Mixing in reservoirs. In H.B. Fischer (ed), Mixing in Inland and Coastal Waters. Academic Press, New York, NY. pp. 150–228.

Johnson, A.H., Bouldin, D.R., Goyette, E.H. & Hedges, A.M., 1976. Phosphorus loss by stream transport from rural watershed: Quantities, processes, and sources. Jour. Envir. Qual. 5:148–157.

Kantha, L.H., Phillips, O.M. & Azad, R.S., 1977. On turbulent entrainment at a density interface. J. Fluid Mech. 79:753–768.

38

Kennedy, R.H., Thornton, K.W. & Carroll, J.H., 1981. Suspended sediment gradients in Lake Red Rock. In Stefan, H., (ed.), Proc. Symp. on Surface Water Impoundments. ASCE Vol 2:1318–1328.

Kennedy, R.H., Thornton, K.W. & Gunkel, R.C., 1982. The establishment of water quality gradients in reservoirs. Can. Wat. Res. J. 7:71–87.

Kennedy, R.H., Gunkel, R.C. & Carlile, J.M., 1983. Riverine influence on the water quality characteristics of West Point Lake. Technical Report, US Army Engineer Waterways Experiment Station, Vicksburg, Miss.

Kennedy, R.H., Montgomery, R.H., James, W.F. & Nix, J., 1983. Phosphorus dynamics in an Arkansas reservoir: The importance of seasonal loading and internal recycling. Miscellaneous Paper E–83–1, US Army Engineer Waterways Experiment Station, Vicksburg, Miss.

Knapp, R.T., 1942. Density currents: Their mixing characteristics and their effect on the turbulence structure of the associated flow. Proc. of the Second Hydraulic Conference. State University of Iowa, Iowa City, pp. 289–306.

Martin, D.B., & Arneson, R.D., 1978. Comparatieve limnology of a deep-discharge reservoir and a surface-discharge lake on the Madison River, Montana. Freshwat. Biol. 8:33–42.

Minshall, G.W., 1978. Autotrophy in stream ecosystems. Bioscience 28:767–771.

Naiman, R.J. & Sedell, J.R., 1980. Relationships between metabolic parameters and stream order in Oregon. Can J. Fish aquat. Sci. 37:834–847.

Naiman, R.J. & Sedell, J.R., 1981. Stream ecosystem research in a watershed perspective. Verh. int. Ver. Limnol. 21:804–811.

Rao, D.V.S., 1977. Effect of physical processes on the production of organic matter in the sea. Symposium on Modeling of Transport Mechanisms in Oceans and Lakes. Manuscript Report Series. Marine Sciences Directorate Department of Fisheries and the Environment, Ottawa. 43:161–169.

Roberts, P.J.W., 1981. Jet entrainment in pumped storage reservoirs. Technical Report E–81–3. US Army Engineer Waterways Experiment Station, CE, Vicksburg, Miss.

Schuiling, R.D., 1976. Source and composition of lake sediments. In Golterman, H.L. (ed.), Interactions between sediment and fresh water. Dr W. Junk Publishers, The Hague. pp. 12–18.

Sharpley, A.N. & Syers, J.K., 1979. Phosphorus inputs into a stream draining an agricultural watershed. II: Amounts contributed and relative significance of runoff types. Wat. Air. Soil Pollut. 11:417–428.

Swanson, C.D. & Bachman, R.W., 1976. A model of algal exports in Iowa streams. Ecology 57:1076–1080.

Thornton, K.W., Kennedy, R.H., Carroll, J.H., Gunkel, R.C. & Ashby, S., 1981. Reservoir sedimentation and water quality-an heuristic model. In Stefan, H., (ed.), Proc. Symp. on Surface Water Impoundments. Vol. I:654–661.

Vannote, R.L., Minshall, G.W., Cummins, K.W., Sedell, J.R. & Cushing, C.E., 1980. The river continuum concept. Can. Fish. aquat. Sci. 37:130–137.

Verhoff, F.H. & Melfi, D.A., 1978. Total phosphorus transport during storm events. J. Env. Engr. ASCE. 104:1021–1023.

Verhoff, F.H., Melfi, D.A. & Yaksich, S.M., 1979. Storm travel distance calculations for total phosphorus and suspended materials in rivers. Wat. Resour. Res. 15:1354–1360.

Walker, W.W., 1981. Empirical methods for predicting eutrophication in impoundments. Phase I: Data base development. Technical Report E–81–9. US Army Engineer Waterways Experiment Station, Vicksburg, Miss.

Wright, J.C., 1967. Effect of impoundments on productivity, water chemistry, and heat budgets of rivers. In: Reservoir Fishery Resources. Am. Fish. Soc., Washington, DC pp. 188–199.

Authors' addresses:
Robert H. Kennedy, Ph D.
Environmental Laboratory
US Army Waterways Experiment Station
Vicksburg, Mississippi
USA

Kent W. Thornton, Ph D.
and Dennis E. Ford, Ph D.
Ford, Thornton, Norton and Associates
Little Rock, Arkansas
USA

CHAPTER 3

Chemistry and microbiology of newly flooded soils: relationship to reservoir-water quality

DOUGLAS GUNNISON, ROBERT M. ENGLER and WILLIAM H. PATRICK, JR.

Abstract. Upon filling, a new reservoir undergoes several years of intensive biological and chemical transformations resulting from the decomposition of flooded organic matter together with reductive reactions of inorganic portions of the soil. Decomposition of flooded substrates provides a source of food for valuable fish species, but also lowers dissolved oxygen. Microbial interactions with flooded soil components result in release of algal growth promoting nutrients plus various metals and sulfide, making achievement of water quality objectives difficult.

The work presented here summarizes the result of approaches developed to determine the effects of microbial processes occurring in flooded soils upon reservoir water quality. Emphasis is placed on the nature of the microbial processes involved, rather than specific microorganisms responsible for the processes. Examples are taken from soil-water interaction studies that simulate newly impounded reservoirs. In addition to an examination of the microbially-mediated release of nitrients and metals from flooded soil, this chapter also describes the effects of temperature and reservoir aging upon oxygen uptake and nutrient regeneration. Emphasis is also given to the unique hydrodynamic properties of reservoirs; these properties determine both the fate and the impact of products released by microbial activity.

Introduction

Upon filling, a new reservoir is subject to several years of intensive biological and chemical changes characteristic of the transition from a terrestrial to an aquatic ecosystem (Wilroy & Ingols, 1976; McLachlan, 1977). Major processes occurring during this phase are generally microbially-mediated and involve the utilization of readily biodegradable remnants of brush, standing trees, and other above-ground organic matter flooded during filling. Degradation of these materials is considered in detail in the Chapter by Godshalk & Barko. However, leaching and microbial decomposition of labile components in the soil litter layer and the A layer soil horizon are also important (Gunnison *et*

al., 1980). In fact, these processes may exert a strong influence on reservoir water quality through depletion of dissolved oxygen and release of nutrients and metals.

During initial filling, the bottom of the reservoir moves through a sequential transition of its own – from dry soil to wet soil to saturated soil to sediment. Accompanying this waterlogging process are shifts in the physiochemical composition and structure of the soil and in the structure and function of the former terrestrial soil microbial community. Such changes are further evidenced by a gradual release from the soil to the water of color contributed by humic substances from organic soils, high levels of dissolved organic carbon, excessive concentrations of dissolved and particulate froms of

Gunnison, D. (ed.) Microbial Processes in Reservoirs.

nitrogen and phosphorus, and oxygen-demanding materials, both particulate and dissolved. If the situation is such as to permit the establishment of anaerobic conditions, accumulation of ammonium-nitrogen, orthophosphate- phosphorus, dissolved iron and manganese, hydrogen sulfide, and methane may occur. Vigorous decompositional processes associated with the first portion of the transition phase can cause the release of large quantities of nutrients that support the growth of aquatic plants plus small particles of detritus that are easily used by large invertebrates and fish. Thus, the transition period is often characterized as a period of high biological productivity, particularly with regard to sport fisheries (Ploskey, 1981).

The last phase of the transition period, often some six to ten years after initial filling, is characterized by a decline in the levels of undersirable materials in the reservoir water column, an increase in overall water quality and a sharp, often precipitous, decrease in biological productivity (Wilroy & Ingols, 1964). Decreased productivity often results from a depleted availability of substrates to the microbial community. Details of the phenomena responsible for this decrease will be considered later in this chapter.

In planning the construction and operation of a new reservoir project, the engineer must consider traditional aspects such as placement, sizing, and appropriate structural features of the dam; however, he must also weigh the potential impact of many in-reservoir environmental factors upon water quality both within and downstream from the new project. Previously unutilized concepts, such as whether to cut or burn standing vegetation, whether to strip, cover, or remove mineral deposits and organically rich soils, etc., now must enter into the planning of a new reservoir project. Certain tools are available to assist the engineer with his decision-making, including: water quality studies done on adjacent reservoirs during the initial fill and transition periods; preimpoundment investigations performed using soil and vegetation samples from the proposed reservoir site (Gunnison et al., 1979; 1984); and mathematical water quality models.

This chapter considers the influence of chemical and microbial processes in newly flooded soils on the water quality of new reservoirs. In so doing, the chapter describes the behind-the-scenes activities that form the real reasons for the engineer's having to consider the nature and composition of soil in a new reservoir site. Emphasis in this chapter is placed on major processes of concern with respect to water quality both within and downstream from the new reservoir. Since only a limited amount of research effort has been devoted to reservoirs, much of what is presented here is derived from recent studies on lakes, ponds, lake sediments, and flooded soils. Process-oriented research, rather than the study of individual organisms or groups of organisms, will be stressed. The effects of decomposition of vegetation upon water quality in reservoirs are also extremely important to consider in the planning and construction of a new reservoir (Chapter 4). Since the present chapter is restricted to the transformation of newly flooded soils into reservoir sediments, the reader may wish to refer to Chapter 7 for a consideration of the relationship of sediments to water quality in established reservoir, i.e., reservoirs that have completed their transition phase.

Chemical, microbial and physical processes

The transformation from dry soil to flooded soil has been described in detail by several workers for the case of rice paddy soils (Tusneem & Patrick, 1971; Ponnamperuma, 1972; Yoshida, 1975). The basic tenet of this chapter is that the same processes are extant in newly flooded reservoir soils as for rice paddy soils; however, in a reservoir, physical factors exert a considerable influence over the situation, often serving to alter the rate of chemical and microbial reactions occurring in the flooded soil and dominating the fate of materials released from the soil into the water column. The reader is encouraged to review the previous discussions by Baxter (Chapter 1) and by Kennedy et al. (Chapter 2) for detailed consideration of the physical processes mentioned here.

While the process occurring in newly flooded reservoir soils are similar to those occurring in rice paddy soils, there are some major differences; these will be examined on page 53. We consider here some of the minor differences.

Water covering rice paddies tends to be shallow. This, in turn, promotes ease of mixing by wind, resulting in waters that are generally well-aerated. Moreover, the proximity of the soil to the water surface permits adequate light penetration, resulting in high photosynthesis by benthic algae – also serving to keep the water aerobic. The shallow depth of the rice paddy results in warmer water and soil during sunlight hours, accelerating those reactions occurring at the soil-water interface. By contrast, much of the flooded soil in a new reservoir is deeply submerged, is not necessarily subject to waters mixed by the wind, and often remains at low temperature, thus slowing microbial activity.

Another important factor distinguishing a newly flooded reservoir soil from a rice paddy is duration of submergence. The permanent flooding of soils in the conservation (storage) pool of the reservoir, as opposed to the intermittant submergence occurring in rice paddies, insures that flooded soil development proceeds beyond the rice paddy stage, approaching what Ponnamperuma (1972) has termed the 'subaquatic soil.' Such a soil has been characterized as (1) being formed from soil components, (2) having normal soil-forming reactions, (3) having a bacterial community similar to terrestrial soils, (4) having a metabolism similar to terrestrial soils, and (5) having areas that differ in texture, composition, clay mineralogy, organic matter content, and redox potential in a manner similar to terrestrial soils (Ponnamperuma, 1972).

This section will present an overview of the changes occurring in newly flooded soils with an emphasis on soil reduction processes from a microbial and biochemical viewpoint. Physicochemical interactions resulting from the soil flooding process will also be considered, particularly where implications for the transformation of soil into sediment are important.

Shifts in microbial community structure and function

Biochemical changes occurring in newly flooded soils may be viewed as a series of stepwise oxidation-reduction reactions driven by organic matter and mediated by microorganisms. The sequential nature of this process is evident from the metabolic characteristics of the microorganisms present, involving initially aerobes, then facultative anaerobes, and finally anaerobes (Takai *et al.*, 1956; Takai & Kamura, 1966; Ponnamperuma, 1972). Activities of the microorganisms, primarily bacteria, are reflected by a change in the pH and a decrease in the oxidation-reduction potential (Eh) as the reduction process proceeds in the flooded soil. Development of the flooded soil environment from the biochemical viewpoint parallels the successive progression of microorganisms as aerobic respiration is followed by anaerobic respiration and fermentation, and finally methanogenesis (Fig. 1). This process has been described in detail by several authors (Tusneem & Patrick, 1971; Ponnamperuma, 1972; Yoshida, 1975) and will be outlined here as a general scenario. The reader should consult these authors for additional details. Many examples of the processes described here are given in the accompanying figures; most of these are taken from our own work using soil-water reaction chambers with continuous flow as described in Gunnison *et al.* (1979, 1980b, 1983) (Fig. 2).

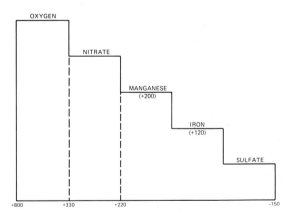

Fig. 1. Sequential reduction of inorganic electron acceptors. Numbers at the base of figure are approximate oxidation-reduction (redox) potentials in mV at which the preceding constituent becomes undetectable in solution. Numbers in parentheses are approximate redox potentials in mV at which the reduced form of the constituent becomes detectable in solution (Data used to construct figure taken from Ponnamperuma, 1972).

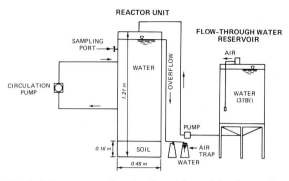

Fig. 2. Soil-water reaction chamber with provision for continuous flow of flood waters. In this system, water within reactor unit is continuously recirculated while inflows are pumped in and outflows are removed by gravity. For details of this system and its operation, see Gunnison *et al.* (1980b).

Aerobic respiration

Upon flooding, consumption of dissolved oxygen within the soil proceeds at a rapid pace. Oxygen demand easily exceeds the capacity of the supply route which requires diffusion of dissolved oxygen into the soil from the overlying water column.

Aerobic microorganisms undergo a 'flush' of activity, converting available organic matter to its inorganic components, carbon dioxide, nitrate, sulfate, and phosphate, and reducing oxygen to water. Supplies of oxygen within the soil are often exhausted within 24–28 hours after flooding, resulting in anaerobic conditions (Takai *et al.*, 1956; Takai & Kamura, 1966; Patrick & Mikkelsen, 1971). Initial supplies of easily utilizable organic matter are also often rapidly depleted, however, and aerobic microbial activity decreases. At this point, the supply of dissolved oxygen to the flooded soil surface may be sufficient to reoxidize the surface, forming what has been termed the 'aerobic-anaerobic double layer' (Patrick & DeLaune, 1972;

DeLaune *et al.*, 1973; Reddy *et al.*, 1980). Once this point has been reached, organic materials required to support the growth of heterotrophic microorganisms in the oxidized surface layer may be supplied by anaerobic microbial activity in the underlying soil. If the supply of these materials is adequate, aerobic heterotrophic activity in the surface layer may continue for a prolonged period. Alternatively, the supply of dissolved oxygen in the overlying water may become exhausted. As levels of dissolved oxygen supplied consumption in the surface layer may be limited to that carried out by autotrophic microorganisms oxidizing reduced chemical species diffusing into the surface layer from the underlying anaerobic soil. Table 1 summarizes the major reactions carried out by these organisms. Eventually, this layer, too, becomes anaerobic and dissolved oxygen is removed from the overlying water column (Fig. 3).

Anaerobic respiration

Anaerobic respiration may be defined as a biological reaction wherein oxidized inorganic compounds serve as electron acceptors and are reduced; this activity is coupled to the energy-yielding oxidation of organic or inorganic compounds (Doetsch & Clark, 1973; Yoshida, 1975). Reduction of inorganic compounds follows a stepwise chain of events in accord with thermodynamic predictions (Fig. 1). Once almost all dissolved oxygen has been consumed, facultatively anaerobic bacteria take over from the aerobic bacteria, and nitrate begins to be reduced. Upon exhaustion of nitrate, manganic manganese is reduced. Next comes ferric iron, followed by sulfate and then carbon dioxide. The following discussion summarizes the salient features of each of the reduction steps.

Table 1. Major oxidation reactions carried out by autotrophic (chemolithotrophic) micro-organisms.

Reaction	Micro-organisms
$NH_4^+ + 1\frac{1}{2}O_2 \rightarrow NO_2^- + 2H^+ + H_2O$	Ammonium oxidizers
$NO_2^- + \frac{1}{2}O_2 \rightarrow NO_3^-$	Nitrite oxidizers
$Fe^{2+} + \frac{1}{4}O_2 + H_2O \rightarrow Fe^{3+} + \frac{1}{2}H_2O$	Iron bacteria
$S^{2-} + 2O_2 \rightarrow SO_4^{2-}$	Sulfur bacteria
$CH_4 + 2O_2 \rightarrow CO_2 + 2H_2O$	Methane oxidizers

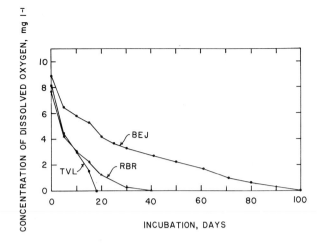

Fig. 3. Removal of dissolved oxygen from the water column overlying newly flooded soils from three reservoir projects. BEJ = B. Everett Jordan Lake, RBR = Richard B. Russell Lake; and TVL = Twin Valley Lake. Studies were done in the reaction chambers depicted in Fig. 2.

Nitrate reduction. Microbial reduction of nitrate can be placed into two categories: assimilatory and dissimilatory. The product of assimilatory nitrate reductions is NH_4^+ which is incorporated into the cell (Payne, 1973).

Dissimilatory reduction of NO_3^- to NH_4^+ has been shown to occur in marine environments (Koike & Hattori, 1978; Sørensen, 1978) and may also be of importance in freshwaters (Knowles, 1979). In dissimilatory NO_3^- reduction to NH_4^+, the reduced product is released rather than being incorporated into the cell, and the process is coupled to energy production (Hasan & Hall, 1975). Organisms carrying out dissimilatory reduction of NO_3^- to NH_4^+ have not been listed, although Sørensen (1978) indicates that *Clostridium perfringens* and *Paracoccus denitrificans* are capable of this process.

In anaerobic environments, respiratory reduction of NO_3^- to N_2 serves as an alternative to 0_2 respiration for denitrifying bacteria and occurs in a zone immediately below the oxidized surface and reduced microniches within the surface zone (Koike & Hattori, 1978; Sørenson, 1978). In this scheme, NH_4^+ is oxidized to NO_2^- or NO_3^- either in the aerobic water column or in the oxidized surface layer; these ionic nitrogen oxides then diffuse down into a zone of lower oxygen tension where they are

subsequently denitrified (Reddy & Patrick, 1975). In both marine and freshwater systems, depletion of dissolved oxygen below approximately $0.2\,mg\,l^{-1}$ favors denitrification (Knowles, 1982). Until nitrate reduction is complete, the Eh is poised in the $+400$ to $+200\,mV$ range (Fig. 1) (Turner & Patrick, 1968). The products of this dissimilatory process are N_2O, NO and N_2.

The initial step in the nitrate reduction process, reduction of NO_3^- to NO_2^-, can be carried out by a variety of bacteria, including *Bacillus, Clostridium,* and *Pseudomonas.* The capacity to carry out the full sequence of nitrate reduction to N_2O and N_2 is apparently also possessed by bacteria ranging from heterotrophs to autotrophs. Examples include species in the genera *Alcaligenes, Agrobacterium, Bacillus, Chromobacterium, Corynebacterium, Hyphomicrobium, Hydrogenemonas, Paracoccus (Micrococcus), Pseudomonas,* and *Thiobacillus* (Knowles, 1982).

Manganese reduction. MnO_2 serves as an electron acceptor in anaerobic respiration at an oxidation level comparable to that of nitrate; however, manganese lacks the effectiveness of nitrate in stabilizing Eh (Patrick & Mikkelsen, 1971; Ponnamperuma, 1972; Brannon *et al.*, 1978). This weakness may be a consequence of several factors, including the limited solubility of MnO_2 and the use of MnO_2 as an electron acceptor by only a small number of bacteria (Ponnamperuma, 1972). The relatively high standard oxidation-reduction potential of the MnO_2-Mn^{2+} redox couple may regulate the Eh of flooded soils as is demonstrated by several studies wherein native or added MnO_2 was found to retard the fall in Eh in flooded soils (Nhung & Ponnamperuma, 1966) and also to prevent the accumulation of Fe^{2+} and other reduction products that normally appear in substantial quantities when MnO_2 reduction has been completed (Ponnamperuma & Castro, 1964; Yamane & Sato, 1968). However, other investigators have not found a relationship between Eh and the MnO_2-Mn^{2+} couple (Bohn, 1968) or have found deviations between the theoretical Eh values for pure systems and those obtained in actual experiments (Bohn, 1970; Gotoh & Patrick, 1972).

That the reduction of MnO_2 is the result of bio-logical rather than chemical activity is shown in several studies wherein: (1) glucose added to soil was observed to stimulate Mn reduction while ad-dition of oxide to glucose-amended soils retarded Mn reduction (Mann & Quastel, 1946) and (2) addition of organic matter to flooded soils in-creased the accumulation of Mn^{2+} (Mandal, 1961; Gotoh & Yamashita, 1966). We have observed similar results using cellulose as a source of organic matter (Fig. 4). Preferential use of NO_3^- as an electron acceptor by facultatively anaerobic bacte-ria is suggested by retardation of MnO_2 reduction which occurs upon addition of NO_3^- to a flooded soil (Takai & Kamura, 1966). However, it is not presently understood whether NO_3^- exerts this effect through chemical oxidation of reduced spe-cies or by a toxic or inhibitory action on the micro-organisms responsible for Mn reduction (Brannon et al., 1978).

Many soil fungi and bacteria are capable of re-ducing the oxidized forms of Mn; one estimate of abundance included 1.1×10^4 fungi and 3.0×10^6 bacteria per gram dry weight of soil (Bautista & Alexander, 1972).

Iron reduction. The concentration of Fe is much greater than NO_3^- of Mn in most soils, and several investigators have found a close relationship be-tween the amount of reducible iron and the Eh and pH of flooded soils (Takai et al.), 1963); Takai & Kamura, 1966). By contrast, the effect of the $Fe(OH)_3$–Fe^{2+} redox couple on flooded soil Eh is not as great as those of NO_3^- or MnO_2, presumably because of the low standard potential of the $Fe(OH_3$–Fe^{2+} system (Ponnamperuma, 1972). Moreover, the effect of adding $Fe(OH)_3$ on the Eh of a flooded soil is not as pronounced as the effect of adding MnO_2 (Nhung & Ponnamperuma, 1966).

Evidence that microbial activity is responsible for Fe^{3+} reduction is provided by the stimulation of iron reduction observed when organic matter is added to a soil or an Fe oxyhydroxide slurry (Man-dal, 1961; Gotoh & Yamashita, 1966; Brannon et al., 1978 – see Fig. 4). In addition, bacteria that actively reduce Fe^{3+} to Fe^{2+} have been isolated (Bromfield, 1954), and Fe^{3+} reduction has been

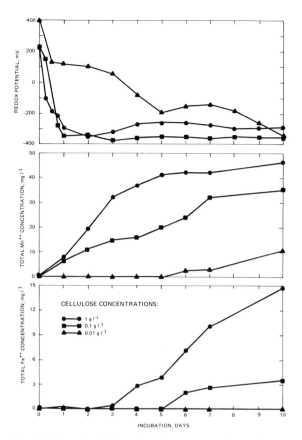

Fig. 4. Effect of addition of varying concentrations of organic matter to sealed suspensions containing particulate forms of manganese dioxide and iron oxyhydroxide and inoculated with a reservoir sediment. Note depression of oxidation-reduction potential (above) and appearance of soluble reduced forms of manganese (middle) and iron (below).

observed to correlate strongly with CO_2 production (Asami & Takai, 1970). Addition of MnO_2 or NO_3^- to a flooded soil system undergoing Fe^{3+} reduction retards the reduction process (Takai & Kamura, 1966). Again, the mechanism of retardation, whether biological or chemical, is not understood, but NO_3^- addition to anaerobic microcosms has been shown to rapidly oxidize Fe^{2+} and Mn^{2+}, providing strong evidence for a chemical oxidation process (Brannon et al., 1978).

The capacity for iron reduction is apparently a common phenomenon among microorganisms. Between 10^4 and 10^6 bacteria per gram of soil can reduce Fe^{3+} to Fe^{2+} with members of the genera *Bacillus, Pseudomonas, Clostridium, Klebsiella,* and *Serratia* having this capacity (Alexander, 1977).

Sulfate reduction. While respiratory sulface reduction is of primary importance for the production of sulfide in the natural ecosystem, a combination of low Eh (–150 mV) and a restricted pH range (6.8–7.1) are required for reduction to occur (Connell & Patrick, 1968). The biochemistry of sulfate metabolism in aquatic ecosystems has been the subject of much recent research (See, for example: Larson & Bella, 1974; Abram & Nedwell, 1978; Jørgensen, 1978; Laanbroek & Pfennig, 1981; Widdel & Pfennig, 1981). At present, a particularly strong controversy exists over the competition between sulfate reducers and methanogens for organic substrates, particularly acetate and hydrogen (Winfrey & Zeikus, 1977; Dwyer & Klug, 1982; Lovely & Klug, 1983; Winfrey & Ward, 1983). Sulfate reduction is throught to be carried out predominantly by members of the genus *Desulfovibrio*, which contains species capable of both autotrophic and heterotrophic growth (Takai, 1952; Taha Mahmoud & Ibrahim, 1967), and by members of the genera *Desulfomonas* and *Desulfotomaculum* (Gottschalk, 1979). Reduction of sulfate to sulfide is inhibited by the addition of nitrate, MnO_2, and $FePO_4$ (Engler & Patrick, 1973). MnO_2, while incapable of competing with SO_4^{2-} as a H acceptor, can oxidize H_2S. Free sulfide in solution readily combines with Fe^{2+} to form insoluble iron sulfides.

Fermentation and gas metabolism

Fermentation in the classical sense is considered as life in the absence of oxygen and is a low energy-yielding process consuming large quantities of sugar, often with a minimum uptake of the energy source carbon (Sokatch, 1969). In practical terms, fermentation refers to the anaerobic conversion of organic compounds to organic acids, alcohols, and carbon dioxide, although some authors consider reduction of sulfate to sulfide and reduction of carbon dioxide to methane as fermentation (Gottschalk, 1979). In previous work, production of organic acids has been noted to occur between 1 and 23 days after flooding, after which the level of these materials undergoes a marked decrease, presumably due to their removal through the formation of carbon dioxide and methane (Gunnison *et al.*, 1979). Organic acids observed in flooded soils have included aliphatic acids, the lower volatile fatty acids – formic, acetic, propionic, butyric, lactic and aromatic acids (Stevenson, 1967; Batistic, 1974). The rate of production, amount, and spectrum of acids produced varies with the soil, the source compounds, and the time after initial flooding (Ponnamperuma, 1972).

Microbial degradation of substrates such as cellulose, rice straw, green manure, and weeds in flooded soil has been examined, but most work has been done with individual anaerobes in pure culture. Much the same can be said for aromatic compounds where certain phases of the decompositional process apparently require the presence of oxygen. Likewise, fermentation of nitrogenous organic compounds in flooded soils has not been extensively studied. The mechanism for this process probably involves deamination followed by an attack on the residual compound to form carbon dioxide and various organic compounds, largely acids (Vogels & Van der Drift, 1976; Alexander, 1978).

Major gas products evolved after flooding include CO_2, N_2, CH_4 and H_2. Carbon dioxide is first evolved from aerobic respiration of carbohydrates. Nitrogen gas results from the nitrification-denitrification coupling process wherein nitrate is produced from ammonium and, upon diffusion into an anaerobic zone, is denitrified to N_2O, NO and N_2 Additional CO_2 and H_2 are produced by fermentation. Once all sulfate has been depleted, normally a rapid process in a freshwater environment because of the limited availability of sulfate, methanogenesis becomes the terminal electron acceptor for carbon respiration. In this instance CO_2 is reduced with either H_2 or an organic compound serving as the hydrogen donor. Alternatively, acetic acid may be decarboxylated (transmethylated), the apparent pathway of choice in flooded soils. Shifts in the pattern of gas evolution with time are temperature sensitive and are evident when the gas headspace above a sealed water column is sampled during incubation of a flooded soil in a closed system (Fig. 5).

46

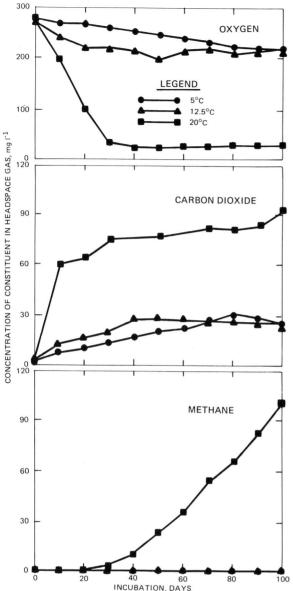

Fig. 5. Effect of temperature on the patterns of oxygen consumption (above) and evolution of carbon dioxide (middle) and methane (below) with time. Measurements were made in the headspace above a flowing water column overlying an alluvial piedmont soil using the reactor unit depicted in Fig. 2. Systems were sealed from atmos, .eric contact at 0 days, but inflowing waters were fully saturated with dissolved oxygen (modified from Gunnison *et al.*, 1983).

The soil reduction process

Upon submergence, the soil microbial community moves through the oxidation-reduction scenario just described. Simultaneously, the flooded soil itself undergoes reduction as oxygen and nitrate disappear and reduced products accumulate either as soluble components of the interstitial water or as precipitated components that become part of the soil itself. Reduction of the soil can be considered the most important chemical change resulting from flooding (Ponnamperuma *et al.*, 1967).

Under emergent conditions, the well-drained aerated soil differs considerably from its submerged counterpart. This soil has a high redox potential and a mineral fraction which is generally well-oxidized and which tends to be light colored. The organic fraction is composed of materials in various stages of aerobic decomposition. Agents of decomposition include primarily bacteria and fungi, although an assemblage of other microorganisms is also present. The soil pore space occupies nearly half its volume and consists of both air and water.

Submergence causes several immediate changes. The air in the pore space is displaced by water. Oxygen disappears within 1–3 days after submergence as carbon dioxide begins to accumulate. As the oxygen becomes depleted, the microbial community shifts its metabolism from aerobic pathways to fermentation and the use of alternate inorganic electron acceptors is initiated. As nitrate is consumed, N_2O, NO, N_2, and NH_4^+ are produced, with the latter compound accumulating in the interstitial water. Organic fermentation products also begin to appear in the interstitial water. As the reduction continues, manganese, iron, and then sulfide are released. Upon formation, sulfide begins to associate with reduced iron resulting in the formation of a dark insoluble precipitate. The precipitate accumulates in the flooded soil causing it to assume a black, gray, or greenish-gray color (Ponnamperuma, 1972). Refractory organic matter may further contribute to the coloration process, causing the flooded soil to become dark gray or even black. Methanogenesis, the terminal electron transfer process in flooded soils and sediments, becomes the final consumer of low molecular weight organic acids and hydrogen generated during the fermentation process (Wolin & Miller, 1982). If methanogenesis is extensive and the overlying water pressure not too great, bubbles of the gas may escape from the soil into the water column (Strayer & Tiedje, 1978).

The final product of the reduction process is a soil which is dark in color, has a low redox potential, is devoid of dissolved oxygen, and contains the reduced components: NH_4^+, H_2S, Mn^{2+}, Fe^{2+}, and CH_4, plus refractory organic matter. The reduced submerged soil may or may not have a thin, light colored oxidized layer at the surface, depending on the aeration status of the overlying water column (Ponnamperuma, 1972).

Other changes in the flooded soil

In addition to the soil reduction process just described, several other changes are apparent in the flooded soil. These include (1) changes in the nature and abundance of organic matter, (2) changes in redox potential, (3) shifts in the mineral equilibria, (4) changes in cation/anion exchange systems, (5) changes in pH, and (6) changes in conductance and ionic strength. These alterations in the organic and physiocochemical character of the flooded soil have been treated in excellent detail in Ponnamperuma's (1972) treatise, and the reader should consult this work for further information on the subject. The significance of these factors for the transformation of the newly flooded soil into a true aquatic sediment was considered in the previous section.

Relationship to reservoir water quality

Interactions between the flooded soil and the overlying water column

We now shift our focus from the chemical and microbial processes occurring within the flooded soil matrix to the relationship between the changes that have occurred within this environment and the variations to be expected in the chemical composition of the overlying water column. There are several mechanisms by which materials released within the flooded soil matrix can find their way into the water column.

Bioturbation, or the disruption of the integrity of the flooded soil surface through biological activities that displace and suspend materials in the surface layers, can be an important factor, although this process will not be considered here.

Suspension of materials in the flooded soil surface by underwater currents can be significant process for moving materials in the soil surface layers into the water column; however, the actual impact of this depends upon the factors influencing the existence and circulation patterns of underwater currents – i.e., reservoir basin morphometry, presence or absence of standing trees, pattern of flows, etc. For details of this process, see Chapters 1 and 2.

Diffusion of materials from the flooded soil into the water column constitutes the third major process wherein a flooded soil affects the water column. This process, in turn, is influenced by the redox status of the flooded soil surface layer, since this layer is the interface between the solid and liquid phases of the water column. When oxidized, the surface layer also serves as the interface between oxidizing and reducing activities, both chemical and microbial. Ammonium, derived from decompositional or dissimilatory reductive processes occurring in underlying anaerobic mud, diffuses upward through the surface layer and is oxidized to nitrite and nitrate on its way into the water column. A similar fate befalls sulfides as they are oxidized to sulfate upon upward diffusion. Carbon dioxide may be released either from the aerobic activities of microbes upon soil organic matter or from anaerobic fermentation processes that release carbon dioxide in the oxidized surface layer and the water column.

The direction of diffusion is, of course, not restricted to the upward direction and, depending on the concentration gradients involved, materals can also move into the flooded soil from the water column. This is particularly true of dissolved oxygen, and the newly flooded soil can serve as a sink for this compond. Nitrate and sulfate can also move downward with the reductive processes of denitrification and dissimilatory nitrate reduction to ammonium and sulfate reduction serving to keep these compounds at low levels in the flooded soil. If vegetative decomposition in the litter layer is active, this process can provide a source of soluble organic matter both to the overlying water column and the underlying soil surface layer.

The net result of vigorous microbial decomposition in flooded soils is to reduce the flooded soil matrix, as described previously. In the case of the new reservoir that has not been cleared, decomposition of herbaceous plants, leaves, grasses, and other above-ground materials can consume so much dissolved oxygen that the flooded soil surface has little chance to become oxidized, except under the most vigorous of mixing conditions. Where wind-induced mixing or inflowing currents are strong enough, turbulent displacement of the uppermost soil layers into the water column may occur with resulting oxidation of the suspended material as described by Gorham (1958). This material may then resettle to form an oxidized surface layer. The new oxidized layer may later disappear from the surface downward, either as a result of oxygen consumption resulting from decomposition of sedimented plankton (Gorham, 1958) or of organic debris settling from flooded plant material above the soil surface. Alternatively, waters of the hypolimnion may themselves become anoxic, preventing the surface layer from reoxidizing as consumption of reducing substances diffusing upward from the underlying anaerobic layers utilize the remaining oxidized substances in the surface layer.

There are various consequences resulting from the development of a reduced surface layer. A reduced surface permits direct release of reduced materials from the soil into the water column. Reduction of the flooded soil surface parallels movement of the boundary layer between the aerobic and anaerobic zones from within the flooded soil to above the surface layer and into the hypolimnion (Ponnamperuma, 1972). In our own experience, the water column overlying the sediment becomes an extension of the sediment interstitial water, although the intensity of reduction may not be identical (Gunnison *et al.*, 1979). Mortimer (1941, 1942), in his study of English lakes that develop thermal stratification, has observed a decline in redox potential in the hypolimnion that may attain values as low as the underlying reduced sediments. The net result of the release of reduced mterials from

flooded soils/sediments and from reduced materials suspended in an anaerobic hypolimnion is that NH_4^+, Mn^{2+} and Fe^{2+} may accumulate, although the latter element may precipitate out upon appearance of appreciable concenrations of sulfide. Release of such materials from a flooded soil into the overlying water column is depicted in Figs. 6, 7, and 8.

Influence of other factors

In the newly-filled reservoir, several additional factors govern the impact that the soil flooding process has on reservoir water quality. Reservoir operation plays a predominant role. Also of major importance are the effects of reservoir inflows and uncleared vegetation, particularly standing trees. Because these factors have already been considered in detail (Chapters 1 and 2), we will only given them brief consideration here.

Waters flowing into the reservoir (inflows) can interact with flooded soils, producing synergistic positive or negative effects. Placement of inflows, or the level at which inflows move into and flow through the reservoir, has several ramifications in this regard. Depending upon their density relative to that of the reservoir water, inflows may move into the empilimnion (overflows), flow into the metalimnion (interflows), or plunge to the bottom (underflows). In the latter case, if the reservoir is equipped with bottom withdrawal facilities, it is possible for an inflow to move along the bottom of the reservoir and flow out the other end. This may have several possible effects on water quality with respect to flooded soil-water column interactions. If the inflow is fully saturated with dissolved oxygen and much lower in temperature than the reservoir hypolimnion, the inflow may serve beneficially by keeping decompositional processes from driving the hypolimnion anaerobic and by sweeping material released from the flooded soil out of the reservoir, thus preventing accumulation of undesirable products that ordinarily would be stored up and then dumped in slug loads. In addition, if the inflows transport large loads of relatively inert sediments into the reservoir, deposition of these sediments over the newly flooded soils may cover the soils, preventing release of undesirable materials

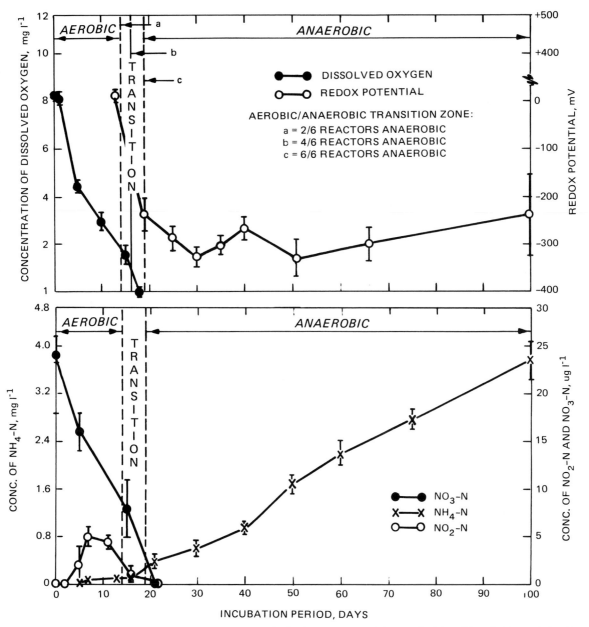

Fig. 6. Dynamics of the various forms of inorganic nitrogen in flowing waters overlying an alluvial prairie soil. Note relationship of disappearance of nitrate-N and appearance of ammonium-N (below) to aeration status of the water column (above). Study was done using the reaction chamber depicted in Fig. 2.

into the hypolimnion. Alternatively, if the inflows are anaerobic and/or contain large loads of biologically available organic matter, they may tend to force the hypolimnion into anaerobiosis, thus intersifying problems resulting from anaerobic releases from flooded soils. In a large reservoir, the problem can develop within the reservoir itself, with the inflow picking up large loads of decomposable organic matter at the head of the reservoir, transporting the maerial downstream, and adding these substances to those already present, creating a situation for intense anaerobiosis near the dam.

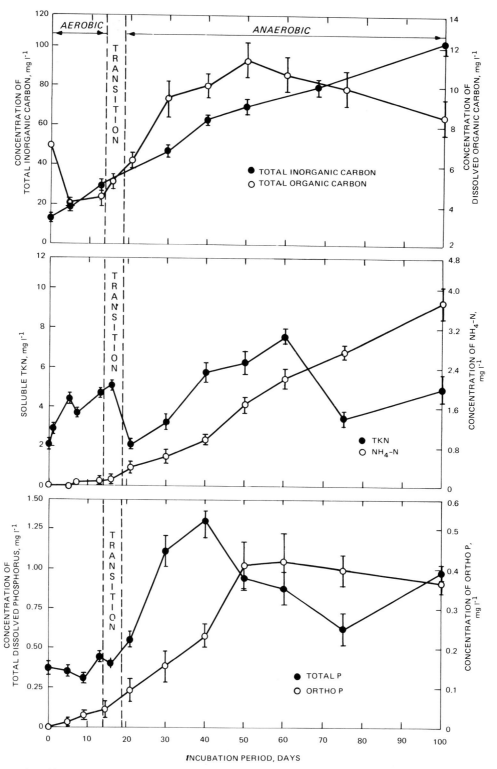

Fig. 7. Release of total inorganic carbon and soluble forms of organic carbon (above), total Kjeldahl nitrogen (TKN) and ammonium nitrogen (NH₄-N) (middle) total phosphorus (total P), and orthophosphate phosphorus (ortho P) (below) into flowing waters overlying an alluvial prairie soil in soil-water reaction chambers (modified from Gunnison *et al.*, 1980a).

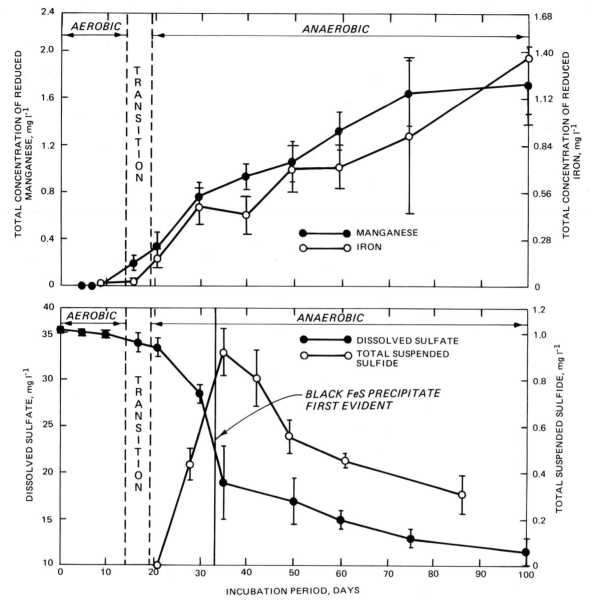

Fig. 8. Release of reduced manganese and iron (iron) and formation of sulfide (below) into flowing waters overlying an alluvial prairie soil in soilwater reaction chambers.

Volume and seasonality of inflows can also influence soil-water interactions. In certain reservoirs, there are little or no inflows during dry seasons. In other cases, inflow volumes and outflow conditions are sufficient to replace the entire reservoir volume several times in one year. All of this can dictate the pooling of soil-water releases or the net flushing of soil-water releases in a given time span.

Uncleared vegetation also influences soil-water column interactions. Standing trees, shrubs, and bushes can impede water movement, permitting establishment of areas or pockets with little or no circulation. The resulting stagnation may encourage development of anaerobic sections where intensive release of substances from soil to water can occur, particularly during periods of low flow; these may later add their materials to the remaining reservoir water when periods of stronger circulation occur. In this regard, reservoir hydrodynamics – the way water moves through the reservoir ecosystem – can exert a moderating influence. As mentioned previously in this book (Chapter 2), water moving through a reservoir tends to follow the Thalweg – the original river course. This, in combination with standing trees, old structures, and the like, tends to isolate areas on the side of the reservoir from the main stream with obvious effects on materials released from flooded soil. Decomposition of above ground vegetation adds materials to the water column and depletes dissolved oxygen levels, the latter process pushing the surrounding water towards anaerobiosis.

Reservoir operational modes further influence the system by determining the rate of flow of water moving through the system, the retention time of water in the system, and the daily and seasonal periodicity of releases. Waters having a longer retention time will have a greater contact period with the newly flooded soil and thus more opportunity to acquire high levels of materials released from soil.

Placement of the withdrawal structure – the location of the intake for reservoir discharges – affects the downstream consequences of soil-water interchanges. Reservoir discharges drawn from the bottom will tend to be cooler and to carry higher levels of materials released from the flooded soil than will waters removed from the surface. Initial discharges made from the anaerobic hypolimnion of a new reservoir will often contain reduced products such as NH_4^+, Mn^{2+}, PO_4^{3-}, and Fe^{2+} or S^{2-} as well as soluble organic matter; these will tend to have undesirable effects upon downstream areas, resulting in high biological and chamical oxygen demands, unaesthetic properties related to odor, taste, and discoloration, and high water treatment costs for the downstream user.

Several approaches to minimize the negative consequences of releases from newly filled reservoirs have been tried, with varying degrees of success. Perhaps the most successful of all procedures in terms of minimizing negative effects of soil flooding on the water quality in new reservoirs is the clearing of all surface vegetation and stripping of the litter and soil A-horizon layers from the soil

surface (Campbell *et al.*, 1976). However, while the water quality is vastly improved, the biological productivity markedly declines owing the lack of shelter and food (Ploskey, 1982). In addition, the cost of this type of undertaking is prohibitive for all but the smallest, flattest reservoir basin, and if filling does not immediately follow clearing, vegetative regrowth may negate many of the benefits gained by clearing.

Aeration or injection of oxygen into the hypolimnion can be effective in preventing development of anaerobic conditions. Again, the cost for applying this procedure to a large reservoir project is often prohibitively high. While the cost of operating such a system can be lowered somewhat by using it only during seasons when the reservoir tends to become anaerobic, the monitoring program required to determine when to initiate and when to terminate the aeration process detracts from this.

Use of flushing or repetitive filling and emptying and use of staged or incremental filling offer unique operational approaches to the flooded soil release problem. In the first case, rapid movement of water into and out of the reservoir system promotes removal of readily soluble materials from the soil surface while simultaneously preventing large concentrations of any one substance from accumulating. In the second case, keeping the reservoir relatively shallow over the area most recently flooded promotes wind mixing of the water, tending to keep aerated waters over areas where the most vigorous decompositional processes are occurring. While both of these methods can be effective, they each tend to delay the initial use of the preservoir for its intended purposes and are, again, difficult to apply to large projects.

Many newer reservoir projects are equipped with selective withdrawal facilities – the capacity to remove water from one or more of various depths in the reservoir. This capability has been used to minimize problems with new reservoir filling and during the transition phase by using surface withdrawal only for perods of anoxia in the hypolimnion. While this prevents release of materials from the hypolimnion, it is also promotes stagnation in the hypolimnion, permitting high levels of soil-released materials to accumulate. Moreover, such withdrawal facilities are costly and not all reservoirs are equipped with them; this is particularly true of projects having a hydropower as a principal function, and hydropower projects often have the greatest problems with negative downstream impacts of releases. This problem is compounded by the rigid operational schedule used for most hydropower projects, which do not allow for maximizing variations in the use of operational procedures to prevent release of poor quality waters that result from soil-water interactions.

Effect of reservoir aging

The primary effect of prolonged submergence on flooded soils is to permit a gradual depletion of microbially-available organic matter and of products of soil origin that are released to the water column. We are referring here to the original soil only and are not giving consideration to deposition of additional sediments or to sedimentation of algae or other biological sources of organic matter that may extend the depletion process.

Upon submergence, the soil and its attendant microflora undergo the sequence of events described previously in this chapter. The reduced soil becomes depleted of most of its soluble organic matter with the bulk of the remaining material being refractory to microbial attack, at least under anaerobic conditions. Iron and sulfur exist in their reduced forms with the resulting ferrous sulfide undergoing diagenetic processes to form more complex versions of this material. Any additional iron plus manganese, and phosphorus that have not been released to the water column are bound to the flooded soil in unavailable form, again barring any substantial change in the soil redox status.

This description, while oversimplified, fairly accurately depicts what may be expected if a soil sample is taken from a potential reservoir site, brought to the laboratory, and subjected to a prolonged period of submergence under continuous flow conditions. An example of this process is shown in Fig. 9 where an alluvial soil, having components of fertile prairie soil origin, was flooded for a one year period. Gradual declines in the releases

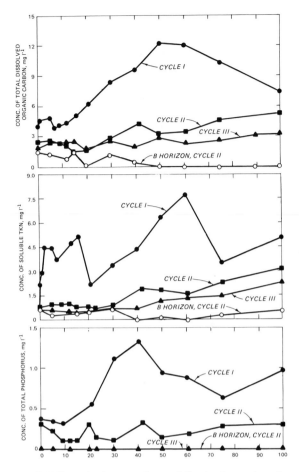

Fig. 9. Effect of aging on release of dissolved organic carbon (above), soluble total Kjeldahl nitrogen (middle) and total phosphorus (below) into flowing waters overlying an alluvial prairie soil. Soil-water reaction units were completely drained of water and then refilled at the end of each cycle and then equilibrated for 30 days prior to start of the next cycle. Cycles I, II, and III were done with the A layer soil horizon. Cycle II of a run using the B layer soil horizon is shown as a reference.

of total phosphorus, total organic carbon, and total Kjeldahl nitrogen within the overlying water column are representative of what occurs in the real reservoir system. In another study, oxygen consumption and releases of carbon dioxide and methane from a flooded alluvial piedmont soil were found to decrease with aging (Fig. 10). Such laboratory studies have the advantage of predicting what may be expected in soil-water interactions in the actual system, but lack the capacity to simulate the effects of the presence of large amounts of vegetation, the deposition of sediments by inflows, and

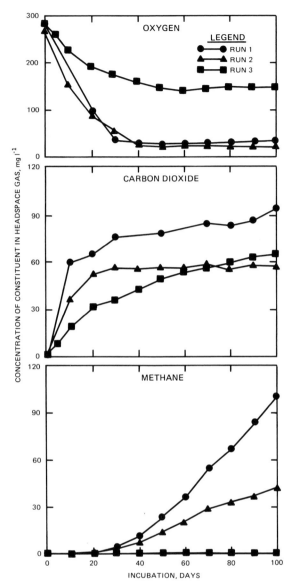

Fig. 10. Changes in the pattern of oxygen consumption (above) and evolution of carbon dioxide (middle) and methane (below) with time and aging in the headspace above flowing waters overlying an alluvial prairie soil. Soil-water reaction units were completely drained of water and then refilled at the end of each run and then equilibrated for 30 days prior to start of the next run. Systems were sealed from atmospheric contact at 0 days, but inflowing waters were fully saturated with dissolved oxygen.

the settling of autochthonous biological materials such as algal blooms in prolonging the duration of the transition phase. In this regard, use of ecological computer models, such as those decribed in Chapter 9, can be of benefit.

The aging process itself has been observed in many reservoirs with descriptions of the process available from several authors (McLachlan, 1977; Sylvester & Seabloom, 1965; Wilroy & Ingols, 1964). The length of the transition phase is site-specific and cannot, at present, be predicted with certainty for a given reservoir, although the process generally requires from six to ten years after initial filling. At the last stage of the transition period, levels of undesirable materials in the reservoir water column decrease overall water quality and biological productivity declines (Wilroy & Ingols, 1964; Ploskey, 1982). Microbial processes in newly flooded soils of reservoirs are of great importance in terms of their influence on the initial water quality. The benefits to be obtained from understanding the means by which to modify or minimize these effects are significant.

Summary and conclusions

Major factors of concern

Upon filling, a new reservoir undergoes several years of biological and chemical changes resulting from the decomposition of flooded organic matter. Inundation of brush, debris, and standing timber is desirable because of the refuge and food these materials provide for fish. However, decay of these substances lowers dissolved oxygen, and microbial activity causes a release of objectionable materials, making achievement of water quality objectives difficult.

This chapter summarizes the results of approaches developed to determine the effects of flooding of soil in new reservoirs as they relate to reservoir water quality. Emphasis is placed on the chemical and microbial activities involved in release of nutrients and contaminants from soils flooded by newly impounded reservoirs.

Investigations conducted to date have demonstrated that processes occurring in newly flooded soils, particularly those having high contents of organic matter, strongly influence water quality. The intensity of these activities is a function of the properties of the soil and this, in turn, is site specific. The following concerns related to soil flooding and decomposiion of organic matter were considered in detail in this chapter.

Depletion of dissolved oxygen

Factors influencing the rate and extent of dissolved oxygen depletion in reservoirs that should be considered in the evaluation of the effects of soil flooding and decomposition of organic matter on reservoir water quality are as follows:

Composition of unremoved soils that have high biological oxygen demands. Decomposition of organic matter in such materials will readily remove oxygen from waters not having continuous sources of reaeration.

Temperature of waters overlying unremoved materials that have a high biochemical oxygen demand (BOD). Rate of depletion is directly related to temperature, and increased temperatures may cause depletion to exceed reaeration.

Release of nutrients and metals

Plant nutrients, including carbon dioxide, inorganic phosphorus, inorganic sulfur, and inorganic nitrogen are released by decompositional processes acting on flooded sols. These nutrients are then available to support the growth of bloom-producing algae with consequent undesirable effects. Likewise, reductive processes release iron and manganese. Except for the following, the factors that influence the rate and extent of nutrient and metal releases are the same as for dissolved oxygen:

Aerobic conditions in areas of nutrient release will favor production of carbon dioxide, sulfate.sulfur, and nitrate-nitrogen.

Anaerobic conditions in areas of nutrient release will favor production of carbon dioxide and methane, ammonium-nitrogen, sulfide-sulfur, orthophosphate-phosphorus, iron, and manganese.

Release of color-producing substances

Soils that contain high amounts of organic matter will tend to release large amounts of color-producing substances upon flooding. These materials are readily leached from soil, and the improvement in water quality, from a coloration standpoint, occurs rapidly with reservoir aging.

Conclusions

Concerns for soil flooding and decomposition of organic matter must include consideration of the possible detrimental effects on water quality of not removing trees, herbaceous plants, forest litter, and soil surface layers. These concerns must be balanced with the understanding that algal blooms supported by nutrients released from newly flooded soils and decomposing forest litter and herbaceous plants, serve as food sources for microorganisms, zooplankton, benthos, and some fish, while both herbs and woody vegetation provide cover for young fish and spawning sites for adult fishes. Removal of these materials can be costly, and therefore, it is important to assess whether the presence of these materials may contribute towards development of prolonged periods of anoxia in the hypolimnion. If a strong potential for hypolimnetic anoxia exists, particularly in a project with hypolimnetic withdrawal, removal of materials in the hypolimnetic region prior to filling may be of greater value than leaving them as a potential contribution to the reservoir fishery. In addition, nutrients and metals released by microbial processes in flooded soils and decomposing organic matter may, in themselves, contribute to poor water quality both within and downstream from the new reservoir.

Reservoir filling practices that mitigate against adverse effects of soil flooding and decomposition of vegetation include seasonal timing of inundation, control of water levels, and the use of staged (or incremental) filling. Timing of filling to mimic spring flooding and control of water levels through staged filling to inundate areas covered by new pool levels at intervals of 2 to 3 years can be effective in increasing fish production although practicality may limit this to large storage reservoirs. Moreover, use of staged filling also keeps newly inundated areas shallow and subject to wind mixing. This will tend to maintain aerobic conditions in areas where the strongest release of nutrients from flooded soil and the most vigorous decomposition is occurring, resulting in improved water quality.

With rapidly increasing public pressure being placed on US reservoirs, it is important to consider ways to provide high water quality when considering the effects of soil flooding and decomposition of vegetation on reservoir water quality. Soil flooding and decomposition of vegetation must be considered as potentially significant sources of oxygen depletion, nutrients, and undesirable compounds, including reduced forms of iron and manganese and toxic substances, plus hydrogen sulfide. These factors must be evaluated and appropriate corrective measures taken in order to ensure minimal negative impacts on reservoir water quality when the new reservoir is filled and operation initiated.

Acknowledgements

This research was supported by the Environmental and Water Quality Operational Studies Program of the US Army Corps of Engineers. Permission was granted by the Chief of Engineers to publish this information.

References

Abram, J.W. & Nedwell, D.B., 1978. Inhibition of methanogenesis by sulfate reducing bacteria competing for transferred hydrogen. Arch. Mikrobiol. 117:89–92.

Alexander, M., 1977. Introduction to soil microbiology, 2nd Edition, John Wiley and Sons, Inc. New York, 467 pp.

Bautista, E.M. & Alexander, M., 1972. Reduction of inorganic compounds by soil microorganisms. Proc. Soil Sci. Soc. Amer. 36:918–920.

Bohn, H.L., 1968. Electromotive force of inert electrodes in soil suspensions. Proc. Soil Sci. Soc. Amer. 32:211–215.

Brannon, J.M., Gunnison, D., Butler, P.L. & Smith, I., Jr., 1978. Mechanisms that regulate the intensity of oxidation-reduction in anaerobic sediments and natural water systems. Technical Report Y–78–11. Environmental Laboratory, US Army Engineer Waterways Experiment Station, Vicksburg, MS.

Campbell, P.G., Bobee, B., Caille, A., Demalsy, M.J., Demalsy, P., Sasseville, J.L., Visser, S.A., Coutoure, P., Lachance, M., Lapointe, R. & Talbot, L., 1976. Effects du décapage de la cuvette d'un réservoir sur la qualité de l'eau emmagasinée: Élaboration d'une méthode d'étude et application au Réservoir de Victoriaville (Rivière Bulstrode, Québec). Rapport scientifique No. 37. Université de Québec, Canada., 238 pp.

Connell, W.E. & Patrick, W.H., Jr., 1968. Sulfate reduction in soil: Effect of redox potential and pH. Science 159:86–87.

DeLaune, R.D., Patrick, W.H., Jr. & Brannon, J.M., 1976. Nutrient transformations in Louisiana salt marsh soils. Sea Grant Publ. No. LSU–T–76–009. Center for Wetland Resources, Louisiana State University, Baton Rouge, LA.

Doetsch, R.N. & Cook, T.M., 1973. Introduction to bacteria and their ecobiology. University Park Press, Baltimore. 371 pp.

Engler, R.M. & Patrick, W.H., Jr., 1973. Sulfate reduction and sulfide oxidation in flooded soil as affected by chemical oxidants. Proc. Soil Sci. Soc. Amer. 37:685–688.

Gorham, E., 1958. Observations on the formation and breakdown of the oxidized microzone at the mud surface in lakes. Limnol. Oceanogr. 3:291–298.

Gottschalk, G., 1979. Bacterial metabolism. Springer-Verlag, New York., 231 pp.

Gunnison, D., Brannon, J.M., Smith, I., Jr., Burton, G.A. & Butler, P.L., 1979. Appendix B. A determination of potential water quality changes in the hypolimnion during the initial impoundment of the proposed Twin Valley Lake. In Water quality evaluation of proposed Twin Valley Lake Wild Rice River, Minnesota. Technical Report EL–79–5. Environmental Laboratory, US Army Engineer Waterways Experiment Station, Vickburg, MS.

Gunnison, D., Brannon, J.M., Smith, I., Jr. & Burton, G.A., 1980a. Changes in respiration and anaerobic nutrient regeneration during the transition phase of reservoir development. In Barica, J. & Mur, L.R., (eds.), Hypertrophic ecosystems. SIL Workshop on hypertrophic ecosystems. Developments in Hydrobiol. 2:151–158.

Gunnison, D., Brannon, J.M., Smith, I., Jr. & Preston, K.M., 1980b. A reaction chamber for study of interactions between sediments and water under conditions of static or continuous flow. Wat. Res. 14:1529–1532.

Gunnison, D., Chen, R.L. & Brannon, J.M., 1983. Relationship of materials in flooded soils and sediments to the water quality of reservoirs – I. Oxygen consumption rates. Wat. Res. 17:1609–1617.

Gunnison, D., Brannon, J.M., Chen, R.L., Smith, I., Jr. & Sturgis, T.C., 1984. Richard B. Russell Dam and Reservoir: Potential water quality effects of initial filling and decomposition of vegetation. Miscellaneous paper MP–E–84–2. Environmental Laboratory, US Army Engineer Waterways Experiment Station, Vicksburg, MS.

Hasan, S.M. & Hall, J.B., 1975. The physiological function of nitrate reduction in Clostridium perfringens. J. gen. Microbiol. 87:120–128.

Jørgensen, B.B., 1978. A comparison of methods for the quantification of bacterial sulfate reduction in coastal marine sediments. II. Calculations from mathematical models. Geomicrobiol. J. 1:29–51.

Knowles, R., 1979. Denitrification, acetylene reduction, and methane metabolism in lake sediment exposed to acetylene. Appl. envir. Microbiol. 38:486–493.

Koike, I. & Hattori, A., 1978. Denitrification and ammonia formation in anaerobic coastal sediments. Appl. envir. Microbiol. 35:278–282.

Laanbroek, H.J. & Pfennig, N., 1981. Oxidation of short-chain fatty acids by sulfate-reducing bacteria in freshwater and marine sediments. Arch. Mikrobiol. 128:330–335.

Lovely, D.R., Dwyer, D.F. & Klug, M.J., 1982. Kinetic analysis of competition between sulfate reducers and methanogens for hydrogen in sediments. Appl. envir. Microbiol. 43:1373–1379.

Lovely, D.R. & Klug, M.J., 1983. Sulfate reducers can outcompete methanogens at freshwater sulfate concentrations. Appl. envir. Microbiol. 45:187–192.

McLachlan, A.J., 1977. The changing role of terrestrial and autochthonous organic matter in newly flooded lakes. Hydrobiologia. 54:251–217.

Mann, P.J.G. & Quastel, J.H., 1946. Manganese metabolism in soils. Nature. 158:154–156.

Mortimer, C.H., 1941. The exchange of dissolved substances between mud and water in lakes. J. Ecol. 29:208–329.

Mortimer, C.H., 1942. The exchange of dissolved substances between mud and water in lakes. J. Ecol. 30:147–201.

Nhung, M.F.M. & Ponnamperuma, F.N., 1966. Effects of calcium carbonate, manganese dioxide, ferric hydroxide, and prolonged flooding on chemical and electrochemical changes and growth of rice in a flooded acid sulfate soil. Soil Sci. 102:29–41.

Patrick, W.H., Jr. & DeLaune, R.D., 1977. Chemical and biological redox systems affecting nutrient availability in the coastal wetlands. Geosci. Man. 18:131–137.

Patrick, W.H., Jr. & Mikkelson, D.S., 1971. Plant nutrient behavior in flooded soil. In Fertilizer technology and use. Soil Sci. Soc. Amer., Madison, Wisconsin.

Patrick, W.H., Jr. & Tusneem, M.E., 1972. Nitrogen loss from flooded soil. Ecology 53:735–737.

Payne, W.J., 1973. Reduction of nitrogenous oxides by microorganisms. Bact. Rev. 37:409–452.

Ploskey, G.R., 1981. Factors affecting fish production and fishing quality in new reservoirs with guidance on timber clearing, basin preparation, and filling. Technical Report E–81–11. Environmental Laboratory, US Army Engineer Waterways Experiment Station, Vicksburg, MS.

Ponnamperuma, F.N., 1972. The chemistry of submerged soils. Adv. Agronomy. 24:29–88.

Ponnamperuma, F.N. & Castro, R.V., 1964. Redox systems in submerged soils. 8th Internat. Congr. Soil Sci., Bucharest, Romania.

Ponnamperuma, F.N., Tianco, E.M. & Loy, T.A., 1967. Redox equilibria in flooded soils. I. The iron oxyhydroxide system. Soil Sci. 103:374–382.

Reddy, K.R. & Patrick, W.H., Jr., 1975. Effect of alternate aerobic and anaerobic conditions on redox potential, organic matter decomposition, and nitrogen loss in a flooded soil. Soil biol. biochem. 7:87–94.

Reddy, K.R., Patrick, W.H., Jr. & Phillips, R.E., 1980. Evaluation of selected processes controlling nitrogen loss in a flooded soil. J. soil Sci. Amer. 44:1241–1246.

Sokatch, J.R., 1969. Bacterial physiology and metabolism. Academic Press, London and New York, 443 pp.

Sørensen, J., 1978. Capacity for denitrification and reduction of nitrate to ammonia in a coastal marine sediment. Appl. envir. Microbiol. 35:301–305.

Stevenson, F.J., 1967. Organic acids in soil. In Paul, E.A. & Peterson, G.H., (eds.), Soil Biochemistry, Marcel Dekker, Inc., NY, 1:119–146.

Strayer, R.F. & Tiedje, J.M., 1978. In situ methane production in a small, hypereutrophic, hardwater lake. Loss of methane from sediments by vertical diffusion and ebullition. Limnol. Oceanogr. 23:1201–1206.

Sylvester, R.O. & Seabloom, R.W., 1965. Influence of site characteristics on the quality of impounded water. J. am. Wat. Wks. Ass. 57:1528–1546.

Takai, Y. & Kamura, T., 1966. The mechanism of reduction in waterlogged paddy soil. Folia microbiologia. 11:304–313.

Takai, Y., Koyama, T, & kamura, T., 1956. Microbial metabolism in reduction process of paddy soils (Part I). Soil Sci. plant nutri. 2:63–66

Turner, F.T. & Patrick, W.H., Jr., 1968. Chemical changes in waterlogged soils as a result of oxygen depletion. 9th Internat. Congr. Soil Sci., Adelaide, Australia. IV:53–65.

Vogels, G.D. & Van Der Drift, C., 1976. Degradation of purines and pyrimidines by microorganisms. Bact. Rev. 40:403–468.

Widdel, F. & Pfennig, N., 1977. A new anaerobic, sporing, acetate-oxidizing, sulfate-reducing bacterium, *Desulfotomaculum* (emend.) *Acetoxidans*. Arch. Mikrobiol. 112:119–122.

Wilroy, R.D. & Ingols, R.S., 1964. Aging of waters in reservoirs of the Piedmont Plateau. J. am. Wat. Wks. Ass. 56:886–890.

Winfrey, M.R. & Zeikus, J.G., 1977. Effect of sulfate on carbon and electron flow during microbial methanogenesis in freshwater sediments. Appl. envir. Microbiol. 33:275–281.

Winfrey, M.R. & Ward, D.M., 1983. Substrates for sulfate reduction and methane production in intertidal sediments. Appl. envir. Microbiol. 45:193–199.

Wolin, M.J. & Miller, T.L., 1982. Interspecies hydrogen transfer: 15 years later. ASM News. 48:561–565.

Yoshida, T., 1975. Microbial metabolism of flooded soils. In Paul, E.A. & McLaren, A.D. (eds.), Soil Biochemistry. Marcel Dekker, Inc., New York, 3:83–122.

Yamane, A. & Sato, K., 1968. Initial rapid drop of oxidation-reduction potential in submerged air-dried soils. Soil Sci. plant nutri. 14:68–72.

Author's address:

Douglas Gunnison and Robert M. Engler
Ecosystem Research and Simulation Division/EL
USAE Waterways Experiment Station
P.O. Box 631
Vicksburg, MS 39180

William H. Patrick, Jr.
Center for Wetland Resources
Louisiana State University
Baton Rouge, LA 70803

CHAPTER 4

Vegetative succession and decomposition in reservoirs

GORDON L. GODSHALK and JOHN W. BARKO

Abstract. Vegetative succession and decomposition in reservoirs are examined. Information is summarized on vegetative tolerance to inundation, subsequent aquatic plant succession, sources of detritus, factors affecting the rate and extent of decomposition, organisms involved in decomposition, the sequence of decomposition, and the effects of decomposition on water quality. The processes of vegetative succession and decomposition in reservoirs are essentialy the same as those occurring in natural lakes. Succession is hastened in reservoirs compared with natural lakes by greater watershed loadings and proportionately greater inputs of refractory detritus. The lifespan of reservoirs is relatively limited and will depend on several factors such as basin morphometry, flushing rate, and site preparation which interact to control rates of production and decomposition.

Introduction

Many reservoirs possess physical and hydraulic features notably different from those of natural lakes (Baxter, 1977; Wetzel, 1984). Moreover, reservoirs (particularly at their inception) are usually subject to relatively greater inputs of refractory detritus of terrestrial origin. Vegetative colonization and succession in reservoirs, dependent primarily upon basin morphometry, can be rapid due to high rates of sedimentation. The nature of detritus, varying with vegetative source, and the extent to which detritus is decomposed have important influences on water quality, and ultimately on reservoir ontogeny. This article examines vegetative succession and decomposition in reservoirs. Whereas our intention is not specifically to contrast these processes in lakes and reservoirs, we attempt elucidation where distinctions do exist.

Tolerance of terrestrial vegetation to inundation

Permanent flooding rapidly devastates most terrestrial vegetation. Death or severe injury to woody vegetation due to flooding is usually attributed directly to soil anoxia, but the accumulation of toxic products of reducing processes (cf. Drew & Lynch, 1980) predominating in flooded soils may also be contributory (reviewed by Gill, 1970). Few tree species or other woody plants are adapted to survive permanently flooded soil conditions because of difficulties in maintaining aerobic root respiration. Even the most flood-tolerant of trees need to be periodically emerged in order to rejuvenate root systems (Crawford, 1982).

Factors determining survival of woody plants to flooding include: specific morphological and physiological adaptations such as increased aerenchymous tissue in roots, adventitious and buttressed roots, and temporary anaerobic respiratory pathways which accumulate malate and other non-toxic compounds instead of ethanol; edaphic factors,

Gunnison, D. (ed.) Microbial Processes in Reservoirs.

especially clay content; flooding characteristics (seasonality, periodicity, duration, maximum and average depths); chemical composition of flood water; and age and size of flooded plants (see Gill, 1970 for specific details). Recent evidence suggests that, among these factors, tree mortality may be most closely related to water depth (Harms *et al.*, 1980). In an investigation of bottomland hardwood survival connected with lock and dam construction on the Upper Mississippi River, it was concluded that mortality, regardless of species, was confined to areas where water levels remained permanently above the root crowns (Green, 1947). Under such conditions, mortality of all temperate woody vegetation occurs in 3 to 4 years (Green, 1947; Crawford, 1982).

Herbaceous plants are generally more tolerant of flooding than are woody plants. In both types of plants, adaptability to flooding usually involves the provision of an adequate root ventilating system (Armstrong, 1975). Flood-tolerant plants are capable of transporting atmospheric oxygen to roots by molecular diffusion (Hutchinson, 1975; Armstrong, 1978) or by pressurized mass flow (Dacey, 1980, 1981). Gas exchange in these plants is enhanced by increased root porosity which further facilitates aerobic functioning (Crawford, 1982). In nearly all flood-tolerant plants, internal ventilation is aided by a network of internal air spaces (Williams & Barber, 1961), the volume of which varies considerably among species (Crawford, 1982, Table 14.5). Morphological and anatomical variations affecting internal ventilation are imporant in determining specific tolerances to flooding; however, the widely differing tolerances to flooding demonstrated by various species with similar morphologies suggest the importance also of metabolic adaptations to anoxia (Gill, 1970).

In general, only flood-tolerant plant species (usually herbaceous, i.e., wetland plants), and then only if not overtopped by flood water (cf. Sale and Wetzel, 1983), will survive permanent flooding. Inundated flood-intolerant vegetation will become available at various rates and in various forms (considered later in text) for microbial decomposition. Reservoirs subject to seasonally fluctuating water levels may continue to experience pulsed loadings of terrestrial vegetative debris. The magnitude of these loadings will vary with the extent and duration of water level fluctuations, and with the composition of vegetation in the periodically flooded watershed. Impoundments with extreme fluctuations in water level will have greater periodic inputs of terrestrial vegetation than occur in lakes or reservoirs that do not cyclically expose and submerse shoreline margins.

Vegetative succession following inundation

The evolutionary development of different aquatic plant life forms (see below, after Sculthorpe, 1967) facilitates variety in the occupation of newly-formed aquatic habitats.

Life form	Example
emergent	cattail
floating-leaved	waterlily
submersed	pondweed
free-floating	duckweed

Emergent, floating leaved, and submersed aquatic plants are rooted and attached to the sediment substratum. Free-floating plants often have roots, but these normally do not function in attachment. Emergent aquatic plants are structurally complex, and although physiologically adapted to constant or periodic standing water, are most similar to terrestrial herbaceous plants. Submersed aquatic plants lack structural rigidity, are dependent upon the buoyancy of the water for their support, and are the most highly adapted to the aquatic environment. Floating-leaved plants are morphologically intermediate between emergent and submersed life forms and possess physiological characteristics of both.

The colonization and establishment of aquatic plants in new impoundments can be very rapid (Eggler, 1961; Little, 1966). However, the extent of vegetative development within reservoirs is ultimately dictated by basin morphometry. Shallow reservoirs or those formed in dendritic river valleys with gently sloping bottoms will generally support greater aquatic plant populations than reservoirs that are deeper, rounder, or have steeper sides (see Baxter, Chapter 1). Submersed aquatic plants, with few exceptions, do not often grow beyond water depths of about 10 m (Wetzel, 1983), and are usually restricted by low light (Spence, 1982) or low temperature (Barko *et al.*, 1982) to somewhat lesser depths. Emergent aquatic plants are generally restricted to water depths of less than 1.5 m due to constraints on ventilation. Free-floating aquatic plants and rooted aquatic plants with floating leaves are subject to displacement or damage by waves and are therefore restricted to sheltered areas.

The majority of aquatic plants are rooted. Their growth and distribution are markedly affected by sediment composition (Pond, 1905; Pearsall, 1920; Misra, 1938; Moyle, 1945; Macan, 1977). In general, aquatic plants grow best on inorganic fine-textured sediments (Barko & Smart, 1983, 1984) which provide adequate anchorage and an excellent source of nutrients for uptake by roots. The provision of nutrients from both sediment and open water (see Denny, 1980) maximizes aquatic plant productivity within constraints posed by other environmental factors such as light, temperature, pH, inorganic carbon availability, etc. (Barko, 1981; Spence, 1982). Given favorable environmental conditions, the development of aquatic vegetation in reservoirs can be diverse and quite excessive (Peltier and Welch, 1970; Kight, 1980).

The succession of aquatic plant communities in natural lakes usually involves the progression over thousands of years from dominance by submersed to floating-leaved to emergent vegetation (Pearsall, 1920; Walker, 1972; Wetzel and Hough, 1973; Wetzel, 1979). The same successional pattern can be expected in reservoirs, but at greatly accelerated rates, often in less than a century due to greater sedimentation (Wetzel, 1979). Aquatic plants themselves contribute to sedimentation by intercepting soluble as well as particulate inputs from the watershed (Wetzel & Allen, 1972; Mickle & Wetzel, 1978a,b, 1979), and by contributing their own remains to the sediment (Wetzel, 1979; Carpenter, 1981a). Aging in lakes and reservoirs is a direct function of sedimentation, which can be greatly accelerated by the expansion of aquatic plant communities (Wetzel, 1979; Carpenter, 1981a, 1983).

Changes in aquatic plant community composition and/or in the mass of intercepted terrestrial inputs have an important influence on the extent of microbial decomposition, and ultimately on reservoir 'life expectancy.' Organic matter derived from emergent aquatic vegetation and in particular from most terrestrial vegetation is relatively resistant to decomposition (Godshalk and Wetzel, 1978a,b: also see later discussion in text). Decomposition is further retarded by increasing anaerobiosis (Rich and Wetzel, 1978). As a consequence sediment organic matter content increases as reservoirs age. Organic sediments are considerably less dense than inorganic sediments, and occupy a much greater volume than the latter (Barko, unpublished). In the final stages of succession, reservoir filling by organic matter will proceed at prodigous rates, eventually leading to encroachment by terrestrial vegetation (Wetzel, 1979).

Sources of detritus

One of the important ways that reservoirs differ from natural lakes is in their usually very large initial loading of refractory organic matter from inundated terrestrial vegetation. The magnitude of various components of detritus, namely trees, marsh plants, phytoplankton, etc., depends on reservoir filling, basin morphology, surface area, volume, and on site preparation.

Organic matter inputs as terrestrial vegetation likely to be flooded can be roughly estimated. A temperate deciduous forest has a dry biomass of 30 kg m^{-2}, a boreal forest 20 kg m^{-2}, and a temperate grassland 2 kg m^{-2} (Whittaker 1975; also Baxter, Chapter 1 and Gunnison *et al.*, Chapter 3). This biomass can be further partitioned into non-woody and woody tissues (Larcher, 1980). A hypothetical reservoir with average characteristics of all United States reservoirs having a usable capacity of 5,000 acre-feet or more (Todd, 1970) would have a surface area of 3,840 ha. If this reservoir was built in any of the three terrestrial communities delineated above without vegetative clearing, very large amounts of terrestrial biomass would be inundated in each case (Table 1).

The hypothetical 'average' reservoir would have a usable volume of $2.8 \times 10^8 \, m^3$ and an average mean depth (computed as volume/surface area) of 7.4 m. Assuming a shoreline development value of 2.0, this hypothetical reservoir would have a shoreline of 43.9 km. If it is further assumed that the littoral zone (i.e. plant-occupied shoreline) extends about 10 m above and below the water line, then a reservoir with sufficient nutrients to be classified as eutrophic (cf. Wetzel, 1983) would have rates of annual autochthonous production similar to those in Table 2.

In comparing these projections (Tables 1, 2), it is apparent that an amount of terrestrial vegetation many times greater than the subsequent annual autochthonous production of the reservoir may be available to undergo decomposition as impoundment takes place. Decomposition of flooded terrestrial vegetation may dominate all heterotrophic processes for several years, even though its decay rate is much lower than those of aquatic plant tissues produced annually in the reservoir (Fig. 1).

A detailed prediction of decomposition in new reservoirs was developed by researchers in the Soviet Union based on batch culture experiments simulating decay of several categories of vegetation

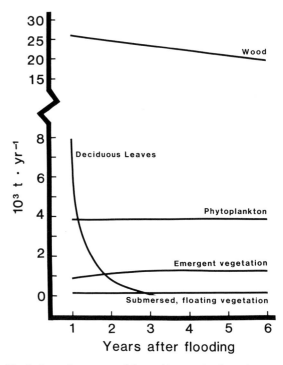

Fig. 1. Annual amounts of decay (thousands of metric tons of carbon) associated with various components of detritus in a hypothetical reservoir inundating uncleared temperate deciduous forest. See text for further explanation.

(Maystrenko & Denisova, 1972). Their predictions regarding the relative importance of various sources of organic matter were very good, but much different than for the hypothetical reservoir con-

Table 1. Types and amounts of terrestrial plant tissue that could be flooded by a hypothetical reservoir (see text) built in different natural communities. Units are metric tons of carbon per whole reservoir.

Terrestrial community	Non-woody tissue	Woody tissue
Temperate deciduous forest	8,650	568,000
Boreal forest	17,300	367,000
Grassland	38,400	–

Table 2. Source and amounts of autochthonous primary production in a hypothetical reservoir (see text). Areal productivity is in metric tons of carbon per hectare per year; total is metric tons per year for the entire reservoir.

Primary producers	Areal productivity	Total
Phytoplankton	1	3,840
Submersed, floating-leaved vegetation	3	66
Emergent, marsh vegetation	19	1,250

sidered above. For example, in the very large Kiev Reservoir, which has a storage volume of $3.7 \times 10^9 \, m^3$, the greatest input of organic carbon was the inflowing river (74% of total), then decomposition of plankton (20.7%), release from flooded soils (9.4%), decomposition of meadow, aquatic, and forest vegetation (3.1, 1.8, 1.1%, respectively), and precipitation (0.5%). (These percentages add to greater than 100 probably because they were estimated independently and contain error). Differences in results can be explained by the higher turnover rate of the Kiev Reservoir water and its large size compared to the hypothetical reservoir. Both of these factors would allow for high phytoplankton production relative to littoral production. A notable point here is that reservoirs are vastly different from each other, and any predictions should be limited to specific systems or reservoir types.

Factors controlling decomposition

Decomposition in aquatic ecosystems of all types has been studied extensively. A great deal is known about relative decay rates of a variety of substrates under an array of environmental conditions. Less is known, however, about the organisms responsible for decay and the ultimate fate of recalcitrant detritus. Various aspects of aquatic decomposition have been recently reviewed (Willoughby, 1974; Saunders, 1976; Saunders et al., 1980; Brinson et al., 1981). Decomposition processes, particularly

weight loss, have been mathematically described by several investigators. Most find a simple exponential decay equation statistically, if not biologically, satisfactory, and the resultant decay coefficients are used routinely to compare rates of decomposition (Table 3). More sophisticated models usually in some way take into account the fact that detritus is not homogeneous, and that rates of decay (i.e., percent loss per day) decrease through time (cf. Godshalk & Wetzel, 1978b; Carpenter, 1981b, 1982). Related considerations are elaborated below.

As with most biological processes, decay accelerates with increasing temperature. The breakdown of detrital organic compounds depends in large part on extracellular enzyme activity, which proceeds faster under warmer conditions. Low temperatures, however, do not halt decomposition processes; decomposition proceeds during the winter months even under ice-covered north temperate lakes (e.g., Boylen & Brock, 1973; Tison et al., 1980). Howard-Williams et al. (1983), in studies of decay of watercress (*Nasturtium officinale* R.Br.), found Q_{10} values (the factor of rate increase with a 10°C rise in temperature) of 1.15 to 1.81 for decay rates between 10 and 20°C. Federle et al. (1982a) measured a Q_{10} of 2.7 for *Carex* decay in an arctic lake, and Carpenter & Adams (1979) estimated a Q_{10} of 3 for *Myriophyllum* decomposing in a Wisconsin lake. Terrestrial leaves decay in streams at rates influenced by temperature (Kaushik & Hynes, 1971; Petersen & Cummins, 1974; Suberkropp et al., 1975).

Table 3. Exponential decay coefficients and other parameters representing rate of weight loss of various substrates during aquatic decomposition. Data compiled from various sources.

Tissue type	Intermediate k^* (range)		Percent loss per year	Mean persistence time (days)[†]
Phytoplankton	0.0159	(0.0055 –0.0320)	99.7	44
Submersed, floating vegetation	0.0080	(0.0051 –0.0370)	94.6	87
Emergent, marsh vegetation	0.0031	(0.0008 –0.0090)	67.7	224
Deciduous leaves	0.0064	(0.0007 –0.0175)	90.3	108
Conifer needles	0.0049	(0.0009 –0.0131)	83.4	141
Wood	0.00013	(0.00006–0.0025)	4.6	5,332

* Percent remaining = (100) e^{-kt}, t = days
[†] 'Biological halflife'

Oxygen content of the water is another important regulator of decomposition. Numerous studies have demonstrated that anaerobic decay rates are significantly less than aerobic decay rates (Mills & Alexander, 1974; Reddy & Patrick, 1975; Godshalk & Wetzel, 1978b). Lack of oxygen affects decomposition also by limiting the activity of some critical enzymes. Lignin, for instance, does not decay appreciably under anoxic conditions due to reduced enzyme activity (Hackett *et al.*, 1977). More importantly, decay in anoxic compared to oxic conditions is not as complete (i.e., organic matter is not reduced to the reactants of photosynthesis: carbon dioxide and water), since the oxidizing power and abundance of alternate electron acceptors (e.g., sulfate and nitrate) are usually inadequate relative to demand (Godshalk & Wetzel, 1978a; Jones & Simon, 1981; King & Klug, 1982). In some lakes anaerobic decomposition predominates (e.g., Molongoski & Klug, 1980b). Decay under anaerobic conditions causes accumulation of a wide variety of partially oxidized chemical compounds. Some of these (e.g. ammonia, fatty acids, carbohydrates) can be used by the biota, but many (e.g. sulfides, tannins, phenols, lignocellulose) are resistant and/or inhibitory to metabolic processes.

Plant materials dried before being subjected to decomposition under water have repeatedly been shown to more quickly release ions, produce dissolved organic matter (DOM), and lose more weight initially than non-dried materials (Brock *et al.*, 1982; Furness & Breen, 1982; Larsen, 1982). However, dried plant materials may show slower and less extensive decay over long time periods than non-dried plant materials (Rogers & Breen, 1982). It is reasonable to expect that organic substrates on wet, but not totally submerged soils, would decompose most rapidly because of their concomitant exposure to dissolution, microbes, and atmospheric oxygen, but this is apparently not so. Studies of decomposition in floodplain ecosystems have demonstrated that leaves disappear faster when submersed than when above the water surface, and that repeated cycles of inundation and desiccation do not accelerate decomposition rate (Brinson, 1977; Day, 1982, 1983; cf., however, Reddy & Patrick, 1975).

Nutrient availability has various effects on the rates and dynamics of decomposition. The influence of nitrogen on decay is best studied and perhaps most variable. Generally, plant tissues with high nitrogen contents decompose quickly and release nitrogen to the environment (Triska *et al.*, 1975; Godshalk & Wetzel, 1978b; Aumen *et al.*, 1983; Melillo *et al.*, 1983). In a detailed study of nitrogen release during decomposition, Howard-Williams *et al.*, (1983) observed a clear succession of forms of dissolved nitrogen produced, corresponding to steps in the nitrogen cycle: first to accumulate was ammonium followed by nitrite, then nitrate. Dissolved organic nitrogen was the dominant form at the end of decomposition. In a rigorous and extensive study of terrestrial decomposition, which probably applies to aquatic decay as well, Zieliński (1980) compared the degradation of 31 species of agricultural plants and determined statistically that the factor most influential in determining decay rates was initial nitrogen content of the tissue. He also observed that tissues, depending on initial nitrogen content, would either release or immobilize nitrogen until the tissue nitrogen content approached a value of ca. 3.5%. Nitrogen that accumulates on detritus is a component of microbial biomass and probably also complexed exoenzymes (Iversen, 1973; Haslam, 1974; Harrison & Mann, 1975a; Suberkropp *et al.*, 1976).

Nitrogen to promote decay may be obtained from the detrital substrate, from the surrounding water, or both. Baker *et al.* (1983) found that wood of red alder (*Alnus rubra*) promoted associated nitrogen fixation during decomposition and, as a result, decayed faster than wood of Douglas fir (*Pseudotsuga menziesii*), which did not promote nitrogen fixation. However, decay of particularly recalcitrant components of tissue, especially if isolated, is not always stimulated by nitrogen additions (Federle & Vestal, 1980b). Polunin (1982) compared the response of microbes on fresh and decomposed litter of *Phragmites*. He found that aerobic respiration of microbes increased on fresh detritus when nitrogen and phosphorus were added, but addition of glucose had no effect. Later, when only the most refractory components of the detritus remained, addition of a readily usable substrate such as glucose caused increased microbial respiration, but added nitrogen and phosphorus did not.

The role of phosphorus and other elements in decomposition has been less well studied than that of nitrogen. In some studies, phosphorus has been found to be important in affecting decomposition, while in other investigations it has not (Kaushik & Hynes, 1971; Nichols & Keeney, 1973; Carpenter & Adams, 1979; Federle & Vestal, 1980b; Federle et al., 1982b). Release and accumulation of many elements including nitrogen and phosphorus by decomposing plant tissue is variable, depending on environmental conditions and plant species (e.g., Mason & Bryant, 1975; Howard-Williams & Howard-Williams, 1978; Davis & van der Valk, 1978).

The substrate itself has much to do with the rate at which it is decomposed. Plant tissues subject to decomposition in aquatic environments are highly diverse with respect to composition, structure, size, and surface area. At one extreme, algal cells are small, contain little structural tissue that is resistant to decay, and are readily decomposed. Phytoplankton cells are generally thought to be decomposed largely while still in the water column, i.e., they do not reach the lake bottom (Kleerekoper, 1953; Golterman, 1972; cf., however, Wetzel et al., 1972, Molongoski & Klug, 1980a). However, Gunnison & Alexander (1975) have shown that algal cells of different species are not equally susceptible to microbial decay. At the other extreme, most vascular plants have greater contents of decay-resistant structural tissues such as cellulose and lignin, an anatomy that includes layers of cuticle or sclerenchyma retarding mechanical and microbial breakdown, and a much lower surface: mass ratio than phytoplanktonic organisms (Godshalk & Wetzel, 1978b; Panwar & Sharma, 1981).

Vascular plants show vast differences in their susceptibility to decomposition. Aquatic plants generally decay quickly if they are soft, flaccid, and pithy (e.g. water lilies) or finely dissected (most submersed aquatic plants). Emergent aquatic plants with significant amounts of cellulose and lignin, tough vascular tissues, and protective outer layers decay more slowly (Godshalk & Wetzel,

1978b; Danell & Sjöberg, 1979). Esteves & Barbieri (1983) and others have demonstrated that leaves generally decay faster than petioles and stems. The underground portions of aquatic plants, which in emergent species often dominate their biomass, decay less rapidly than the aboveground plant parts (Hackney & de la Cruz, 1980; Howard-Williams et al., 1983)

Terrestrial plants are more resistant to decomposition than are most aquatic plants. Tree leaves, an important source of energy for many aquatic ecosystems, have been studied extensively with regard to their decay in streams (e.g., Kaushik & Hynes, 1968, 1971; Petersen & Cummins, 1974) and lakes (e.g., Gasith & Lawacz, 1976; Reed, 1979; Hanlon, 1982). Decay of tree leaves is quite variable. Some leaves decay more readily than flaccid aquatic plants while others are more resistant than leaves of emergent aquatic or herbaceous terrestrial vegetation.

Relatively little is known of the decomposition of woody tissues, even though in some aquatic systems wood may be a much larger component of the detrital mass than other plant tissues (Ball et al., 1975; Anderson et al., 1978). Wood decays extremely slowly. Features unique to decomposition of wood reflect this tissue's high refractility under water, large particle size, and variable composition. Whereas tree leaves may have a fiber (cellulose plus lignin) content of 20 to 40% and a carbon:nitrogen ratio of 25 to 100:1, wood may be 70 to 80% fiber and have a carbon:nitrogen ratio of 300 to 1000:1 (Anderson et al., 1978). This has a strong bearing on degradability, since the carbon:nitrogen ratios are well outside the optimum levels for microbial utilization (~10:1) (Alexander, 1977). Attack on wood, whether by microbes or animals, in freshwater systems is largely limited to exposed surfaces. The surface area: volume ratio of detrital wood likely to be in a reservoir is typically much lower than for particles of herbaceous tissues and varies with tree size and overall maturity of the forest (Anderson et al., 1978). Bark decomposition in laboratory situations varies broadly even within a single species, depending on where on the tree and the time of year samples are collected (Solbraa, 1979).

Mechanical factors are also important in regulating decomposition in aquatic ecosystems. Moving waters provide abrasion and breakage and also replenish oxygen and nutrients (Witkamp & Frank, 1969; Hodkinson, 1975; Gasith & Lawacz, 1976; Federle *et al.*, 1982a). Finer detrital particles are generally more accessible to attack by microorganisms because of their relatively larger surface area (Fenchel, 1970). However, particles that are very fine may tend to pack, impeding decay. Sedimentation of fine inorganic particles (silts and clays, common in many reservoirs) may inhibit decomposition by burying detritus, isolating it from oxygen and nutrients, sequestering nutrients by adsorption, and packing the particles more tightly. Reice (1974, 1977) suggests that sediment particle size influences rates of leaf decomposition in streams. Decay proceeds faster when detritus is associated with surficial bottom sediments than when it is suspended in open water (Brock *et al.*, 1982) or buried in deeper sediments (Hackney & de la Cruz, 1980).

Decomposer organisms

Several types of organisms are involved in aquatic decomposition. Bacteria and fungi are probably the most important and certainly the best known. Protozoans may also be important, but very little research has been performed to elucidate their role in decomposition (cf. however, Fenchel, 1975; Harrison & Mann, 1975b). The impact of multicellular animals on decomposition processes in aquatic habitats is uncertain despite extensive research efforts on their interaction with and uses of detritus.

Bacteria are the most versatile decomposers, both in terms of habitats exploited and substrates metabolized. Bacteria can be divided into two groups: attached and free-living (not attached). The interrelationships and interchangeability between these two groups are not clear. The two groups are also separated functionally by their differing specializations of nutritional sources. Free-living bacteria take up DOM from the open water (Kang & Seki, 1983; Kato & Sakamoto, 1983), and attached bacteria use exoenzymes to mineralize detrital particles to which they are attached (Bobbie *et al.*, 1978; Tanaka & Tezuka, 1982).

Bacteria differ in abundance and activity from lake to lake and even among regions in the same lake as a result of environmental factors. Eutrophic waters generally have more bacteria (as determined by direct counts), more bacterial ATP, and greater respiratory activity (Jones *et al.*, 1979). Jones has also shown (1979; Jones & Simon, 1981; cf. also Jüttner & Schröder, 1982) that microbial metabolic pathways vary along gradients of oxidation-reduction potential in lakes. Aerobic respiratory systems, depending on the tricarboxylic acid cycle, dominate in shallow waters while anaerobic fermentation pathways prevail in the deoxygenated hypolimnion and sediments. Drabkova (1983) reported that the quality of bacteria present in lake sediments is not necessarily proportional to sediment organic matter content. Relatively little has been done to isolate and characterize the many kinds of bacteria responsible for the detrital transformations observed in the multitude of decomposition studies performed.

Fungi are obligate aerobes and, therefore, definitely limited in their distribution in most aquatic systems. They are, however, most effective at mineralizing the very recalcitrant compounds of plant detritus, cellulose and especially lignin. Mineralization of wood is almost entirely dependent on the activity of the fungi. In a comparative study using antibacterial and antifungal agents, Kaushik & Hynes (1971) found that decay of leaves in streams was not affected by removal of bacteria, but was significantly retarded if fungi were eliminated from the microbial community. In contrast, Mason (1976) found that bacteria and fungi were equally important, each accounting for ca. 25% of total weight loss; the other half of the weight lost was due to activities of organisms not affected by the antibiotic treatments, presumably protozoans, nematodes, and periphytic algae.

Long-term studies of the microflora associated with detrital particles document a succession of microbial colonization that correlates well with the biochemical changes occurring during decomposition (e.g. Maystrenko *et al.*, 1969; Oláh, 1972; Úlehlová, 1976; Morrison *et al.*, 1977; Chamier & Dixon, 1982; Federle & Vestal, 1982). Bacteria are typically first to colonize, and their numbers increase rapidly after a few days. A diversity of bacteria is present, but gram-negative rods predominate (Bastardo, 1979). Fungi are usually slower to accumulate, but in several studies are reported to be the first colonizers (e.g. Suberkropp & Klug, 1976; Lee *et al.*, 1980). Higher temperature and low light may promote relatively greater fungal activity (Egglishaw, 1972; Federle *et al.*, 1982a). The rate of biologically-mediated weight loss (i.e., non-leaching losses) reaches its peak when both fungi and bacteria are abundant on the detritus (Fig. 2). Concomitantly, early high activities of saccharide-degrading enzymes decline in favor of increasing cellulolytic activities (Bastardo, 1979). Rate of weight loss correlates with the amount of ATP and protein accumulating on the detritus and with activities of alkaline phosphatase and cellulase (Federle & Vestal, 1980a; Federle *et al.*, 1982a). Subsequent sloughing of fungi accounts for secondary declines in detrital nitrogen and slowing of decay. Further decomposition is primarily due to bacterial processes alone. Peaks in microbial proteolytic activities have been observed both initially and later in the succession of decomposers (Maystrenko *et al.*, 1969; Bastardo, 1979). Other nitrogen-metabolizing bacteria succeed as well during decomposition, at least in laboratory batch cultures (Maystrenko *et al.*, 1969).

The overall effectiveness of animals at transforming plant detritus has received considerable attention. While many animals, and entire food webs, depend either directly or indirectly on detrital plant tissue as an energy source, it is probably rare for a major proportion of the extant detritus to be consumed by animals as opposed to being metabolized by microorganisms. Laboratory experiments often associate animals with significant losses of herbaceous plant tissue (e.g. Cummins *et al.*, 1973; Carpenter, 1982). However, field studies using enclosures or exclosures usually indicate that the impact of animal feeding on loss of leaf detritus is negligible (Mathews & Kowalczewski, 1969; Kaushik & Hynes, 1971; Pieczyńska, 1972; Dickinson & Maggs, 1974; Brock *et al.*, 1982; Hanlon,

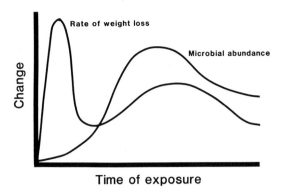

Fig. 2. Generalized trends in rate of weight loss and microbial abundance on detritus in a reservoir. Weight loss is rapid at first because of abiotic leaching; the second smaller peak in weight loss corresponds to maximum microbial abundance and activity later in decomposition.

1982; however, cf. Sedell *et al.*, 1975; Iversen, 1975; Danell & Sjöberg, 1979). The reason most commonly used to explain low consumption of detritus by animals is that the substrates are either not nutritive or not palatable until they have been well colonized by microorganisms (Petersen & Cummins, 1974; Barlöcher & Kendrick, 1975; Hill & Webster, 1982). In this regard, the mechanical breakdown of large particles by animals is important in increasing surface area for microbial colonization. Many studies have concentrated on the role of aquatic invertebrates in the decomposition of inundated wood, yet the direct quantitative importance of animals in wood degradation appears to be small compared with that of microorganisms (Petr, 1970; McLachlan, 1970; Anderson *et al.*, 1978).

Sequence of decomposition

In spite of the myriad of controlling factors and the variety of conditions under which decomposition proceeds, a quite typical sequence of events occurs during aquatic decomposition. As decay begins there is release from fresh detritus of much readily soluble material. This material is generally considered to consist of non-cell wall cytoplasmic constituents and is usually rich in cations (Na^+, K^+, Ca^{++}, and Mg^{++}), nitrogen and phosphorus, and labile organic compounds (sugars, fatty acids, amino acids). As cell membranes lose their integrity much of the contents of the cells may be merely dissolved or 'leached' out of the tissue. Such release may begin by autolysis even before the death of the tissue (Golterman, 1973; Otsuki & Wetzel, 1974). Howard-Williams *et al.* (1978) postulate that epiphytic growth on submersed macrophyte leaves may pre-condition them to rapid leaching and breakdown. This early stage of rapid material release from fresh detritus by leaching can account for significant weight loss (more than one third) and production of DOM in a matter of hours or days (Figs. 2, 3) (Cummins *et al.*, 1972; Wetzel & Manny, 1972; Petersen & Cummins, 1974; Polunin, 1982; Esteves & Barbieri, 1983); Howard-Williams *et al.*, 1983). Even newly flooded wood undergoes significant leaching, with detrimental effects on water quality in reservoirs (Sylvester & Seabloom, 1965). Maystrenko *et al.* (1968) discovered in laboratory experiments that wood samples released more material by leaching than by microbial activity for the first five months of decay. Enough labile organic matter is released during early decomposition of woody tissues to promote high biochemical oxygen demand (BOD) and depletion of dissolved oxygen for well over 100 days (Maystrenko *et al.*, 1968; McKeown *et al.*, 1968; Servizi *et al.*, 1971).

It is during this period of most rapid weight loss that detrital nitrogen content usually begins to decline. Decomposing plant tissue high in nitrogen (i.e. has a low C:N ratio) generally loses nitrogen to the surrounding water for the duration of exposure. Tissues that contain little nitrogen (high C:N) lose readily soluble (non-cell wall) nitrogen during the beginning of decomposition, then accumulate nitrogen for a relative long period, and then when only the most resistant tissues remain, lose nitrogen again (Kaushik & Hynes, 1971; Suberkropp *et al.*, 1976; Godshalk & Wetzel, 1978b; Sardana & Mehrotra, 1981; Tiwari & Mishra, 1983; Marinucci *et al.*, 1983).

Total oxygen consumption (i.e., biochemical oxygen demand (BOD) during decomposition depends on the amount of labile material subject to oxidation and the intensity of microbial activity.

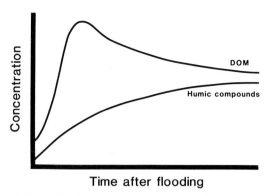

Fig. 3. Generalized trends in concentrations of total dissolved organic matter (DOM) and the fraction composed of humic compounds in reservoir water. DOM peaks initially because of leaching, then declines as it is metabolized. Later, DOM is dominated by the large, refractory molecules of humic substances.

Commonly, especially with restricted circulation and aeration, BOD is high and dissolved oxygen is totally consumed during early stages of decay (Fig. 4). Slower anaerobic metabolic pathways dominate during subsequent decay until oxygen becomes available again, as by lake mixing (Godshalk & Wetzel, 1977). Deep reservoirs are subject to intense thermal stratification resulting in large areas of lake bottom becoming anaerobic, and to detrital loss by permanent sedimentation; waters of shallow reservoirs are more frequently and thoroughly mixed, aerating bottom waters, distributing nutrients, and resuspending small detrital particles, allowing continued decay, or loss associated with reservoir flushing (Gunnison *et al.*, Chapter 3).

Plant tissue is greatly modified during decomposition by microbial activity and environmental conditions, but the ultimate character of detritus is highly dependent on plant species (Boon & Haverkamp, 1982; Boon *et al.*, 1983). Very resistant particles persist in the sediments and are buried by new particles. In the ultimate stages of decay, refractory components dominate. The rates of further weight loss by microbial degradation may be dependent on the supplies of oxygen and nitrogen in the media surrounding the detritus (Figs. 2, 4). As decay continues, fibrous compounds become more concentrated in the detritus. There may even be an increase in the absolute amount of lignin-like mate-

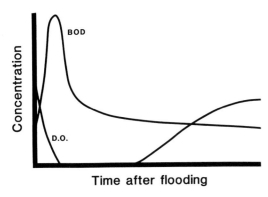

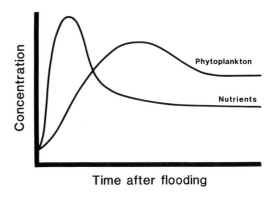

Fig. 4. Generalized trends in oxygen concentration (D.O.) and short term oxygen consumption (BOD). Consumption is rapid initially due to metabolism of leached labile DOM (cf. Fig. 2). As a result, oxygen is depleted.

Fig. 5. Generalized trends in concentrations of major nutrients released during decomposition in a reservoir and response of phytoplankton to nutrient supply. Initial peak in nutrient release results from leaching of newly flooded soils and terrestrial vegetation.

rial present. This increase is usually attributed to the formation by a variety of biogeochemical pathways of complex substances from reactants in the detritus, its associated microflora, and the surrounding environment. Analysis by pyrolysis mass spectrometry has shown that specific recalcitrant compounds such as paracoumeric esters and N-acetyl aminosugars accumulate in older detritus (Boon *et al.*, 1982, 1983).

Effects of decomposition on water quality

Massive releases of nutrients from inundated herbaceous vegetation and from flooded soil (Gunnison *et al.*, 1980; Gunnison, 1981) give rise initially to relatively high microbial activity and associated phytoplankton productivity following impoundment (Fig. 5) (Rodhe, 1964; Ostrofsky, 1978; Ostrofsky & Duthie, 1980). This phenomenon, associated primarily with the direct leaching of phosphorus (Grimard & Jones, 1982) is usually short lived (<3 years), but in lakes filled over a period of several years, may persist somewhat longer (Petr, 1975). The gradual release of nutrients from submerged trees represents a form of internal nutrient loading which may contribute for many years to localized eutrophy (Hendricks & Silvey, 1977).

Aquatic plants, with roots in the sediment and foliage in the water, effectively link the sediment with the overlying water. Many nutrients, but not all, incorporated into aquatic plant tissues are obtained via root uptake from the sediment (Denny, 1980; Barko and Smart, 1981; Barko, 1982; Smart & Barko, 1984). The release of these nutrients from aquatic plants during decomposition represents an important mode of sediment nutrient cycling and another form of internal nutrient loading to aquatic systems (Prentki *et al.*, 1979; Carpenter, 1980; Jacoby *et al.*, 1982; Welch *et al.*, 1982).

Losses of nutrients from aquatic plants occur primarily during senescence and tissue decay, involving leaching, autolysis, and microbial decomposition. Such losses can be very rapid in aquatic systems. In studies involving a variety of submersed aquatic plant species, tissue decay typically resulted in 50–100% losses of nitrogen, phosphorus, and potassium within about 50 days (Kistritz, 1978; Sudo *et al.*, 1978; Howard-Williams & Davies, 1979; Hill, 1979). Since aquatic plants concentrate major nutrients (greater than ten fold) relative to concentrations in sediments (Gopal & Kulshreshtha, 1980), tissue decay can have a major impact on water chemistry. For example, in Lake Wingra, Wisconsin, decay of *Myriophyllum spicatum* alone was estimated to account for about half of the observed flux of dissolved total phosphorus and dissolved organic matter from the littoral zone to the pelagial zone (Carpenter, 1980). The release of organic leachates and inorganic nutrients from aquatic vegetation during decay stimulates microbial metabolism (Carpenter *et al.*, 1979) and, either directly or indirectly, the growth of algae (Rho & Gunner, 1978; Landers, 1982).

Internal nutrient loading is likely to increase in importance in reservoirs as areas potentially colonizable by aquatic plants expand due to accumulation of sediments and as the reservoirs become more eutrophic. The associated growth of algae, often excessive in the open water (phytoplankton) or directly attached to aquatic plants (epiphytes) may promote declines in some submersed aquatic vegetation due to shading (Jupp & Spence, 1977; Phillips *et al.*, 1978; Sand-Jensen & Søndergaard, 1981). However, many submersed aquatic plants, notably those which form dense foliar canopies near the water surface such as *Hydrilla* and *Myriophyllum* (Haller & Sutton, 1975; Titus & Adams, 1979), are relatively insensitive to shading, and can grow prolifically even in turbid eutrophic aquatic systems.

It has been suggested that the development of submersed aquatic vegetation closes a positive feedback loop, promoting increased phytoplankton production, increased sedimentation, and thus increased area for even further vegetative expansion (Wetzel & Hough, 1973; Carpenter, 1983). Internal loading by aquatic vegetation is potentially self-accelerating, with concomitant stimulatory effects on sedimentation, associated vegetative succession, and reservoir filling (cf. Wetzel, 1979; Carpenter, 1983; also see previous section on vegetative succession).

In addition to the sedimentation of refractory detrital particles which contributes to the filling of reservoir basins, many organic products of partial decay accumulate in the water, especially in the anoxic hypolimnion (Fig. 3); these may also be displaced in high concentrations downstream from reservoirs with hypolimnetic withdrawal. These substances, collectively called humic compounds, are high molecular weight polyphenolic macromolecules of variable and complex structure (see reviews by Schnitzer & Kahn, 1972; Jackson, 1975; Gjessing, 1976; Stevenson, 1982; see also Christman & Gjessing, 1983). There is considerable evidence that they are produced directly as a result of leaching and decomposition processes acting on both terrestrial and aquatic vegetation. It is widely believed that humic substances are specifically derived from the lignin component of plants (e.g., Christman & Oglesby, 1971; Matsumoto, 1982; Ertel *et al.*, 1984) and modified by biotic and abiotic processes (Larson & Hufnal, 1980; Zeikus, 1981).

Humic compounds reduce water quality in several ways. Most noticeably, they give the water a color that varies from light yellow to red to dark brown. Color originates from the large molecules that have many linked aromatic rings, and is affected by environmental parameters such as pH, temperature, and dissolved minerals (Christman & Ghassemi, 1966; Ghassemi & Christman, 1968; Stewart & Wetzel, 1981). Humic substances have been implicated in taste and odor problems of reservoirs used as water supplies, especially those in which existing vegetation and organic soils were not removed prior to inundation (Allen, 1960).

Humic compounds are extremely recalcitrant, and they persist and accumulate in many aquatic ecosystems (e.g., Barber, 1968; Williams *et al.*, 1969). Alexander (1975) has discussed the several ways that ecosystem metabolism is drastically influenced by the presence of humic substances. They may rapidly absorb much of the light entering the water surface. Because they are chemically 'sticky,' humic compounds are thought to compete with organisms by sequestering nutrients (Jackson & Schindler, 1975; Jackson & Hecky, 1980; Stewart & Wetzel, 1982) and by complexing with and inactivating enzymes (cf. Ladd & Butler, 1975). Microbial activity in the presence of these complex materials is often inhibited (e.g. Mahadevan & Muthukumar, 1980; Geller, 1983). Humic substances are not inert, and their alteration and degradation in both lakes and flowing waters has been observed (Gjessing & Samdal, 1968; Gjessing, 1970; de Haan, 1972, 1974; Stewart & Wetzel, 1981).

Summary

Dynamics of decomposition in reservoirs are not greatly different from those of natural lakes except with regard to rates. Sudden initial availability for decomposition of large amounts of refractory detritus, mostly wood, slows the decomposition process and makes it less complete because of oxygen depletion. Rapid leaching and mineralization of nutrients from flooded soils and vegetation release large amounts of nutrients which are quickly taken up by phytoplankton and littoral plants, accelerating new production. As allochthonous and autochthonous detritus accumulates, at rates greatly in excess of those occurring during aging of natural lakes, the importance of various primary producers shifts away from easily decayed phytoplankton to more refractory vascular plants of the littoral zone, further increasing the accrual of organic matter in the reservoir basin. This process of reservoir filling is not different from that hypothesized for most natural lakes, except that it occurs much faster in reservoirs, and it is often augmented by large inputs of inorganic sediments. Ultimately, the lifespan of the reservoir depends on several factors such as basin morphometry, flushing rate, and site preparation which interact to control the rates of production and decomposition in the impoundment.

References

Alexander, M., 1975. Environmental and microbiological problems arising from recalcitrant molecules. Microb. Ecol. 2:17–27.

Alexander, M., 1977. Introduction to Soil Microbiology. Wiley, New York, NY. 467 pp.

Allen, E.J., 1960. Taste and odor problems in new reservoirs in wooded areas. J. am. Wat. Wks Ass. 52:1027–1032.

Anderson, N.H., Sedell, J.R., Roberts, L.M. & Triska, F.J., 1978. The role of aquatic invertebrates in processing of wood debris in coniferous forest streams. Am. Midl. Nat. 100:64–82.

Armstrong, W., 1975. Waterlogged soils. In Etherington, J.R., (ed.), Environment and Plant Ecology. J. Wiley & Sons, NY: 181–218.

Armstrong, W., 1978. Root aeration in the wetland condition. In Hook, D.D. & Crawford, R.M.M., (eds.), Plant Life in Anaerobic Environments. Ann Arbor Science Publishers, Ann Arbor, MI, pp. 269–297.

Aumen, N.G., Bottomley, P.J., Ward, G.M. & Gregory, S.V., 1983. Microbial decomposition of wood in streams: distribution of microflora and factors affecting (^{14}C) lignocellulose mineralization. Appl. envir. Microbiol. 46:1409–1416.

Baker, J.H., Morita, R.Y. & Anderson, N.H., 1983. Bacterial activity associated with the decomposition of woody substrates in a stream sediment. Appl. envir. Microbiol. 45:516–521.

Ball, J., Weldon, C. & Crocker, B., 1975. Effects of original vegetation on reservoir water quality. Texas Water Resources Institute Report, National Technical Information Services, Springfield, VA, 120 pp.

Barber, R.T., 1968. Dissolved organic carbon from deep waters resists microbial oxidation. Nature 220:274–275.

Barko, J.W., 1981. The influence of selected environmental factors on submersed macrophytes-a summary. In Stephan, H.G., (ed.), Proceedings of the Symposium on Surface Water Impoundments ASCE, June 2–5, 1980, Minneapolis, Minnesota: American Society of Civil Engineers, NY:1378–1382.

Barko, J.W., 1982. Influence of potassium source (sediment vs. open water) and sediment composition on the growth and nutrition of a submersed freshwater macrophyte (*Hydrilla verticillata* (L.F.) Royle). Aquat. Bot. 12:157–172.

Barko, J.W. & Smart, R.M., 1981. Sediment-based nutrition of submersed macrophytes. Aquat. Bot. 10:339–352.

Barko, J.W. & Smart, R.M., 1983. Effects of organic matter additions to sediment on the growth of aquatic plants. J. Ecol. 71:161–175.

Barko, J.W. & Smart, R.M. 1984. Ecology of aquatic plant species: effects of sediment composition. In Proceedings, 18th Annual Meeting, Aquatic Plant Research Program. Misc. Paper A-84-4. USAE Waterways Experiment Station, CE, Vicksburg, MS 39180, USA. pp. 148–153.

Barko, J.W., Hardin, D.G. & Matthews, M.S., 1982. Growth and morphology of submersed freshwater macrophytes in relation to light and temperature. Can. J. Bot. 60:877–887.

Barlöcher, F. & Kendrick, B., 1975. Leaf-conditioning by microorganisms. Oecologia 20:359–362.

Bastardo, H., 1979. Laboratory studies on decomposition of littoral plants. Pol. Arch. Hydrobiol. 26:267–299.

Baxter, R.M., 1977. Environmental effects of dams and impoundments. Ann. Rev. Ecol. Syst. 8:255–283.

Bobbie, R.J., Morrison, S.J. & White, D.C., 1978. Effects of substrate biodegradability on the mass and activity of the associated estuarine microbiota. Appl. envir. Microbiol. 35:179–184.

Boon, J.J. & Haverkamp, J., 1982. Pyrolysis mass spectrometry of intact and decomposed leaves of *Nuphar variegatum* and *Zostera marina*, and some archeological eelgrass samples. Hydrobiol. Bull. 16:71–82.

Boon, J.J., Wetzel, R.G. & Godshalk, G.L., 1982. Pyrolysis mass spectrometry of some *Scirpus* species and their decomposition products. Limnol. Oceanogr. 27:839–848.

Boon, J.J., Windig, W., Wetzel, R.G. & Godshalk, G.L., 1983. The analytical pyrolysis of particulate residues of decomposing *Myriophyllum heterophyllum*. Aquat. Bot. 15:307–320.

Boylen, C.W. & Brock, T.D., 1973. Bacterial decomposition processes in Lake Wingra sediments during winter. Limnol. Oceanogr. 18:628–634.

Brinson, M.M., 1977. Decomposition and nutrient exchange of

litter in an alluvial swamp forest. Ecology 58:601–609.

Brinson, M.M., Lugo, A.E. & Brown, S., 1981. Primary productivity, decomposition and consumer activity in freshwater wetlands. Ann. Rev. Ecol. Syst. 12:123–161.

Brock, T.C.M., Huijbregts, C.A.M., van de Steeg-Huberts, M.J.H.A. & Vlassak, M.A., 1982. In situ studies on the breakdown of *Nymphoides peltata* (Gmel.) O. Kuntze (Menyanthaceae); some methodological aspects of the litter bag technique. Hydrobiol. Bull. 16:35–49.

Carpenter, S.R., 1980. Enrichment of Lake Wingra, Wisconsin, by submersed macrophyte decay. Ecology 61:1145–1155.

Carpenter, S.R., 1981a. Submersed vegetation: an internal factor in lake ecosystem succession. Am. Nat. 118:372–383.

Carpenter, S.R., 1981b. Decay of heterogenous detritus: a general model. J. theor. Biol. 89:539–547.

Carpenter, S.R., 1982. Comparisons of equations for decay of leaf litter in tree-hole ecosystems. Oikos 39:17–22.

Carpenter, S.R., 1983. Submersed macrophyte community structure and internal loading: relationship to lake ecosystem productivity and succession. In Taggart, J., (ed.), Lake Restoration, Protection, and Management. US Environmental Protection Agency, Washington, DC, pp. 105–111.

Carpenter, S.R. & Adams, M.S., 1979. Effects of nutrients and temperature on decomposition of *Myriophyllum spicatum* L. in a hard-water eutrophic lake. Limnol. Oceanogr. 24:520–528.

Carpenter, S.R., Gurevitch, A. & Adams, M.S., 1979. Factors causing elevated biological oxygen demand in the littoral zone of Lake Wingra, Wis. Hydrobiologia 67:3–9.

Chamier, A.-C. & Dixon, P.A., 1982. Pectinases in leaf degradation by aquatic hyphomycetes. I: the field study. Oecologia 52:109–115.

Christman, R.F. & Gjessing, E.T., (eds.), 1983. Aquatic and terrestrial humic materials. Ann Arbor Science, Ann Arbor, MI. 538 pp.

Christman, R.F. & Ghassemi, M., 1966. Chemical nature of organic color in water. J. am. Wat. Wks Ass. 58:723–741.

Christman, R.F. & Oglesby, R.T., 1971. Microbiological degradation and the formation of humus. In Sarkanen, K.V. & Ludwig, C.H., (eds.), Lignins. Occurrence, Formation, Structure and Reactions. Wiley-Interscience, NY, pp. 769–795.

Crawford, R.M.M., 1982. Physiological responses to flooding. In Lange, O.L., Nobel, P.S., Osmond, C.B. & Ziegler, H., (eds.), Physiological Plant Ecology. II: Water Relations and Carbon Assimilation. Springer-Verlag, NY, pp. 453–477.

Cummins, K.W., Klug, M.J., Wetzel, R.G., Petersen, R.C., Suberkropp, K., Manny, B.A., Wuycheck, J.C. & Howard, F.O., 1972. Organic enrichment with leaf leachate in experimental lotic ecosystems. BioScience 22:719–722.

Cummins, K.W., Petersen, R.C., Howard, F.O., Wuycheck, J.C. & Holt, V.I., 1973. The utilization of leaf litter by stream detritivores. Ecology 54:336–345.

Dacey, J.W.H., 1980. Internal winds in water lilies: an adaptation for life in anaerobic sediments. Science 210:1017–1019.

Dacey, J.W.H., 1981. Pressurized ventilation in the yellow waterlily. Ecology 62:1137–1147.

Danell, K. & Sjöberg, K., 1979. Decomposition of *Carex* and *Equisetum* in a northern Swedish lake: dry weight loss and colonization by macroinvertebrates. J. Ecol. 67:191–200.

Davis, C.B. & van de Valk, A.G., 1978. The decomposition of standing and fallen litter of *Typha glauca* and *Scirpus fluviatilis*. Can. J. Bot. 56:662–675.

Day, F.P., Jr., 1982. Litter decomposition rates in the seasonally flooded Great Dismal Swamp. Ecology 63:670–678.

Day, F.P., Jr., 1983. Effects of flooding on leaf litter decomposition in microcosms. Oecologia 56:180–184.

Denny, P., 1980. Solute movement in submerged angiosperms. Biol. Rev. 55:65–92.

Dickinson, C.H. & Maggs, G.H., 1974. Aspects of the decomposition of *Sphagnum* leaves in an ombrophilous mire. New Phytol. 73:1249–1257.

Drabkova, V.G., 1983. Bacterial decomposition of organic matter in lacustrine sediments. Hydrobiologia 103:99–102.

Drew, M.C. & Lynch, J.M., 1980. Soil anaerobiosis, microorganisms, and root function. Ann. Rev. Phytopathol. 18:37–66.

Eggler, W.A., 1961. The vegetation of Lake Chicot, Louisiana, after eighteen years impoundment. Southwest Nat. 6:175–183.

Egglishaw, H.J., 1972. An experimental study of the breakdown of cellulose in fast-flowing streams. Mem. Ist. ital. Idrobiol. 29 Suppl., pp. 405–428.

Ertel, J.R., Hedges, J.I. & Perdue, E.M., 1984. Lignin signature of aquatic humic substances. Science 223:485–487.

Esteves, F.A. & Barbieri, R., 1983. Dry weight and chemical changes during decomposition of tropical macrophytes in Lobo Reservoir – São Paulo, Brazil. Aquat. Bot. 16:285–295.

Federle, T.W. & Vestal, J.R., 1980a. Microbial colonization and decomposition of *Carex* litter in an arctic lake. Appl. envir. Microbiol. 39:888–893.

Federle, T.W. & Vestal, J.R., 1980b. Lignocellulose mineralization by arctic lake sediments in response to nutrient manipulation. Appl. envir. Microbiol. 40:32–39.

Federle, T.W. & Vestal, J.R., 1982. Evidence of microbial succession on decaying leaf litter in an arctic lake. Can. J. Microbiol. 28:686–695.

Federle, T.W., McKinley, V.L. & Vestal, J.R., 1982a. Physical determinants of microbial colonization and decomposition of plant litter in an arctic lake. Microb. Ecol. 8:127–138.

Federle, T.W., McKinley, V.L. & Vestal, J.R., 1982b. Effects of nutrient enrichment on the colonization and decomposition of plant detritus by the microbiota of an arctic lake. Can. J. Microbiol. 28:1199–1205.

Fenchel, T., 1970. Studies on the decomposition of organic detritus derived from the turtle grass *Thalassia testudinum*. Limnol. Oceanogr. 15:14–20.

Fenchel, T., 1975. The quantitative importance of the benthic microfauna of an arctic tundra pond. Hydrobiologia 46:445–464.

Furness, H.D. & Breen, C.M., 1982. Decomposition of *Cynodon dactylon* (L.) Pers. in the seasonally flooded areas of the Pongolo river floodplain: pattern and significance of dry matter and nutrient loss. Hydrobiologia 97:119–126.

Gasith, A. & Lawacz, W., 1976. Breakdown of leaf litter in the littoral zone of a eutrophic lake. Ekol. Pol. 24:421–430.

Geller, A., 1983. Degradability of dissolved organic lake water compounds in cultures of natural bacterial communities. Arch. Hydrobiol. 99:60–79.

Ghassemi, M. & Christman, R.F., 1968. Properties of the yellow organic acids of natural waters. Limnol. Oceanogr. 13:583–597.

Gill, C.J., 1970. The flooding tolerance of woody species-a review. For. Abstr. 31:671–688.

Gjessing, E.T., 1970. Reduction of aquatic humus in streams. Vatten 26:14–23.

Gjessing, E.T., 1976. Physical and chemical characteristics of aquatic humus. Ann Arbor Science Publishers Inc., Ann Arbor, MI. 120 pp.

Gjessing, E.T. & Samdal, J.E., 1968. Humic substances in water and the effect of impoundment. J. am. Wat. Wks Ass. 60:451–454.

Godshalk, G.L. & Wetzel, R.G., 1977. Decomposition of macrophytes and the metabolism of organic matter in sediments. In H.L. Golterman (ed.), Interactions Between Sediments and Fresh Water. Dr W. Junk Publishers, The Hague, pp. 258–264.

Godshalk, G.L. & Wetzel, R.G., 1978a. Decomposition of aquatic angiosperms. I. Dissolved components. Aquat. Bot. 5:281–300.

Godshalk, G.L. & Wetzel, R.G., 1978b. Decomposition of aquatic angiosperms. II. Particulate components. Aquat. Bot. 5:301–327.

Golterman, H.L., 1972. The role of phytoplankton in detritus formation. Mem. Ist. ital. Idrobiol. 29 Suppl., pp. 89–103.

Golterman, H.L., 1973. Vertical movement of phosphate in freshwater. In Griffith, E.J., Beeton, A., Spencer, J.M. & Marshall, D.T., (eds.), Environmental Phosphorus Handbook. J. Wiley & Sons, NY, pp. 509–538.

Gopal, B. & Kulshreshtha, M., 1980. Role of aquatic macrophytes as reservoir of nutrients and in their cycling. Int. J. Ecol. Environ. Sci. 6:145–152.

Green, W.E., 1947. Effect of water impoundment on tree mortality and growth. J. For. 45:118–120.

Grimard, Y. & Jones, H.G., 1982. Trophic upsurge in new reservoirs: a model for total phosphorus concentrations. Can. J. Fish. aquat. Sci. 39:1473–1483.

Gunnison, D., 1981. Microbial processes in recently impounded reservoirs. ASM News 47:527–531.

Gunnison, D. & Alexander, M., 1975. Resistance and susceptibility of algae to decomposition by natural microbial communities. Limnol. Oceanogr. 20:64–70.

Gunnison, D., Brannon, J.M., Smith, I., Jr. & Burton, G.A., 1980. Changes in respiration and anaerobic nutrient regeneration during the transition phase of reservoir development. In Barica, J. & Mur, L.R., (eds.), Developments in Hydrobiology, Proceedings of the Workshop on Hypertrophic Ecosystems, Vaxjo, Sweden, 10–14 September 1979. Dr W. Junk Publishers, The Hague, pp. 151–158.

Hackett, W.F., Connors, W.J., Kirk, T.K. & Zeikus, J.G.,

1977. Microbial decomposition of synthetic ^{14}C-labeled lignins in nature: lignin biodegradation in a variety of natural materials. Appl. envir. Microbiol. 33:43–51.

Hackney, C.T. & de la Cruz, A.A., 1980. In situ decomposition of roots and rhizomes of two tidal marsh plants. Ecology 61:226–231.

Haller, W.T. & Sutton, D.L., 1975. Community structure and competition between *Hydrilla* and *Vallisneria*. Hyacinth Contr. J. 13:48–50.

Haan de, H., 1972. Some structural and ecological studies on soluble humic compounds from Tjeukemeer. Verh. int. Ver. Limnol. 18:685–695.

Haan de., H., 1974. Effect of a fulvic acid fraction on the growth a *Pseudomonas* from Tjeukemeer (The Netherlands). Freshwat. Biol. 4:301–310.

Hanlon, R.D.G., 1982. The breakdown and decomposition of allochthonous and autochthonous plant litter in an oligotrophic lake (Llyn Frongoch). Hydrobiologia 88:281–288.

Harms, W.R., Schreuder, H.T., Hook, D.D., Brown, C.L. & Shropshire, F.W., 1980. The effects of flooding on the swamp forest in Lake Ocklawaha, Florida. Ecology 61:1412–1421.

Harrison, P.G. & Mann, K.H, 1975a. Chemical changes during the seasonal cycle of growth and decay in eelgrass (*Zostera marina*) on the Atlantic coast of Canada. J. Fish Res. Bd Can. 32:615–621.

Harrison, P.G. & Mann, K.H., 1975b. Detritus formation from eelgrass (*Zostera marina* L.): the relative effects of fragmentation, leaching, and decay. Limnol. Oceanogr. 20:924–934.

Haslam, E., 1974. Polyphenol-protein interactions. Biochem. J. 139:285–288.

Hendricks, A.C. & Silvey, J.K.G., 1977. A biological and chemical comparison of various areas of a reservoir. Wat. Res. 11:429–438.

Hill, B.H., 1979. Uptake and release of nutrients by aquatic macrophytes. Aquat. Bot. 7:87–93.

Hill, B.H. & Webster, J.R., 1982. Aquatic macrophyte breakdown in an Appalachian river. Hydrobiologia 89:53–59.

Hodkinson, I.D., 1975. Dry weight loss and chemical changes in vascular plant litter of terrestrial origin, occurring in a beaver pond ecosystem. J. Ecol. 63:131–142.

Howard-Williams, C. & Davies, B.R., 1979. The rates of dry matter and nutrient loss from decomposing *Potamogeton pectinatus* in a brackish south-temperate coastal lake. Freshwat. Biol. 9:13–21.

Howard-Williams, C. & Howard-Williams, W., 1978. Nutrient leaching from the swamp vegetation of Lake Chilwa, a shallow African lake. Aquat. Bot. 4:257–267.

Howard-Williams, C., Davies, B.R. & Cross, R.H.M., 1978. The influence of periphyton on the surface structure of a *Potamogeton pectinatus* L. leaf (an hypothesis). Aquat. Bot. 5:87–91.

Howard-Williams, C., Pickmere, S. & Davies, J., 1983. Decay rates and nitrogen dynamics of decomposing watercress (*Nasturtium officinale* R. Br.). Hydrobiologia 99:207–214.

Hutchinson, G.E., 1975. A treatise on limnology. 3. Aquatic

macrophytes and attached algae. J. Wiley & Sons, NY, 660 pp.

Iversen, T.M., 1973. Decomposition of autumn-shed beech leaves in a springbrook and its significance for the fauna. Arch. Hydrobiol. 72:305–312.

Iversen, T.M., 1975. Disappearance of autumn shed beech leaves placed in bags in small streams. Verh. int. Ver. Limnol. 19:1687–1692.

Jackson, T.A., 1975. Humic matter in natural waters and sediments. Soil Sci. 119:56–64.

Jackson, T.A. & Schindler, D.W., 1975. The biogeochemistry of phosphorus in an experimental lake environment: evidence for the formation of humic-metal-phosphate complexes. Verh. int. Ver. Limnol. 19:211–221.

Jackson, T.A. & Hecky, R.E., 1980. Depression of primary productivity by humic matter in lake and reservoir waters of the boreal forest zone. Can. J. Fish. aquat. Sci. 37:2300–2317.

Jacoby, J.M., Lynch, D.D., Welch, E.B. & Perkins, M.A., 1982. Internal phosphorus loading in a shallow eutrophic lake. Wat. Res. 16:911–919.

Jones, J.G., 1979. Microbial activity in lake sediments with particular reference to electrode potential gradients. J. gen. Microbiol. 115:19–26.

Jones, J.G. & Simon, B.M., 1981. Differences in microbial decomposition processes in profundal and littoral lake sediments, with particular reference to the nitrogen cycle. J. gen. Microbiol. 123:297–312.

Jones, J.G., Orlandi, M.J.L.G. & Simon, B.M., 1979. A microbiological study of sediments from the Cumbrian lakes. J. gen. Microbiol. 115:37–48.

Jupp, B.P. & Spence, D.H.N., 1977. Limitations on macrophytes in a eutrophic lake, Loch Leven. I. Effects of phytoplankton. J. Ecol.65:175–186.

Jüttner, F. & Schröder, R., 1982. Microbially derived volatile organic compounds in the recent sediment of the *Phragmites australis* bed of the Bodensee (Lake Constance). Arch. Hydrobiol. 94:172–181.

Kang, H. & Seki, H., 1983. The gram-stain characteristics of the bacterial community as a function of the dynamics of organic debris in a mesotrophic irrigation pond. Arch. Hydrobiol. 98:39–58.

Kato, K. & Sakamoto, M., 1983. The function of the free-living bacterial fraction in the organic matter metabolism of a mesotrophic lake. Arch. Hydrobiol. 97:289–302.

Kaushik, N.K. & Hynes, H.B.N., 1968. Experimental study on the role of autumn-shed leaves in aquatic environments. J. Ecol. 56:229–243.

Kaushik, N.K. & Hynes, H.B.N., 1971. The fate of the dead leaves that fall into streams. Arch Hydrobiol. 68:465–515.

Kight, J., 1980. USAE Division/District presentations; aquatic plant problems operations activities: South Atlantic Division, Mobile, Lake Seminole Reservoir. In Proceedings, 14th Annual Meeting, Aquatic Plant Control Research Planning and Operations Review, 26–29 November 1979, Lake Eufaula, Oklahoma, US Waterways Experiment Station, Vicksburg, MS. pp. 57–61.

King, G.M. & Klug, M.J., 1982. Comparative aspects of sulfur mineralization in sediments of a eutrophic lake basin. Appl. envir. Microbiol. 43:1406–1412.

Kistritz, R.U., 1978. Recycling of nutrients in an enclosed aquatic community of decomposing macrophytes (*Myriophyllum spicatum*). Oikos 30:561–569.

Kleerekoper, H., 1953. The mineralization of plankton. J. Fish Res. Bd Can. 10:283–291.

Ladd, J.N. & Butler, J.H.A., 1975. Humus-enzyme systems and synthetic, organic polymer-enzyme analogs. In Paul, E.A. & McLaren, A.D., (eds.), Soil Biochem., 4:143–194.

Landers, D.H., 1982. Effects of naturally senescing aquatic macrophytes on nutrient chemistry and chlorophyll *a* of surrounding waters. Limnol. Oceanogr. 27:428–439.

Larcher, W., 1980. Physiological plant ecology, 2d ed. Springer-Verlag, Berlin. 303 pp.

Larsen, V.J., 1982. The effects of pre-drying and fragmentation on the leaching of nutrient elements and organic matter from *Phragmites australis* (Cav.) Trin. litter. Aquat. Bot. 14:29–39.

Larson, R.A. & Hufnal, J.M., Jr., 1980. Oxidative polymerization of dissolved phenols by soluble and insoluble inorganic species. Limnol. Oceanogr. 25: 505–512.

Lee, C., Howarth, R.W. & Howes, B.L., 1980. Sterols in decomposing *Spartina alterniflora* and the use of ergosterol in estimating the contribution of fungi to detrital nitrogen. Limnol. Oceanogr. 25:290–303.

Little, E.C.S., 1966. The invasion of man-made lakes by plants. In R.H. Lowe-McConnell (ed), Man-made Lakes. Academic Press, Lond., pp. 75–86.

Macan, T.T., 1977. Changes in the vegetation of a moorland fishpond in twenty-one years. J. Ecol. 65:95–106.

Mahadevan, A. & Muthukumar, G., 1980. Aquatic microbiology with reference to tannin degradation. Hydrobiologia 72:73–79.

Marinucci, A.C., Hobbie, J.E. & Helfrich, J.V.K., 1983. Effect of litter nitrogen on decomposition and microbial biomass in *Spartina alterniflora*. Microb. Ecol. 9:27–40.

Mason, C.F., 1976. Relative importance of fungi and bacteria in the decomposition of *Phragmites* leaves. Hydrobiologia 51:65–69.

Mason, C.F. & Bryant, R.J., 1975. Production, nutrient content and decomposition of *Phragmites communis* Trin. and *Typha angustifolia* L. J. Ecol. 63:71–95.

Mathews, C.P. & Kowalczewski, A., 1969. The disappearance of leaf litter and its contribution to production in the River Thames. J. Ecol. 57: 543–552.

Matsumoto, G., 1982. Comparative study on organic constituents in polluted and unpolluted inland aquatic environments – III. Phenols and aromatic acids in polluted and unpolluted waters. Wat. Res. 16:551–557.

Maystrenko, Yu.G. & Denisova, A.I., 1972. Method of forecasting the content of organic and biogenic substances in the water of existing and planned reservoirs. Soviet Hydrology: Selected Papers 6:515–540.

Maystrenko, Yu. G., Denisova, A.I. & Yenaki, G.A., 1968. Forest vegetation as a source of biogenic and organic substances in natural inland waters. Gidrobiologicheskiy Zhurnal 4:12–19. (Draft Translation 608, US Army Cold Regions Research and Engineering Laboratory, Hanover, New Hamshire, March, 1977)

Maystrenko, Yu. G., Denisova, A.I., Bagnyuk, V.M. & Aryamova, Zh. M., 1969. Role of higher aquatic vegetation in the accumulation of organic and biogenic substances in inland waters. Gidrobiologicheskiy Zhurnal 5:28–39. (Draft Translation 607, US Army Cold Regions Research and Engineering Laboratory, Hanover, New Hampshire, March, 1977).

McKeown, J.J., Benedict, A.H. & Locke, G.M., 1968. Studies on the behavior of benthal deposits of wood origin. J. Wat. Pollut. Cont. Fed., Res. Suppl. 40:R333–R353.

McLachlan, A.J., 1970. Submerged trees as a substrate for benthic fauna in the recently created Lake Kariba (Central Africa). J. appl. Ecol. 7:253–266.

Melillo, J.M., Naiman, R.J., Aber, J.D. & Eshleman, K.N., 1983. The influence of substrate quality and stream size on wood decomposition dynamics. Oecologia 58:281–285.

Mickle, A.M. & Wetzel, R.G., 1978a. Effectiveness of submersed angiosperm-epiphyte complexes on exchange of nutrients and organic carbon in littoral systems. I. Inorganic nutrients. Aquat. Bot. 4:303–316.

Mickle, A.M. & Wetzel, R.G., 1978b. Effectiveness of submerged angiosperm epiphyte complexes on exchanges of nutrients and organic carbon in littoral systems. II. Dissolved organic carbon. Aquat. Bot. 4:317–329.

Mickle, A.M. & Wetzel, R.G., 1979. Effectiveness of submersed angiosperm epiphyte complexes on exchange of nutrients and organic carbon in littoral systems. III. Refractory organic carbon. Aquat. Bot. 6:339–355.

Mills, A.L. & Alexander, M., 1974. Microbial decomposition of species of freshwater planktonic algae. J. envir. Qual. 3:423–428.

Misra, R.D., 1938. Edaphic factors in the distribution of aquatic plants in the English Lakes. J. Ecol. 26:411–451.

Molongoski, J.J. & Klug, M.J., 1980a. Quantification and characterization of sedimenting particulate organic matter in a shallow hypereutrophic lake. Freshwat. Biol. 10:497–506.

Molongoski, J.J. & Klug, M.J., 1980b. Anaerobic metabolism of the particulate organic matter in the sediments of a hypereutrophic lake. Freshwat. Biol. 10:507–518.

Morrison, S.J., King, J.D., Bobbie, R.J., Bechtold, R.E. & White, D.C., 1977. Evidence for microfloral succession on allochthonous plant litter in Apalachicola Bay, Florida, USA. Mar. Biol. 41:229–240.

Moyle, J.B., 1945. Some chemical factors influencing the distribution of aquatic plants in Minnesota. Am. Midl. Nat. 34:402–420.

Nichols, D.S. & Keeney, D.R., 1973. Nitrogen and phosphorus release from decaying water milfoil. Hydrobiologia 42:509–525.

Oláh, J., 1972. Leaching, colonization and stabilization during detritus formation. Mem. Ist. ital. Idrobiol. 29 Suppl., pp. 105–127.

Ostrofsky, M.L., 1978. Trophic changes in reservoirs: an hypothesis using phosphorus budget models. Int. Rev. Gesamten Hydrobiol. 64:481–499.

Ostrofsky, M.L. & Duthie, H.C., 1980. Trophic upsurge and the relationship between phytoplankton biomass and productivity in Smallwood Reservoir, Canada. Can. J. Bot. 58:1174–1180.

Otsuki, A.K. & Wetzel, R.G., 1974. Release of dissolved organic matter by autolysis of a submersed macrophyte, *Scirpus subterminalis*. Limnol. Oceanogr. 19:842–845.

Panwar, M.R.S. & Sharma, P.D., 1981. Possible factors in tardy decomposition of *Scirpus tuberosus* leaves by fungi. Acta bot. Ind. 9:213–217.

Pearsall, W.H., 1920. The aquatic vegetation of the English Lakes. J. Ecol., 8:163–199.

Peltier, W.H. & Welch, E.B., 1970. Factors affecting growth of rooted aquatic plants in a reservoir. Weed Sci. 18:7–9.

Petersen, R.C. & Cummins, K.W., 1974. Leaf processing in a woodland stream. Freshwat. Biol. 4:343–368.

Petr, T., 1970. Macroinvertebrates of flooded trees in the manmade Volta Lake (Ghana) with special references to the burrowing mayfly *Povilla adusta* Navas. Hydrobiologia 36:373–398.

Petr, T., 1975. On some factors associated with the initial high fish catches in new African man-made lakes. Arch. Hydrobiol. 75:32–49.

Phillips, G.L., Eminson, D. & Moss, B., 1978. A mechanism to account for macrophyte decline in progressively eutrophicated freshwaters. Aquat. Bot. 4:103–106.

Pieczyńska, E., 1972. Production and decomposition in the eulittoral zone of lakes. In Kajak, Z. & Hillbricht-Ilkowska, A., (eds.), Productivity Problems of Freshwaters. PWN Polish Scientific Publishers, Warsaw, pp. 271–285.

Polunin, N.V.C., 1982. Processes contributing to the decay of reed (*Phragmites australis*) litter in fresh water. Arch. Hydrobiol. 94:182–209.

Pond, R.N., 1905. The relation of aquatic plants to the substratum. Rept. US Fish. Comm. 19:483–526.

Prentki, R.T., Adams, M.S., Carpenter, S.R., Gasith, A., Smith, C.S. & Weiler, P.R., 1979. The role of submersed weedbeds in internal loading and interception of allochthonous materials in Lake Wingra, Wisconsin, USA Arc. Hydrobiol. Suppl. 57:221–250.

Reddy, K.R. & Patrick, W.H., Jr., 1975. Effect of alternate aerobic and anaerobic conditions on redox potential, organic matter decomposition and nitrogen loss in a flooded soil. Soil Biol. Biochem. 7:87–94.

Reed, F.C., 1979. Decomposition of *Acer rubrum* leaves at three depths in a eutrophic Ohio lake. Hydrobiologia 64:195–197.

Reice, S.R., 1974. Environmental patchiness and the breakdown of leaf litter in a woodland stream. Ecology, 55:1271–1282.

Reice, S.R., 1977. The role of animal associations and current velocity in sediment-specific leaf litter decomposition. Oikos 29:357–365.

Rho, J. & Gunner, H.B., 1978. Microfloral response to aquatic weed decomposition. Wat. Res. 12:165–170.

Rich, P.H. & Wetzel, R.G., 1978. Detritus in the lake ecosystem. Am. Nat. 112:57–71.

Rodhe, W., 1964. Effects of impoundment on water chemistry and plankton in Lake Ransaren (Swedish Lappland). Verh. int. Ver. Limnol. 15:437–443.

Rogers, K.H. & Breen, C.M., 1982. Decomposition of *Potamogeton crispus* L.: the effects of drying on the pattern of mass and nutrient loss. Aquat. Bot. 12:1–12.

Sale, P.J.M. & Wetzel, R.G., 1983. Growth and metabolism of *Typha* species in relation to cutting treatments. Aquat. Bot. 15:321–334.

Sand-Jensen, K. & Søndergaard, 1981. Phytoplankton and epiphyte development and their shading effect on submersed macrophytes in lakes of different nutrient status. Int. Rev. Gesamten Hydrobiol. 66:529–552.

Sardana, R.K. & Mehrotra, R.S., 1981. Decomposition studies on three submerged hydrophytes in the Brahmsarovar Tank of Kurukshetra, India. Trop. Ecol. 22:187–193.

Saunders, G.W., 1976. Decomposition in freshwater. In Anderson, J.M. & Macfadyen, A., (eds.), The Role of Terrestrial and Aquatic Organisms in Decomposition Processes. Blackwell Scientific Publications, Oxford, pp. 341–373.

Saunders, G.W., Cummins, K.W., Gak, D.Z., Pieczyńska, E., Straškrabová, V. & Wetzel, R.G., 1980. Organic matter and decomposers. In Le Cren, E.D. & Lowe-McConnell, R.H., (eds.), The Functioning of Freshwater Ecosystems. Cambridge Univ. Press, Cambridge, pp. 341–392.

Schnitzer, M. & Kahn, S.U., 1972. Humic substances in the environment. Marcel Dekker, Inc., NY, 327 pp.

Sculthorpe, C.D., 1967. The biology of aquatic vascular plants. St. Martin's Press NY 610 pp.

Sedell, J.R., Triska, F.J. & Triska, N.S., 1975. The processing of conifer and hardwood leaves in two coniferous forest streams: I. Weight loss and associated invertebrates. Verh. int. Ver. Limnol. 19:1617–1627.

Servizi, J.A., Martens, D.W. & Gordon, R.W., 1971. Toxicity and oxygen demand of decaying bark. J. Wat. Pollut. Cont. Fed. 43:278–292.

Smart, R.M. & Barko, J.W., 1984. Laboratory culture of submersed freshwater macrophytes on natural sediments (submitted to Aquat. Bot.).

Solbraa, K., 1979. Composting of bark. II. Laboratory experiments. Medd. Norsk Inst. Skogforsk 34:335–386.

Spence, D.H.N., 1982. The zonation of plants in freshwater lakes. Adv. ecol. Res. 12:37–125.

Stevenson, F.J., 1982. Humus chemistry. Wiley-Interscience, NY. 443 pp.

Stewart, A.J. & Wetzel, R.G., 1981. Dissolved humic materials: photodegradation, sediment effects, and reactivity with phosphate and calcium carbonate precipitation. Arch. Hydrobiol. 92:265–286.

Stewart, A.J. & Wetzel, R.G., 1982. Influence of dissolved humic materials on carbon assimilation and alkaline phosphatase activity in natural algal-bacterial assemblages. Freshwat. Biol. 12:369–380.

Suberkropp, K. & Klug, M.J., 1976. Fungi and bacteria associated with leaves during processing in a woodland stream. Ecology 57:707–719.

Suberkropp, K., Godshalk, G.L. & Klug, M.J., 1976. Changes in the chemical composition of leaves during processing in a woodland stream. Ecology 57:720–727.

Suberkropp, K., Klug, M.J. & Cummins, K.W., 1975. Community processing of leaf litter in woodland streams. Verh. int. Ver. Limnol. 19:1653–1658.

Sudo, R., Ohtake, H., Aiba, S. & Mori, T., 1978. Some ecological observation on the decomposition of periphytic algae and aquatic plants. Wat. Res. 12:179–184.

Sylvester, R.O. & Seabloom, R.W., 1965. Influence of site characteristics on quality of impounded water. J. am. Wat. Wks Ass. 57:1528–1546.

Tanaka, Y. & Tezuka, Y., 1982. Dynamics of detritus-attached and free-living bacteria during decomposition of *Phragmites communis* powder in seawater. Jap. J. Ecol. 32:151–158.

Tison, D.L., Pope, D.H. & Boylen, C.W., 1980. Influence of seasonal temperature on the temperature optima of bacteria in sediments of Lake George, New York. Appl envir. Microbiol. 39:675–677.

Titus, J.E. & Adams, M.S., 1979. Coexistence and the comparative light relations of the submersed macrophytes *Myriophyllum spicatum* L. and *Vallisneria americana* Michx. Oecologia 40:273–286.

Tiwari, B.K. & Mishra, R.R., 1983. Dry weight loss and changes in chemical composition of pine (*Pinus kesiya* Royle) needles and teak (*Tectona grandis* L.) leaves during processing in a freshwater lake. Hydrobiologia 98:249–256.

Todd, D.K., (ed.), 1970. The water encyclopedia. Water Information Center, Port Washington, NY. 559 pp.

Triska, F.J., Sedell, J.R. & Buckley, B., 1975. The processing of conifer and hardwood leaves in two coniferous forest streams: II. Biochemical and nutrient changes. Verh. int. Ver. Limnol. 19:1628–1639.

Úlehlová, B., 1976. Microbial decomposers and decomposition processes in wetlands. Studie ČSAV 17, Československe Akad. Věd., Prague. 112 pp.

Walker, D., 1972. Direction and rate in some British postglacial hydroseres. In Walker, D. & West, R.G., (eds.), Studies in the Vegetational History of the British Isles. Cambridge University Press, Cambridge, pp. 117–139.

Welch, E.B., Michaud, J.P. & Perkins, M.A., 1982. Alum control of internal phosphorus loading in a shallow lake. Wat. Res. Bull. 18:929–936.

Wetzel, R.G., 1979. The role of the littoral zone and detritus in lake metabolism. Arch. Hydrobiol. Beih. Ergebn. Limnol. 13:145–161.

Wetzel, R.G., 1983. Limnology. W.B. Saunders, Philadelphia, PA. 767 pp.

Wetzel, R.G., 1984. Reservoir ecosystems: conclusions and speculations. In Thornton, K., (ed.), Perspectives in Reservoir Limnology. J. Wiley & Sons, NY (In press.)

Wetzel, R.G. & Allen, H.L., 1972. Functions and interactions of dissolved organic matter and the littoral zone in lake metabolism and eutrophication. In Kajak, Z. & Hillbricht-Ilkowska, A., (eds.), Productivity Problems of Freshwaters. Warszawa-Krakow. Proceedings of the IBP-UNESCO Symposium on Productivity Problems of Freshwaters 1970. Kazimierz Dolny, Poland, pp. 333–347.

Wetzel, R.G. & Hough, R.A., 1973. Productivity and role of aquatic macrophytes in lakes. An assessment. Pol. Arch. Hydrobiol. 20:9–19.

Wetzel, R.G. & Manny, B.A., 1972. Decomposition of dissolved organic carbon and nitrogen compounds from leaves in an experimental hard-water stream. Limnol. Oceanogr. 17:927–931.

Wetzel, R.G., Rich, P.H., Miller, M.C. & Allen, H.L., 1972. Metabolism of dissolved and particulate detrital carbon in a temperate hard-water lake. Mem. Ist. ital. Idrobiol. 29 Suppl., pp. 185–243.

Whittaker, R.H., 1975. Communities and ecosystems, 2d ed. Macmillan Publishing Co., Inc., NY. 387 pp.

Williams, P.M., Oeschger, H. & Kinney, P., 1969. Natural radiocarbon activity of the dissolved organic carbon in the north-east Pacific Ocean. Nature 224:256–258.

Williams, W.T. & Barber, D.A., 1961. The functional significance of aerenchyma in plants. Symp. Soc. exp. Biol. 15:132–144.

Willoughby, L.G., 1974. Decomposition of litter in fresh water. In Dickinson, C.H. & Pugh, G.J.F., (eds.), Biology of Plant Litter Decomposition, 2. Academic Press, Lond., pp. 659–721.

Witkamp, M. & Frank, M.L., 1969. Loss of weight, ^{60}Co, and ^{137}Cs from tree litter in three subsystems of a watershed. Envir. Sci. Technol. 3:1195–1198.

Zeikus, J.G., 1981. Lignin metabolism and the carbon cycle. Polymer biosynthesis, biodegradation, and environmental recalcitrance. Adv. microb. Ecol. 5:211–243.

Zieliński, J., 1980. The effect of nitrogen content on the rate of organic matter decomposition. Pol. ecol. Stud. 6:167–182.

Gordon L. Godshalk
Department of Biological Sciences,
University of Southern Mississippi,
Hattiesburg, Mississippi 39406-5018, USA

John W. Barko
Environmental Laboratory,
Waterways Experiment Station,
Vicksburg, Mississippi 39180, USA

CHAPTER 5

Microbiological water quality of reservoirs

G. ALLEN BURTON, JR.

Abstract. Reservoir microbiological water quality is a complex subject, affected by numerous environmental variables and management decisions. Indicator and pathogenic microorganisms vary in their occurrence, distribution, and density relaionships to each other. Traditional indicators, criteria, and monitoring practices have been shown through numerous studies to be inadequate in regard to detection of pathogen occurrence and waterborne disease outbreaks in recreational waters. Based on watershed and reservoir characteristics, optimal indicators and the likelihood of certain pathogen types occurring can be predicted. Key environmental factors which must be considered include: spatial and temporal varition; survival rates of indicators and pathogens; survival in water versus sediments; and stormwater and feeder stream inputs. Microorganism enumeration factors of importance include: choice of proper indicators and proper enumeration methods; awareness of methodological pitfalls; stressed organism recovery; and sampling of appropriate sites at appropriate times with adequate replicates. Relating monitoring data on bacterial indicator levels to potential for waterborne disease outbreaks requires decisions on acceptable risk and disease potential. Considering the above environmental and enumeration factors and exposure likelihood allows for adequate evaluations of reservoir water quality.

Introduction

The microbial community of reservoirs is comprised of resident microorganisms and those input through runoff and discharges. A small portion of this diverse community is of human health concern since recreational users are exposed and potential victims of waterborne disease. During the period from 1971 to 1980, 315 waterborne disease outbreaks were reported in the United States, resulting in 78,000 illnesses. In most of the outbreaks the waters were contaminated with chemicals or pathogenic microorganisms, with drinking of these water being implicated as the illness source. From 1972 to 1980, there were 38 reported outbreaks of giardiasis, affecting 30,000 people. Viruses were implicated in 12 outbreaks and 5000 cases from 1978 to 1981. Reported illnesses are not always linked to ingestion of contaminated water, and many cases are not reported at all; thus the true number of illnesses due to waterborne pathogens is likely underestimated. It has been estimated that only one-fifth of the actual waterborne illnesses are reported (Federal Register, 1983).

A summary of waterborne infectious diseases which may occur in North American impoundments is given in Tabel 1. The list includes etiologic agents which may produce disease from water contact activities and serves as a general guide to potential disease transmission in reservoirs.

The constant potential for outbreaks of waterborne disease from potable water supplies makes

Gunnison, D. (ed.) Microbial Processes in Reservoirs.
© 1985, Dr W. Junk Publishers, Dordrecht, Boston, Lancaster. ISBN 90 6193 751 5.

routine monitoring for contamination necessary. However, attempting to routinely assess contamination in natural aquatic systems is of lesser importance. The occurrence of microbial pathogens is difficult to determine in reservoirs due to a multitude of factors. Pathogens are usually present at low levels so it is necessary to monitor waters using 'indicator' organisms, e.g., fecal or total col-

Table 1. Potential Agents of Waterborne Disease (modified from Pipes, 1978).

Disease	Agent
Bacterial	
Diarrhea	Enterotoxigenic *E. coli*
	Campylobacter jejuni
	Other agents in this list
Salmonelliosis	*Salmonella* spp.
Yersiniosis	*Yersinia enterocolitica*
Shigellosis	*Shigella* spp.
Leptospirosis	*Leptospira* spp.
Typhoid fever	*S. typhi*
Tularemia	*Francisella tularensis*
Melioidosis	*Pseduomonas pseudomallei*
Otitis externa	*P. aeruginosa*
	Staphylococcus aureus
Pustular dermatitis	*P. aeruginosa*
Folliculitis (dermatitis)	*P. foliculitis*
Wound infections	*Aeromonas hydrophila*
	Staphylococcus aureus
	Other agents
Legionelliosis*	*Legionella pneumophila*
Viral	
Gastroenteritis	Parvovirus-like agents, e.g., Norwalk
	Reovirus, e.g., Rotavirus
	Enteroviruses, e.g., Coxsackie A and B, Echovirus
	Unknown agents
Hepatitis	Hepatitis A
Parasitic and protozoan	
Amebic dysentery	*Entamoeba hystolytica*
Giardiasis	*Giardia lamblia*
Primary Amebic	*Naegleria fowleri* and
Meningioencephalitis	*Acanthamoeba*
Ascariosis	*Ascaris lumbricoides*
Trichuriosis	*Trichuris trichura*
Balantidial dysentery	*Balantidium coli*
Coccidiosis	*Isopora* spp.
Swimmer's itch	*Schistosomes*

* Has not been implicated in waterborne disease outbreak from recreational waters.

iforms. The basic assumption of this approach is that the presence of bacterial indicator organisms is associated with the presence of microbial pathogens.

The traditional indicator group for all microbial pathogens has been the coliform bacteria. Their ease of enumeration, crude relationship to pathogen and disease occurrence, and lack of better indicators led to incorporation of coliform levels into federal criteria and state water quality standards. During the last two decades numerous shortcomings of coliforms as indicators of waterborne disease and possible alternatives have been identified, making their continued use unfavorable. The literature on this subject is extensive and has been reviewed by several authors (Dufour, 1984; LeChevallier & McFeters, 1984; Burton, 1982; Cabelli, 1983, 1980, 1979, 1976). Various indicator and pathogenic microorganisms, reservoir sampling considerations, and monitoring/management strategies will be discussed in the following sections.

Coliform and pathogenic microorganisms relationships

As a result of the widespread presence of 'total coliform' bacteria in nature, the use of this coliform group as an indicator is generally discouraged, except in finished drinking water. Presence of total coliform bacteria in drinking water has greater meaning since it indicates inadequate treatment. The presence of fecal coliforms in water suggests either animal or human wastes have contaminated the system and their associated pathogens may also be present (McKee & Wolf, 1963; Moore, 1959).

Development of indicator standards, prediction of risk of waterborne disease, and assessment of pathogen levels requires knowledge of an indicator-pathogen relationship. For a given concentration of indicator organisms, there should be a related concentration of pathogens under a known set of conditions. This hypothesis is based on the assumption that there are relatively constant levels of pathogens present in sewage which may be true to some extent in large municipal sewage systems;

but as the number of individuals contributing to the waste becomes smaller, the indicator-to-pathogen ratio variance increases. So a waste discharge into a reservoir from a healthy recreational user may be completely free of pathogens, or contain a high density of virulent pathogens, if the user is infected (Cabelli, 1979).

Correlations between fecal coliform (FC) densities and the presence of *Salmonella* in recreational waters has been reported (Geldreich, 1970). Many times, however, no correlation has been observed between FC and *Salmonella* or other pathogens (Payment *et al.*, 1982; Cabelli, 1979). Enteric pathogens have been found when low and acceptable levels of indicator bacteria were present (Kraus, 1977; Dutka, 1973). Moreover, no studies have shown FC to indicate the presence of pathogens such as *Yersinia enterocolitica* and *Campylobacter jejuni*. These bacteria have been implicated in waterborne disease from waters which have been found to be relatively free from FC (Schiemann, 1978; Ghirelli & Marker, 1977).

Fecal coliforms do not serve as indicators of bacterial pathogens which appear to be ubiquitous in freshwaters. *Pseudomonas aeruginosa, Aeromonas hydrophila, Klebsiella,* and *Legionella* are widespread and have been implicated in human disease. *P. aeruginosa* and *A. hydrophila* usually do not pose a health threat unless present in high numbers, yet are frequently found in conjunction with fecal pollution (Cabelli, 1980). Pathogenic *Klebsiella* and *Legionella* species have yet to be implicated in waterborne disease from recreational waters, but their occassional high densities makes them of concern.

Fecal coliform indicator validity is especially tenuous in predicting health hazards resulting from presence of pathogenic protozoa and viruses. Numerous studies have demonstrated the existence of enteric viruses in waters containing acceptable levels of FC (Gerba, 1980). No normal viral flora exists in humans, and the relationship to FC levels is hampered by some of the same problems as bacterial pathogens, e.g., variable percentage of infected excreters, subclinical infections, inadequate enumeration methods, varying survival rates (IAWPRC, 1983; Melnick & Gerba, 1980; Pipes,

1978). Large variations in the numbers of coliphage and enteroviruses in sewage and in the coliphage:coliform ratio have been reported (Pretorius, 1962). The ratio obviously changes with environments and time; furthermore, improved isolation techniques for viruses recently cast question on earlier reported ratios. As discussed later, virus survival is reportedly longer than FC, thus a ratio would change with time and distance from the point of discharge.

In studies of southern waters, *Naegleria* and *Acanthamoeba* appear to be ubiquitous. Relatively high densities have been reported in warm waters, such as thermal discharges from power plants and hot springs (Sykora *et al.*, 1983; Duma, 1980; O'Dell, 1979). These pathogens have not been associated with fecal pollution; therefore, use of the FC as an indicator of their presence is invalid.

Another protozoan pathogen, *Giardia lamblia,* has been reported with increasing frequency as an etiologic agent in waterborne disease (Jakubowski & Hoff, 1979). It is found in waters which are relatively free of FC and its ability to encyst allows extended survival and increased resistence as compared to FC (Craun, 1979).

The relationship of FC densities in the water column to pathogen densities in the sediments is also unclear. Sediment has been shown to harbor significantly higher numbers of bacteria, viruses, and protozoa, both pathogenic and nonpathogenic (Pellet *et al.*, 1983; Grimes, 1980; Gerba *et al.*, 1979; Matson *et al.*, 1978; Winslow, 1976; Van Donsel & Geldreich, 1971). One study of marine sediment found sediment FC concentrations correlated with sediment enteric virus concentrations (LaBelle *et al.*, 1980). When large numbers of enteric organisms are present in the sediment it suggests that there is or has been some degree of contamination of the overlying water (Allen *et al.*, 1953). Nearshore sediments at reservoir swimming areas have shown excessive FC concentrations (Winslow, 1976) with 100- to 1000-fold more than in overlying waters (Grimes, 1980, 1975; Van Donsel & Geldreich, 1971). *Salmonella* levels in sediments have been reported to be from 46 to 90 percent higher than in overlying waters (Goyal *et al.*, 1977; Van Donsel & Geldreich, 1971; Hendricks, 1971). Survival of indicator bacteria and pathogens is greater in sediments than overlying waters (Chan *et al.*, 1979). This factor combined with sediment spatial variation and water spatial/temporal variation discourages establishing FC correlations between sediment and water levels.

There is no doubt that the present indicator system works to some degree. Fecal coliforms do indicate fecal pollution, and the present acceptable levels are low enough that pathogens such as *Salmonella* will probably not be present at critical levels. Fecal indicators and pathogens are removed effectively with proper water treatment, thus the coliform indicator system works well with drinking water supplies. In recreational waters where there is little control over treatment, natural and man-made inputs, or water quality, the indicator relationship breaks down. It may be concluded that as indicator density increases there is a deterioration of water quality, but not necessarily an increase in health hazards (Pipes, 1978). Conversely, as indicator density decreases there is an improvement in aesthetic water quality, but not necessarily a decrease in health hazards.

Indicator and pathogen survival

Numerous studies have measured the survival rates of fecal coliform and pathogenic bacteria in water (McFeters *et al.*, 1974; McFeters & Stuart, 1972; Carter *et al.*, 1967). When enteric microorganisms, e.g., coliforms and pathogens, enter nutrient-poor aquatic systems they become physiologically stressed (Bissonnette, 1975). Factors affecting their survival rates are many, including; sunlight, temperature, nutrients, osmotic change, pH, protozoa, phage, and metal and organic toxicants (Kapuscinski & Mitchell, 1983; Faust *et al.*, 1975; Mitchell *et al.*, 1967; Van Donsel *et al.*, 1967). Since such a multitude of environmental variables influence survival, it is not surprising that reported survival rates vary substantially. In general it appears that indicator organisms die quicker in marine waters than in freshwater environments (Dufour, 1984; Chamberlin & Mitchell, 1978). Survival of 90 to 99 percent of indicator and *Salmonella* spp. in

freshwaters is usually less than four days (Rudolphs *et al.*, 1950), however, there are conflicting reports. The majority of studies have shown *E. coil* to be a good indicator because it survives as long or longer than *Salmonella* spp. A large percentage of the discharged fecal bacteria die-off within hours in aquatic systems, however at high concentrations of organisms, a significant number may still remain. The inconsistency of findings can be attributed not only to varying environmental factors, but also to varied strain characteristics, experimental designs, and enumeration methods.

Survival of enteric bacteria has usually been fitted to Chick's Law, a first order decay rate; i.e., $dC/dt = -kC$, where C is the bacterial density at time t, k is the die-off rate, and d is the difference from beginning to end. A popular model, which contains the usual first order decay for coliforms, is QUAL-II. The validity of first order decay rates has been questioned and numerous cases presented where they did not describe die-off kinetics (Chamberlin & Mitchell, 1978; Velz, 1970). A more common die-off rate is non-linear with time and is more accurately described with a two component decay term (Velz, 1970):

$$Ct = C_o e^{-kt} + C_o' e^{-k't}$$

where: Ct = coliform concentration at time t;

C_o = coliform concentration at time zero;

C_o' = resistant coliform population concentration at time zero (end of first rate);

k = decay rate of initial die-off period;

k' = decay rate of second die-off period.

Models for determining enteric bacteria density and transport in reservoirs have been developed (McDonald & Kay, 1981; Kay & McDonald, 1980; Thorton *et al.*, 1980). Assuming logarithmetic decay rates, a simple model based on distance rather than time dependence was superior, but critical variable coefficients were site dependent (Kay & McDonald, 1980). Critical factors which predominate in survival rates are temperature, sunlight, and turbidity. Die-off in surface waters have been reported to be faster than in deeper waters, resulting from light attenuation (Kapuscinski & Mitchell, 1983; Geldreich et al., 1980; Chamberlin & Mitchell, 1978). Chamberlin & Mitchell (1978) presented a light dependent die-off rate coefficient, which varied with water depth. Coliform densities follow turbidity plumes when storm waters enter reservoir headwaters, with die-off depending principally on temperature (Thorton et al., 1980). Enteric microorganisms die-off faster in warmer waters, such as summer, than in cooler waters (Fattal et al., 1983).

Fecal streptococci (FS) have been shown in most studies to survive longer than FC in surface waters, groundwaters, and sediments (Fattal et al., 1983; Keswick et al., 1982; Miescier & Cabelli, 1982; Sayler et al., 1975). Enterococci, a subgroup of FC (Streptococcus haecalis, S. faceium, and group Q streptococci), have been reported to survive longer than other FS (Geldreich et al., 1980; McFeters et al., 1974). Their survival time is more indicative of virus survival, thus have been suggested as superior indicators of waterborne disease (Cabelli et al., 1983; Cabelli, 1983, 1981; Fattal et al., 1983; Keswick et al., 1982; Miescier & Cabelli, 1982).

Die-off of other pathogens may vary considerably from FC survival. Many of the pathogens have been observed to survive longer than FC or Salmonella, e.g., Mycobacteria, Aeromonas, Pseudomonas, Giardia lamblia, Naegleria fowleri, enterovirus, parasitic ova, Yersinia enterocolitica (Grabow et al., 1983; Keswick et al., 1982; O'Malley et al., 1982; Melnick & Gerba, 1980; Schillinger & McFeters, 1978; Berg, 1978, 1973; McFeters & Stuart, 1972). Y. enterocolitica survives long periods in cold waters which are low in nutrients (Schillinger & McFeters, 1978). Campylobacter jejuni survived from 4 days to 5 weeks in 4° C Colorado water, and from 2 to 4 days at 25° C (Blaser et al., 1980). High densities of heterotrophic and coliform bacteria are inhibitory to Yersinia spp., Shigella, Leptospira, and enteric viruses (Schiemann, 1978; Highsmith et al., 1977; Geldreich, 1972). Giardia, parasitic ova, and enteric viruses survive adverse conditions such as water treatment better than coliform indicators. This characteristic of greater persistence and the lower

number of organisms required for an infective dose makes the occurrence of these organisms critical (Pipes, 1978). Viruses may persist for several weeks to months in cold water environments (Melnick & Gerba, 1980; Bitton, 1978). Enteroviruses survive longer than FC in secondary wastewater; however, there have been reports of E. coli die-off similar or longer than some viruses (Kott, 1981; Grabow et al., 1980). In general, viruses appear to persist longer than FC and enteric pathogenic bacteria (Melnick & Gerba, 1980; Berg, 1978). Viruses, like enteric bacteria, vary in their survival time in aquatic environments and thus add another complicating factor in predicting water quality (Colwell & Foster, 1980). Factors affecting their survival are similar in significance to those affecting coliform survival, e.g., temperature, sunlight, organics, and particulate matter (Melnick & Gerba, 1980).

Sediments greatly extend the survival of most microbial organisms of health significance, as indicated by much higher numbers. Studies of survival in sediments are few (Chan et al., 1979; Van Donsel & Geldreich, 1971), probably due to methodology difficulties. Van Donsel & Geldreich (1971) reported a 90 percent die-off in seven days of both FC and Salmonella spp. in various sediments, which is much longer than in water. Studies by the author (unpublished data) comparing survival of S. newport, E. coli, P. aeruginosa, and K. pneumoniae, in five different freshwater sediments (four of which were reservoirs) showed E. coli and Salmonella to have comparable die-off rates, with 2 to 5 orders of magnitude in 2 weeks, whereas P. aeruginosa and K. pneumoniae decreased only 1 to 2 orders of magnitude. At initial concentrations of 10^8 viable cells per milliliter, such as are found in feces, these pathogens could survive in sediments for months. This increased survival partially accounts for the higher numbers of indicators and pathogens in sediments. Fecal streptococci were observed to survive longer than FC in sediment (Sayler et al., 1975) as they usually do in water. FC are perhaps more indicative of recent contamination in sediments because their numbers near sewage outfalls are usually less than total coliforms, fecal streptococci, enteroviruses, and amoebae (Schaiberger et al., 1982; O'Malley et al., 1982).

Potential indicator organisms

Several indicators of water quality, other than FC, have been suggested in recent years (Berg, 1978; McFeters *et al.*, 1978; Hoadley, 1977; Dutka, 1973;). These include *E. coli*, enterococci (Cabelli, 1983; Dufour, 1983), *Clostridium perfringens* (Bisson & Cabelli, 1980), *K. pneumoniae* (Vlassoff, 1977), *A. hydrophila* (Rippey & Cabelli, 1980), *P. aeruginosa* (McFeters *et al.*, 1978), bifidobacteria (Levin, 1977), *Candida albicans* (Haas & Engelbrecht, 1980; Buck, 1977), coprostanol (Walker *et al.*, 1982; Yae *et al.*, 1982), *Mycobacteria* (Grabow *et al.*, 1983; Haas & Engelbrecht, 1980), *Rhodococcus* (Oragai & Mara, 1983), and coliphage (Kott, 1981).

E. coli is used as the primary fecal indicator in most European countries. It has the advantage of being specific for warm-blooded animals and is not found in nature as are some FC, e.g., *Klebsiella, Enterobacter,* and *Citrobacter.* For this reason it serves as a better indicator of recent fecal pullution than FC. Several studies by the US Environmental Protection Agency (EPA) at marine and freshwater beaches found better correlations between *E. coli* and enterococci with waterborne illness (Gastroenteritis), than other indicators (Dufour, 1984; Cabelli, 1983). In addition, simple methods exist for *E. coli* identification (Dufour *et al.*, 1981), thus continued use of the FC indicator system is of questionable wisdom.

Enterococci are more specific for human wastes. Since they have been reported to survive longer than other FS, FC, and *E. coli,* and are similar to enterovirus in survival periods, they are optimal to FC as indicators. Enterococci have been reported to vary little during waste treatment and decrease less after chlorination than other indicators, similar to enteroviruses (Miescier & Cabelli, 1982). As previously mentioned, EPA studies observed good correlations between enterococci and gastroenteritis incidence among swimmers at recreational beaches. Although it is superior to fecal coliforms as an indicator, it has two shortcomings as an indicator. First, it has been isolated from plants and insects (Geldreich & Kenner, 1969; Mundt, 1962, a, b), thus its source may not always be human feces.

Second, it is a poor indicator of pathogens from animal feces. As yet, epidemiological tests have not been conducted at beaches which are impacted by animal wastes, but not human wastes. Until this factor is studied, its widespread acceptance as an overall indicator of fecal pollution will be in question.

When FS levels are combined with FC information, the FC/FS ratio can be used to crudely identify whether or not the fecal source is human or animal. In general, FC/FS ratios of 4 or greater indicate human feces, whereas ratios of less than 0.7 are indicative of animal wastes. However, as mentioned previously, FS typically survive longer than FC, thus the ratio will change with time. Their use is discouraged unless derived near a contamination source, within hours of discharge (Geldreich, 1976; Geldreich & Kenner, 1969).

Perhaps the most important group of organisms of health significance in reservoir water quality is the viruses, yet they are the most difficult to detect. To date there have not been any methods developed for enterovirus enumeration which are practical for widespread use, e.g., short-term and relatively simple. Recently EPA published a methods manual for viruses (Berg *et al.*, 1984). Numerous methods do exist for recovery from waters and sediments (Gerba & Goyal, 1982): however, recovery rates and interlaboratory comparisons show substantial variation. There is agreement that a significant number of waterborne cases of gastroenteritis are probably caused by enteroviruses (Pipes, 1978). Enteroviruses have been identified in some waterborne disease outbreaks (Kaplan *et al.*, 1982; Wilson *et al.*, 1982; Baron *et al.*, 1982; D'Alessio *et al.*, 1981). It has been suggested that the etiologic agent responsible for the gastroenteritis in the EPA series of studies was of viral origin (Cabelli, 1981). The good correlation of enterococci to gastroenteritis, in the EPA studies, suggests it is a good indicator of enterovirus.

There has been much discussion on the use of coliphage as a virus indicator (Kott, 1981; Colwell & Foster, 1980). Several characteristics of coliphage qualify its use as an indicator: (1) prevalent in sewage at densities approximately 3 orders of magnitude greater than enterovirus densities, (2) short-term, inexpensive, reliable isolation methods, (3) possess chlorine resistance equivalent to or greater than enteroviruses, and (4) survive as long or longer than enteroviruses (Colwell & Foster, 1980). However, as with present indicators, coliphage possess some shortcomings: (1) not present at consistent densities in fecal material, (2) survival rates compared to many viruses of sanitary significance are not known, and (3) survival rates occasionally vary (Cabelli, 1980). A comparison of a short-term (6 h) test between several city laboratories across the United States revealed significant correlations (r^2 = .69) between coliforms (total and fecal) and coliphage in natural waters (Wentsel *et al.*, 1982). Considering the numerous studies which have reported poor correlations between FC levels and enteroviruses, the results of this study are unique. Since conflicting conclusions between studies are the rule, rather than the exception in studies of indicators, pathogens, and waterborne disease, one must look for general trends among conclusions. With refinement in methodologies and test designs and increased understanding of microorganism interactions with aquatic systems and human infection, perhaps more consistent conclusions will increase. At this point in time however, it appears that no one indicator organism exists which can been used with confidence all the time in all aquatic environments.

Factors associated with detection of contamination in reservoirs

Detecting microbial contamination of reservoirs depends on the sampling strategy and enumeration methods used. The true water quality of a reservoir cannot be ascertained with confidence unless valid systems are used which acknowledge and compensate for problems of detection. The following discussion will highlight critical factors in reservoir sampling strategies and microbial enumeration problems.

The hydrology of a reservoir is complex, with varying components such as currents, feeder streams, mixing, bank exposure, and retention time. Generally, impoundment of streams results in improved bacteriological quality because self-purification occurs; increased hydraulic retention time in the reservoirs encourages die-off of pathogens. Since microorganisms are planktonic, they drift with the currents and settle with particulates. When human and animal wastes are discharged into a feeder stream or reservoir, they remain nearshore at higher densities unless mixed by wind, stream inflows, or reservoir mixing, e.g., destratification. Stratification dynamics play a key role in bacterial distribution. In nearshore shallow areas, vertical distribution is fairly uniform due to mixing and uniform temperatures (Geldreich & Kenner, 1980). But during summer, the remaining reservoir often undergoes thermal stratification into three layers, resulting in current restriction and inhibiting nutrient and bacterial distribution. During stratification most mixing is horizontal. Currents are greater in the old river channel and in open waters, and less in coves and isolated portions of the reservoirs (Wetzel, 1975). Knowledge of these currents aids in determining the ability of areas to dilute contamination.

Bacteria in the water column will predominate in layers above the thermocline where biological productivity predominates (Wetzel, 1975). Sedimentation of clay, silt, and organic particulate matter results in high levels of bacteria and viruses at the sediment-water interface. This is a result of microorganism tendency to attach to particulate matter at densities approximating 3,000 to 15,000 organisms per ml (Tsernoglou & Anthony, 1971). Sediments provide protection, nutrients, and extended survival to high concentrations of microorganisms. In fall and winter, cold water inflows and cooler epilimnetic temperatures in the reservoir result in destratification, which results in relatively complete mixing, therefore a more uniform dispersion of bacteria. When destratification occurs, nutrients and microorganisms trapped in the hypolimnion circulate and a temporary deterioration in water quality may occur.

Water quality in reservoirs directly interacts with hydrologic and watershed characteristics to affect the microbiological status of the system. During summer stratification, warm water runoff from the watershed and feeder streams tend to remain in the upper layers of the reservoir. However, cooler inflows will plunge to bottom hypolimnetic waters of a similar density. These inflows may contain organic matter, nutrients, animal and human wastes, and other contaminants, which contribute to increased densities of micro-organisms at various depths (Wetzel, 1975). Therefore in a reservoir which has bottom withdrawal, high densities of contaminants in the hypolimnion would be released downstream. Conversely, high contaminant densities in the epilimnion would be retained in the reservoir for longer periods. This factor must be considered when predicting die-off rates and retention time of contaminants.

Thermocline survival is also affected by light and algal populations. Increased die-off at shallow depths has been attributed to ultraviolet light (Kapuscinski & Mitchell, 1983; Geldreich et al., 1980). Extracellular products excreted by algae also promote growth and survival of aquatic bacteria. Thus, algal blooms may allow indicator or pathogenic bacterial densities to increase in the absence of fecal pollution (McFeters et al., 1978).

Storm events have been shown to play a major role in the water quality of impoundments. Increased flows from feeder streams carry the majority of the annual supply of nutrients during high flows (Nix et al., 1975). Also associated with increased flow and turbidity are high levels of indicator organisms. Indicator densities in swimming areas may increase dramatically to unsafe levels following rainfall as result of runoff (Hendry & Toth, 1982; Horak, 1974; Geldreich, 1972). Sources of these fecal wastes and pathogens can be farm animals, wildlife, pets in urban and recreational areas, inadequate waste treatment systems, and septic tanks.

Increased bacterial densities during high flows are a function of watershed area, land use, duration and intensity of the rainfall. Indicator bacteria generally exhibit the 'first-flush' phenomenon, i.e., like many nutrients and chemical species they are 'flushed' through the stream during the initial increased storm flow. Increasing stream flows, in the absence of storm runoff, have been shown to raise FC levels greater than ten-fold between reservoirs. These FC increases obviously came from exposed stream banks (McDonald et al., 1982; McDonald & Kay, 1981). In large reservoirs, high FC densities in streams are diluted out once the flow reaches the reservoir, with sedimentation, dispersion, and die-off predominating (Thornton et al., 1980; Geldreich et al., 1980). Storm flows enter reservoirs in turbid plumes, proceeding through reservoirs as overflows, interflows, or underflows, depending on relative densities of stream and reservoir waters. As previously mentioned, models have been developed predicting coliform densities and transport distances (McDonald & Kay, 1981; Thornton et al., 1980; Kay & McDonald, 1980). Good correlations for decreasing bacterial densities versus time have been observed focusing on temperature, turbidity, the speed at which a turbid storm plume proceeds, and the distance the plume covers.

Sediments are perhaps the most important yet most under-utilized source of information on microbiological quality of impoundments. As previously pointed out, densities of indicator organisms and pathogens in sediments are often several orders of magnitude higher than in overlying waters and remain relatively stable over time, unlike microbial water densities. Microbial sediment densities are dependent on numerous factors; nearby contamination sources and grain size are most significant. Clays and silts of smaller grain sizes usually have more organic matter and increased surface area, thus higher microbial numbers with longer survival than in sandy sediments (Chan et al., 1979; Gerba & McLeod, 1976; Weiss, 1951). Hydrology and recreational use determine the health significance of contaminated sediments. In areas where sediments are resuspended due to runoff, boats, swimming, wading, and water turbulence, high densities of organisms may be recirculated into the water column. Sediments in areas with stronger currents or in pelagic zones away from shores, seldom possess high indicator or pathogenic microorganisms. Finally, sediments possess spatial variation as do waters. This variability will change from site to site, and must be considered when analyzing results.

All of the above factors must be considered when sampling for potential waterborne disease situations. When hydrologic information is combined with knowledge of watershed characteristics and water quality, sound sampling strategies can be developed.

Methods for collecting water samples are well established (Bordner & Winter, 1978). When sampling sediments in swimming areas, the sample can often be collected by hand by scooping the upper few centimeters into sterile containers. Deeper samples may be obtained using a Van Donsel-Geldreich sampler or more available Eckman or Ponar dredges (Bordner & Winter, 1978). A virus sediment sampler was recently proposed which reportedly recovered 54 percent more virus particles than from dredge samples, whereby the top layer was removed with water suction (Metcalf & Melnick, 1983).

Numerous methods exist which allow easy enumeration of most microbial indicators and bacterial pathogens which possibly could occur in reservoirs. Care must be taken, however, in choosing which methods and materials to use because each has limitations, recovery rates vary substantially, and therefore, limit data comparison.

Of the two predominate standard methods for indicator enumeration, the membrane filtration (MF) method is more popular than the most-probable-number (MPN) method due to its simplicity. The MPN, however, allows greater coliform recovery from chlorinated waters (Schiemann *et al.*, 1978). Numerous studies have compared these two techniques, both of which have advantages and shortcomings.

There are several aspects of the common indicator enumeration methods which frequently cause problems or produce invalid data. Among the major problems are: (1) occurrence of false positive and false negative results, (2) varying rates of recovery between methods and among different brands and batches of membrane filters, (3) effects of turbidity, and (4) inability to recover stressed microorganisms. In addition to these problems, varying methological conditions, and operator, spatial and temporal variances combine to produce the commonly observed widely variable indicator data.

False positives have been noted frequently by investigators attempting to isolate FC from en-

vironmental samples. Some bacteria frequently contributing to this problem are *Aeromonas, Klebsiella, Enterobacter, Citrobacter,* and *Serratia* species. This problem is the result of both the crude definition of FC and the inability of the popular FC isolation media to inhibit nonfecal bacteria.

False negatives are perhaps a greater problem. Many waters will contain such high numbers of bacteria capable of growth on indicator isolation media that they inhibit indicator growth. *P. aeruginosa* and *A. hydrophila* have been shown to interfere with *E. coli* metallic sheen on mEndo media, as did high coliform numbers (Burlingame *et al.*, 1984). This is common in turbid samples which also clog membrane pores during filtration and impede media contact with the bacteria. In addition viable colony plate counts will underestimate true numbers due to sorption of numerous organisms to each particle. Prefiltering can remove sorbed bacteria along with particulates, thus invalidating quantitative measurements. The MPN has been suggested as a superior method for turbid samples with some modifications (LeChevallier *et al.*, 1981). The numerous problems associated with estimating coliform densities in turbid samples with high noncoliform densities have been described in numerous studies (LeChevallier *et al.* 1981; Herson & Victoreen, 1980).

Another significant factor contributing to false negatives is the inability of commonly used methods to recover stressed or injured bacteria. This area has received considerable attention in recent years in regards to coliform enumeration. Several modified methods and new media have been proposed over standard methods for recovery of injured indicators (LeChevallier & McFeters, 1984; LeChevallier *et al.*, 1984; McFeters & Camper, 1983; Bordner, 1977; Bissonnette *et al.*, 1975). Standard Methods have addressed the problem with suggested method and media alternatives (American Publich Health Association (APHA), 1981). Among the modifications and media which have reported promising results are: (1) pre-incubation of coliform and *E. coli* plates at 25 to 35° C for 2 to 6 hours before incubating at 44.5 C, (2) deleting rosolic acid from mFC media, (3) mT7 media for TC and FC, (4) A–1 media for FC, (5) mTEC media for *E. coli,* and (6) FS pre-enrichment for 2 h at 35°C followed by plating on mE media for 48 h at 35° C (LeChevallier *et al.*, 1984; Pagel *et al.*, 1982; Grabow *et al.*, 1981; APHA, 1981; Standridge & Delfino, 1981; Presswood, 1978; Bordner *et al.*, 1977; Green *et al.*, 1977; Lin, 1976).

Since enteric organisms are physiologically adapted to the warm, nutrient-rich guts of animals, dilute aquatic environments cause stress, injury, and eventual death. In drinking water systems, injured total coliform recovery allows a better indication of treatment efficiency (LeChevallier & McFeters, 1984). Indicator bacteria in wastewater effluents treated with chlorine which are not killed are often injured. Numerous microbial processes are affected by the chlorine, with uptake of organic substrates being a primary target of injury (McFeters & Camper, 1983). However, in ambient water samples, e.g., recreational reservoirs, recovery of injured indicator bacteria is of questionable significance. Assuming equal injury to pathogens, as to the measured indicator bacteria, it is questionable whether their subsequent intake by a swimmer would result in infection. Pathogen virulence might be lost upon injury or the injured pathogen might not survive contact with bile and an acidic gastric environment. Recovery of stressed indicators would, however, allow increased safety margins in recreational waters. Since FC and *E. coli* typically die-off faster than enteroviruses, recovering total numbers of the indicator (injured and noninjured) might provide better correlations with virus presence. These possibilities have not been investigated to date, thus the health significance of stressed cells is unknown.

Another complicating problem in enumeration is variation between brands and lots of membrane filters and selective media. Numerous studies and reviews have focused on these problems (Mathewson *et al.*, 1983; Lorenz *et al.*, 1982; Bordner *et al.*, 1977).

Other factors which have been shown to affect enumeration results are holding time of samples and diluent type and temperature, however conflicting results have been reported. In general it is suggested that holding times of iced samples be as short as possible and that cold samples be diluted in cold diluents (APHA, 1981; McFeters *et al.*, 1982).

Several rapid coliform enumeration methods have been proposed recently, allowing results within one working day. These vary in approach, including; microcolonies, radiometric, serologic, electrochemical, enzymic, chromatographic, and chemiluminescent (Cundell, 1981).

Human health criteria

Present criteria and standards for fecal coliform bacteria are based on weak epidemiological studies, arbitrary safety factors, and other questionable assumptions. These weaknesses have been discussed by several authors (Dufour, 1984; Cabelli, 1983).

As mentioned earlier, the EPA conducted several studies which evaluated various microbial indicators of waterborne disease at two freshwater and four marine recreational beaches (Cabelli 1983; Dufour 1984). Several bacterial indicators were evaluated at marine beaches. Correlation coefficients, on results grouped by summers, showed enterococci to have the best relationship to gastroenteritis (r = .75), followed by *E. coli* (r = .52). When results were grouped by separate studies enterococci was the superior indicator (r = .96) and eight other indicators possessed similar coefficients (r = .5 to .6). In freshwater studies, *E. coli* was slightly superior to enterococci based on correlation coefficients (r = .80 vs. r = .74). The confidence bands around the enterococci were only acceptable at a narrow range.

The EPA studies were an improvement in design over earlier investigations supporting present criteria. It appears that *E. coli* and enterococci are superior to FC as indicators of gastroenteritis risk from swimming at recreational beaches. Whether or not the correlations observed are applicable nationwide is uncertain. As yet, similar studies have not been conducted on reservoirs impacted by animal wastes. The EPA sites were principally impacted by municipal wastes, therefore, the bacterial flora, e.g., indicators and pathogens, may be different from agricultural/wilderness areas. The enterococci method used mE media (Levin *et al.*,

1975) which selects for two predominantly human-specific species; *S. faecalis* and *S. faecium*.

Based on the EPA studies and numerous studies documenting the lack of FC correlation with pathogen levels or disease outbreaks, new criteria are warranted. At this point *E. coli* and enterococci are the best candidates for reliable indicators of water-borne disease potential, whether it be of viral or bacterial origin. Other potential indicators such as coliphage, should be tested epidemilogicaly before their validity can be supported. EPA has recently announced in the Federal Register that new bacteriological criteria are being considered for recreational waters, based on the above studies (Federal Register, 1984). The FC criteria may very well be discarded or its use limited in the near future.

In freshwater recreational areas, the present FC criteria/standards are seldom used to initiate regulatory or enforcement actions when violations occur. As the standards are usually written, several samples must be collected a month in order to obtain the percent compliance, e.g., no more than 10 percent exceed 400 FC/100 ml. Since State pollution agencies typically collect only one sample per month, a violation of the standard cannot be technically ascertained. Municipal waste-treatment facilities which are discharging FC levels above permit limits are seldom the target of enforcement actions unless other permit limits, e.g., nutrients, solids, BOD, are also in violation.

The US Corps of Engineers (CE) maintains a monitoring system for indicator bacteria on their reservoirs. During the swimming season, monitoring is usually increased to weekly sampling at beach areas. If excessive levels of indicator bacteria are found, the problem is investigated and repeat samples collected. Occassionally elevated levels have been recorded following storm events at which time the CE may temporarily close swimming beaches.

In some parts of the country, local municipalities monitor reservoir swimming areas for indicator bacteria. This has resulted in swimming restrictions and/or temporary closing following storm events because of elevated bacterial levels.

Exposure to pathogens

The dose of pathogenic microorganisms required to produce illness varies with the age and health of an individual, type of pathogen, metabolic condition and virulence of the microorganism. Studies of infective dose requirements have typically been conducted in laboratory situations using freshly cultured organisms. However, the relationship of these results to reservoir exposure is questionable. Virulence in an organism can easily change and be lost outside an ideal environment, particularly in a stressful aquatic environment. A 1 percent attack rate for gastroenteritis was associated with *E. coli* or enterococci densities of approximately 10/100 ml in the EPA studies (Cabelli, 1983). This indicates that low densities of the unknown etiologic agents (probably viral, Cabelli, 1981) may cause infection. Densities as low as 10 organisms/100 ml may require secondary waste treatment, if dilution and transport are ignored (Miescier & Cabelli, 1982).

Other studies have shown varying infective doses for *Salmonella* ranging from 10^4 to 10^{11} cells and viruses from 1 to 10^4 virions (Bonde, 1981; Dudley *et al.*, 1976; Mechalas *et al.*, 1972, Hornick *et al.*, 1970). A standard of 1 virus per 10 gallons of recreational water has been proposed, but this standard lacks epidemiologic foundation (Melnick, 1976). This criterion may apply to susceptible individuals (IAWPRC, 1983), however a more likely level of significance would be 10 to 100 virions. Due to greater resistance and survival of viruses relative to FC and *Salmonella,* the probability of a viral infection increases more rapidly than does the risk from *Salmonella* (Mechalas *et al.*, 1972). A worst case situation of waters containing 25 percent treated wastewater could contain approximately 2.5 virus particles/10 ml (US EPA, 1978). Another study compared results of virus concentrations in rivers and lakes ranging from .001 to 2.8 virus particles/10 ml (Melnick & Gerba, 1980), similar to the EPA worst case prediction. This would suggest that the majority of recreational waters have little chance of containing virus concentrations in 10 ml capable of infection.

Eye, ear, nose, and throat ailments represent more than half of all illnesses recorded among swimmers, gastrointestinal disturbances up to 20 percent and skin irritation the remainder (Stevenson, 1953). The following discussion gives the more significant waterborne diseases with a brief description of when they may occur in reservoirs.

Infections of the skin and eyes may be caused by *Staphylococcus aureus, A. hydrophila, P. aeruginosa,* and schistosome species. As mentioned previously, these organisms are not always present in conjunction with high FC levels. *A. hydrophila* may be present at high densities in warm waters which are relatively nutrient-rich (Rippey & Cabelli, 1980). Temperature and phytoplankton populations have been considered the most important factors determining population densities (Hazen & Esch, 1983; Colwell & Foster, 1980). Particularly high densities have been noted in thermal discharge waters (Fliermans *et al.*, 1977). *P. aeruginosa* (PA) is found in nutrient-rich and oligotrophic waters (Carson *et al.*, 1973; Nemedi & Layni, 1971). A PA/FC ratio has been suggested as a pollution index (Cabelli *et al.*, 1976). Various species of snail-transmitted bird schistosomes produce dermatitis, including human 'swimmer's itch', in recreational waters.

Gastroenteritis, e.g., diarrhea, associated with waterborne infections may be caused by several organisms including: *Salmonella, Shigella, Yersinia, Campylobacter, Giardia,* and viruses. Viruses probably predominate as etiologic agents with several types being implicated. As noted, their detection is relatively complicated and their incidence of occurrence uncertain. Several outbreaks in recreational waters have occurred with norwalk, hepatitis, and adenoviruses being implicated (Kaplan *et al.*, 1982; Wilson *et al.*, 1982; D'Alessio *et al.*, 1981) without correlation to FC levels. *Y. enterocolitica, C. jejuni,* and *G. lamblia* are the dominating causes of enteric infections in some areas (Harter *et al.*, 1982; Sands *et al.*, 1981; Luechtefeld *et al.*, 1981; CDC, 1979 ab; Pai *et al.*, 1979). *Y. enterocolitica* is particularly common in Europe and Canada, and is increasing across the United States (Bottone, 1977). Small animals, rodents, and pigs have been shown to be possible reservoirs of *Yersinia* (Kaneki & Hashimoto, 1981; Kapperud, 1975). Infections have occurred from waters con-

taining low FC levels. Since *Yersinia* is not routinely measured in US laboratories, its true incidence may be underestimated. Likewise, *C. jejuni* infection reports are increasing, with cattle, pigs, poultry, migratory waterfowl and other birds known as possible reservoirs of this organism (Luechtefeld *et al.*, 1981 1980; CDC, 1979a). *Campylobacter* is a major cause of waterborne diarrhea in some areas (Taylor *et al.*, 1982). From 3 to 11 percent of diarrheal patients in Europe and North America were reported as *C. jejuni* infections. It is also not routinely measured in clinical laboratories therefore true incidence rates are unknown. Improved methods for isolation from water may soon provide better information on its aquatic distribution. The final common diarrheal-causing agent is *Giardia lamblia. G. lamblia's* ability to encyst enables it to survive long periods in the environment and disinfection treatments and is the leading diarrheal agent in some States (CDC, 1979b; Craun, 1976).

Two protozoa, *N. fowleri* and *Acanthamoeba,* which may produce death upon infection have been recently found to be widespread in natural waters (Wellings *et al.*, 1979, 1977; O'Dell, 1977). Fortunately, the incidence of infection is extremely low. Cases of *N. fowleri* infection (primary amoebic meningioencephalitis) have been sporadic and appear to be related to warmer weather periods (Duma, 1980). Organism occurrence has been correlated with warmer water temperatures and overwinters in the sediments (Sykova *et al.*, 1983; Duma, 1980). Isolation methods are presently too complex for most routine monitoring programs.

Concern over *L. pneumophila* has increased in recent years with several deaths and illnesses from infection by aerosolization of potable water and with the realization that *Legionella* is widespread in many natural waters (Dufour & Jakubowski, 1982; Fliermans *et al.*, 1981; Fraser & McDade, 1979). Improved methods of isolation and enumeration by immunofluorescence (DFA method) have shown concentrations of 10^3 to 10^4/ml in lake waters and its association with algae and amoebae (Dufour & Jakubowski, 1982; Tison *et al.*, 1980; Rowbotham, 1980). Since *Legionella* has not been implicated in waterborne disease outbreaks and requires more involved isolation techniques, routine monitoring is impractical. It is effectively controlled, as are most other pathogens, by chlorine treatment.

Reservoir monitoring and management

Well-informed management strategies with respect to microbiological water quality are possible with a well-planned monitoring system. Stratification, agriculture, recreational uses, and rainfall patterns change with the seasons, thus changing types and densities of enteric microorganisms. These factors should drive sampling strategies.

Sampling must be geared toward potential problem areas where pathogen-human contact may occur. When sampling outside shallow areas, vertical samples should be collected down to the thermocline. In shallow swimming areas, emphasis should be placed on increased numbers of horizontal samples and sediment samples. In the initial phase of a monitoring system and during pre-impoundment stages, potential contamination and use areas should be sampled through one year to identify sites of importance as they relate to seasonal changes.

Sample sites should include the mouths of feeder streams; point source discharges; areas subject to agricultural, urban, or recreational runoff; drinking water supply intakes; and swimming areas. These sites should be initially monitored during storm events and storm-flow plumes sampled through the reservoir to determine the degree of contamination to be expected from runoff and feeder streams. Watersheds in which erosion is a problem will likely lead to waters of high turbidity, thus higher bacterial numbers and survival. The choice of microorganisms to monitor in the reservoir will depend on regulatory criteria and watershed use. If studies reveal significant human waste inputs, initial surveys should include indicators, e.g., *E. coli*, enterococci, and pathogens, e.g., *Salmonella*. Agricultural and wildlife dominated watersheds would suggest initial monitoring of FS, *E. coli*, enterococci, *Salmonella*, *Y. enterocolitica*, *C. jejuni*, and *Leptospira*. If more advance identi-

fication facilities are available, obviously other organisms should be surveyed, e.g., enteroviruses, *G. lamblia,* schistosomes, parasitic ova. Pathogen monitoring would only be practical, from a resource aspect, during an initial watershed/reservoir survey.

The confidence one can place on monitoring data depends on the degree of known variability associated with the results. This is only possible by collecting replicate samples, conducting replicate microbial counts on each sample, and calculating confidence levels. Taking one 'grab' sample from an area usually is insufficient, providing relatively meaningless results because the distribution of enteric organisms through the water varies significantly horizontally, vertically, and temporally. Variations in densities as high as three orders of magnitude over a few centimeters have been noted (Thornton *et al.*, 1980). One study analyzed different components comprising variance in bacterial numbers (Maul & Block, 1983). The coefficient of variations for each factor were as follows: daily fluctuation (temporal), 51.9 percent; diurnal, 32.8 percent; method, 5.4 percent; sample to sample, 5.1 percent; and subsamples 4.8 percent. As a general rule Maul suggested 3 to 6 samples optimal. Typical log mean standard deviations of 0.4 have been reported in freshwater FC data (Dufour, 1984), and 0.7 in marine water (Cabelli, 1983; Fuhs, 1975); however, this will depend on good pollutant dispersion and a multitude of environmental factors which vary with each site. After an adequate data base of microbiological levels have been collected for a reservoir, and the degree of spatial and temporal fluctuations have been analyzed, it may be possible to reduce numbers of sample stations, frequency of sampling, and replicate numbers. After the initial reservoir survey, microbiological sampling need only be during recreational use periods, except at drinking water supplies.

A key factor in the potential for waterborne disease transmission from enteric organisms is the siting of swimming areas. Beaches which are established in areas where water currents are prevalent will deter the possibility of significant fecal pollution in the swimming area. Swimming areas should have sandy sediments to deter pathogen build-up, survival, and resuspension. Boat ramps should be located well away from swimming areas because the water turbulence resuspends finegrained sediments/microorganisms into the water column (Gucinski, 1983; Alvarez, 1981; Horak, 1974). Initial surveys should indentify potential sources of pathogen contamination, thus allowing proper siting of beaches.

Conclusions

Reservoir microbiological water quality, from a human health perspective, has not been adequately defined. Many questions exist in this area, including: indicator and pathogenic microorganism relationships; adequacy of the commonly used indicator system; spatial distribution and temporal occurrence of indicators/pathogens: health significance of microbial densities, stressed pathogens, and criteria. In the past twenty years these questions have been recognized and many are being answered.

The fecal coliform bacteria are indicators of fecal pollution but usually are not correlated with waterborne disease outbreaks. Numerous indicator microorganisms have been suggested which are suitable for particular situations; however, all have shortcomings as indicator. The most promising indicators of gastroenteritis appear to be *Escherichia coli* and enterococci. Establishing national criteria is difficult but a necessary safequard; however, the relationship of a national criteria for indicator bacteria to a waterborne disease outbreak may, in many cases, be tenuous. This is a result of several reasons, including: environmental variances between reservoirs which result in different pathogen die-off rates; pathogens which may be injured, stressed, or dormant may have decreased virulence and thus their potential for causing disease questionable; etiologic agents vary; and no indicator exists for all agents of waterborne disease, particularly those which are ubiquitous to aquatic environments.

Reservoir distribution of microorganisms of health significance is dependent primarily on waste input, input location, currents/mixing, stratifica-

tion, retention time, and survival rates. Survival rates are governed mainly by temperature, light, and turbidity/sediments which obviously vary between reservoirs and through time, as does the indicator relationship to the numerous agents of waterborne disease.

The risk of waterborne disease occurring in North American reservoirs is slight, with proper waste treatment and sanitary practices. Determining what level of risk is acceptable is necessary so that monitoring data can be interpreted and appropriate actions taken. Presently, FC data seldom aid in determinations of good or poor water quality. Improvements in our understanding of indicator organism:waterborne illness relationships will allow meaningful water quality assessments. Informed reservoir planning and management can reduce potential risk of waterborne disease to non-detectable levels, simply by establishing a monitoring system which is geared toward critical areas/periods and measures appropriate indicators with reliable methods.

Acknowledgements

Many tanks to G.A. McFeters, A.P. Dufour, and R.H. Bordner for supplying significant current and in press literature.

References

Allen, L.A., Grindley, J. & Brooks, E., 1953. Some chemical and bacterial characteristics of bottom deposits from lakes and estuaries. J. Hygiene 51:185–192.

Alvarez, R.J., 1981. Microbiological quality of selected recreational waters. Envir. Res. 26:372–380.

American Public Health Association, 1981. Standard methods for the examination of water and wasterwater, 15th ed. American Public Health Association, N.Y.:887–993.

Baron, R.C., Murphy, F.D., Greenber, H.B., Davis, C.E., Bregman, D.J., Gary, G.W., Hughes, J.M. & Schenberger, L.B., 1982. Norwalk gastrointestinal illness: an outbreak associated with swimming in a recreational lake and secondary person-to-person transmission. Am. J. Epidemiol. 115:163–172.

Berg, G., 1973. Removal of viruses from sewage, effluents, and waters. A review. Bull. World Health Org. 49:451–460.

Berg, G. (ed.), 1978. Indicators of Viruses in Water and Food. Ann Arbor Science Press, Ann Arbor.

Ann Arbor Science Press, Ann Arbor.

Berg, G., Dahling, D.R., Berman, D. & Hurst, C., 1984. Manual of methods for virology. US Environmental Protection Agency, Cincinnati, Ohio. EPA–600/4–84–013.

Bisson, J.W. & Cabelli, V.J., 1980. *Clostridium perfringens* as a water quality indicator. J. Wat. Pollut. Cont. Fed. 52:241–248.

Bissonnette, G.K., Jezeski, J.J., McFeters, G.A. & Stuart, D.G., 1975. Influence of environmental stress on enumeration of indicator bacteria from natural waters. Appl. Microbiol. 29:186–194.

Bitton, G., 1978. Survival of enteric viruses. In R. Mitchell (ed.), Water Pollution Microbiology, Volume 2. John Wiley & Sons, NY:273–299.

Blaser, M.J., Hardestry, H.C., Powers, B. & Wang, W-L.L., 1980. Survival of *Campylobacter fetus* subsp. *jejuni* in biological milieus. J. clin. Microbiol. 11:309–313.

Bonde, G.J., 1981. *Salmonella* and other pathogenic bacteria. The Sci. total Envir. 18:1–11.

Bordner, R.H., 1977. The membrane filter dilemma. In Bordner, R.H., Frith, C.F. & Winter, J.A. (eds.), Proceedings of the Symposium on the Recovery of Indicator Organisms Employing Membrane Filters. US Environmental Protection Agency, Cincinnati, Ohio. EPA–600/9–77–024. 188 pp.

Bordner, R.H. & Winter, J.A. (eds.), 1978. Microbiological methods for monitoring the environment, water, and wastes. US Environmental Protection Agency, Cincinnati, Ohio. EPA–600/8–78–017. 338 pp.

Bottone, E.J., 1977. *Yersinia enterocolitica*: a panoramic view of a charismatic micro-organism. Crit. Rev. clin. lab. Sci. 5:211–241.

Burlingame, G.A., McElhaney, J., Bennett, M. & Pipes, W.O., 1984. Bacterial interference with coliform colony sheen production on membrane filters. Appl. environ. Microbiol. 47:56–60.

Burton, G.A., 1982. Microbiological water quality of impoundments: A literature review. US Army Engineer Waterways Experiment Station, Vicksburg, Mississippi, 53 pp.

Cabelli, V.J., 1976. Indicators of recreational water quality. In Hoadley, A.W. & Dutka, B.J. (eds.), Bacterial Indicators/ Health Hazards Associated with Water. Am. Soc. Rest. Materials, Philadelphia: 222–238 pp.

Cabelli, V.J., 1979. Evaluation of recreational water quality, the EPA approach. In A. James & L. Evison (eds.), Biological Indicators of Water Quality. John Wiley & Sons, NY: 14.1–14.23 pp.

Cabelli, V.J., 1980. What do water quality indicators indicate? In Colwell, R.R. & Foster, J. (eds.), Aquatic Microbial Ecology. Univ. of Maryland: 305–336 pp.

Cabelli, V.J., 1981. Epidemiology of enteric viral infections. In Goddard, M. & Butler, M. (eds.), Viruses and Wasterwater Treatment. Pergamon Press, NY: p. 291.

Cabelli, V.J., 1983. Health effects criteria for marine recreational waters. US Environmental Protection Agency, Cincinnati, Ohio. EPA–600/1–80–031. 98 pp.

Cabelli, V.J., Dufour, A.P., McCabe, L.J. & Levin, M.A., 1983. A marine recreational water quality criterion consistent with indicator concepts and risk analysis. J. Wat. Pollut. Cont. Fed. 55:1306–1314.

Cabelli, V.J., Kennedy, H. & Levin, M.A., 1976. *Pseudomonas aeruginosa* – fecal coliform relationships in estuarine and fresh recreational waters. J. Wat. Pollut. Contr. Fed. 48:367–376.

Carson, L.A., Favero, M.S., Bond, W.W. & Petersen, M.J., 1973. Morphological, biochemical, and growth characteristics of *Pseudomonas cepacia* from distilled water. Appl. Microbiol. 25:476–483.

Carter, HH., Whaley, R.C. & Carpenter, J.H., 1967. The bactericidal effects of seawater under natural conditions. J. Wat. Pollut. Cont. Fed. 39:1184.

Center for Disease Control, 1979a. *Campylobacter* enteritis – Iowa. Morb. Mortal. weekly Rep. 28:565–566.

Center for Disease Control, 1979b. Intestinal parasite surveillance, annual summary 1978. Atlanta.

Chamberlin, C.E. & Mitchell, R., 1978. A decay model for enteric bacteria in natural waters. In Mitchell, R. (ed.), Water Pollution Microbiology, Volume 2. John Wiley & Sons, NY: 325–348.

Chan, K–Y., Wong, S.H. & Mak, C.Y., 1979. Effects of bottom sediments on the survival of *Enterobacter aerogenes* in seawater. Mar. pollut. Bull. 10:205–210.

Colwell, R.R. & Foster, J. (eds.), 1980. Aquatic Microbial Ecology. University of Maryland, College Park: 460 pp.

Craun, G.F., 1976. Microbiology-waterborne outbreaks. J. Wat. Pollut. Cont. Fed. 48:1378–1379.

Craun, G.F., 1979. Waterborne giardiasis in the United States: a review. Am. J. publ. Health 69:817–819.

Cundell, A.M., 1981. Rapid counting methods for coliform bacteria. Adv. appl. Microbiol. 27:169–183.

D'Alessio, D.J., Minor, T.E., Allen, C.I., Tsiatis, A.A. & Nelson, D.B., 1981. A study of the proportions of swimmers among well controls and children with enterovirus-like illness shedding or not shedding an enterovirus. Am. J. Epidemiol. 113:533–541.

Dudley, R.H., Hekimian, K.K. & Mechalas, B.J., 1976. A scientific basis for determining recreational water quality criteria. J. Wat. Pollut. Cont. Fed. 48:2761–2777.

Dufour, A.P., 1984. Health effects criteria for fresh recreational waters. US Environmental Protection Agency, Cincinnati, Ohio. EPA–600/1–84–004.

Dufour, A.P. & Jakubowski, W., 1982. Drinking water and Legionaires disease. J. am. Wat. Wks. Ass. 74:631–637.

Dufour, A.P., Strickland, E.R. & Cabelli, V.J., 1981. Membrane filter method for enumerating *Escherichia coli*. Appl. envir. Microbiol. 41:1152–1158.

Duma, R.J., 1980. Study of pathogenic free-living amebas in fresh-water lakes in Virginia. US Environmental Protection Agency, Washington, DC EPA–600/S1–80–037. 5 pp.

Dutka, B.J., 1973. Coliforms are an inadequate index of water quality. J. envir. Health 36:39–46.

Fattal, B., Vasl, R.J., Katzenelson, E. & Shuval, H.I., 1983.

Survival of bacterial indicator organisms and enteric viruses in the Mediterranean coastal waters off Tel-Aviv. Wat. Res. 17:397–402.

Faust, M.A., Aotaky, A.E. & Hargadon, M.I., 1975. Effect of physical parameters on the *in situ* survival of *Escherichia coli* MC–6 in an estuarine environment. Appl. Microbiol. 30:800–806.

Federal Register 48:45502–45520, October 5, 1983.

Federal Register 49:21987, May 24, 1984.

Fliermans, C.B., Cherry, W.B., Orrison, L.H., Smith, S.J., Tison, D.L. & Pope, D.H., 1981. Ecological distribution of *Legionella pneumophila*. Appl. envir. Microbiol. 41:9–16.

Fraser, D.W. & McDade, J.E., 1979. Legionellosis. Scien. Am. 241:82–99.

Fuhs, G.W., 1975. A probabilistic model of bathing beach safety. The Sci. total Envir. 4:165–175.

Geldreich, E.E., 1970. Applying bacteriological parameters to recreational water quality. J. am. Wat. Wks. Ass. 62:113–120.

Geldreich, E.E., 1972a. Water-borne pathogens. In R. Mitchell (ed.), Water Pollution Microbiology, Volume 1. John Wiley & Sons, Inc., NY: 207–241.

Geldreich, E.E., 1972b. Buffalo Lake recreational water quality: a study in bacteriological data interpretation. Wat. Res. 6:913–924.

Geldreich, E.E., 1976. Fecal coliform and fecal streptococcus density relationships in waste discharges and receiving waters. Crit. Rev. Microbiol. 6:349–369.

Geldreich, E.E. & Kenner, B.A., 1969. Concepts of fecal streptococci in stream pollution. J. Wat. Pollut. Cont. Fed. 41:12336–12352.

Geldreich, E.E., Nash, H.D., Spino, D.F. & Reasoner, D.J., 1980. Bacterial dynamics in a water supply reservoir; a case study. J. am. Wat. Wks. Ass. 72:31–40.

Gerba. C.P., 1980. Indicator bacteria and the occurrence of viruses in marine waters. In Colwell, R.R. & Foster, J. (eds.), Aquatic Microbial Ecology. Univ. of Maryland: 348–355.

Gerba, C.P. & Goyal, S.M., 1982. Methods in Environmental Virology. Microbiology Series, Volume 7. Marcel Dekker, Inc., NY: 400 pp.

Gerba, C.P. & McLeod, J.S., 1976. Effect of sediments on the survival of *Escherichia coli* in marine waters. Appl. envir. Microbiol. 32:114–120.

Gerba, C.P., Smith, E.M., Schaiberger, G.E. & Edmond, T.D., 1979. Field evaluation of methods for the detection of enteric viruses in marine sediments. In Litchfield, C.D. & Seyfried, P.L. (eds.), Methodology for Biomass Determinations and Microbial Activities in Sediment. Am. Soc. Test. Mat., Pennsylvania: 64–74.

Ghirelli, R.P. & Marker, L., 1977. Bacterial water quality in wilderness areas. Calif. Wat. Resources Center. NTIS #PB–279–690.

Goyal, S.M., Gerba, C.P. & Melnick, J.L., 1977. Occurrence and distribution of bacterial indicators and pathogens in canal communities along Texas coasts. Appl. Microbiol. 34:139–149.

Grabow, W.O.K., Gauss-Muller, V., Prozesky, O.W. & Deinhardt, F., 1983. Inactivation of Hepatitis A virus and indicator organisms in water by free chlorine residuals. Appl. envir. Microbiol. 46:619–624.

Grabow, W.O.K., Hilner, C.A. & Coubrough, P., 1981. Evaluation of standard and modified M–FC, MacConkey, and Teepol media for membrane filtration counting of fecal coliforms in water. Appl. envir. Microbiol. 42:192–199.

Green, B.L., Clausen, E.M. & Litsky, W., 1977. Two-temperature membrane filter method for enumerating fecal coliform bacteria from chlorinated effluents. Appl. envir. Microbiol. 33:1259–1264.

Grimes, D.J., 1975. Release of sediment-bound fecal coliforms by dredging. Appl. Microbiol. 29:109–111.

Grimes, D.J., 1980. Bacteriological water quality effect of hydraulically dredging contaminated upper Mississippi River bottom sediment. Appl. envir. Microbiol. 39:782–789.

Gucinski, H., 1982. Sediment suspension and resuspension from small craft induced turbulence. US Environmental Protection Agency, Cincinnati, Ohio. EPA–600/S3–82–084. 2 pp.

Haas, C.N. & Engelbrecht, R.A., 1980. Chlorine dynamics during inactivation of coliforms, acid-fast bacteria and yeasts. Wat. Res. 14:1749–1758.

Harter, L., Frost, F.F. & Jakubowski, W., 1982. *Giardia* prevalence among 1-to-3 year-old children in two Washington State counties. Am. J. publ. Health 72:386–388.

Hazen, T.C. & Esch, G.W., 1983. Effect of effluent from a nitrogen fertilizer factory and a pulp mill on the distribution and abundance of *Aeromonas hydrophila* in Albemarle Sound, North Carolina. Appl. envir. Microbiol. 45:31–42.

Hendricks, C.W., 1971. Increased recovery rates of *Salmonellae* from stream bottom sediments versus surface waters. Appl. Microbiol. 21:379–380.

Hendry, G.S. & Toth, A., 1982. Some effects of land use on bacteriological water quality in a recreational lake. Wat. Res. 16:105–112.

Herson, D.S. & Victoreen, H.T., 1980. Hindrance of coliform recovery by turbidity and non-coliforms. US Environmental Protection Agency, Cincinatti, Ohio. EPA–600/2–80–097. 66 pp.

Highsmith, A.K., Feeley, J.C., Skaliy, P., Wells, J.G. & Wood, B.T., 1977. Isolation of *Yersinia enterocolitica* from well water and growth in distilled water. Appl. envir. Microbiol. 34:745–750.

Hoadley, A.W. & Dutka, B.J. (eds.), 1977. Bacterial Indicators Health Hazards Associated with Water. Am. Soc. Test. Mat., Philadelphia: 356 pp.

Horak, W.F., 1974. A bacterial water quality investigation of Canyon Lake, Arizona. M.S. thesis. University of Arizona, Tucson. 54 pp.

Hornick, R.B., Greisman, S.E., Woodward, T.E., Dupont, H.L., Dawkins, A.T. & Snyder, M.J., 1970. Typhoid fever: pathogenesis and immunologic control. N. Eng. J. Med. 283–686.

IAWPRC study group on water virology, 1983. The health significance of viruses in water. Wat. Res. 17:121–132.

Jakubowski, W. & Hoff, J.C. (eds.), 1979. Waterborne transmission of giardiasis, proceedings of a symposium. US Environmental Protection Agency, Cincinnati, Ohio. EPA–600/9–79–001. 306 pp.

Kaneko, K.I. & Hashimoto, N., 1981. Occurrence of *Yersinia enterocolitica* in wild animals. Appl. envir. Microbiol. 41:635–638.

Kaplan, J.E., Feldman, R., Campbell, D.S., Lookabaugh, C. & Gary, G.W., 1982. The frequency of a Norwalk-like pattern of illness in outbreaks of acute gastroenteritis. Am J. publ. Health 72:1329–1332.

Kapperud, G., 1975. *Yersinia enterocolitica* in small rodents from Norway, Sweden, and Finland. Acta pathol. Microbiol. scand. Sect. B 83:335–342.

Kapuscinski, R.B. & Mitchell, R., 1983. Sunlight-induced mortality of viruses and *Escherichia coli* in coastal seawater. Envir. Sci. Technol. 17:1–6.

Kay, D. & McDonald, A., 1980. Reduction of coliform bacteria in two upland reservoirs: the significance of distance decay relationship. Wat. Res. 14:305–318.

Keswick, B.H., Gerba, C.P., Secor, S.L. & Cech, I., 1982. Survival of enteric viruses and indicator bacteria in groundwater. J. envir. Sci. Health A17:903–912.

Kott, Y., 1981. Viruses and bacteriophages. The Sci. total Envir. 18:13–23.

Kraus, M.P., 1977. Bacterial indicators and potential health hazard of aquatic viruses. In Hoadley, A.W. & Dutka, B.J. (eds.), Bacterial Indicators/Health Hazards Associated with Water. Am. Soc. Test. Materials. Philadelphia: 196–217.

LaBelle, R.L., Gerba, C.P., Goyal, S.M., Melnick, J.L., Cech, I. & Bogdan, G.F., 1980. Relationships between environmental factors, bacterial indicators, and the occurrence of enteric viruses in estuarine sediments. Appl. envir. Microbiol. 39:588–596.

LeChevallier, M.W. & McFeters, G.A., 1984. Recent advances in coliform methodology for water analysis. J. envir. Health 47:5–9.

LeChevallier, M.W., Evans, T.M. & Seidler, R.J., 1981. Effect of turbidity on chlorination efficiency and bacterial persistence in drinking water. Appl envir. Microbiol. 42:159–167.

LeChevallier, M.W., Jakanoski, P.E., Camper, A.K. & McFeters, G.A., 1984. Evaluation of m–T7 agar as a fecal coliform medium. Appl. envir. Microbiol. 48:371–385.

Levin, M.A., 1977. Bifidobacteria as water quality indicators. In Hoadley, A.W. & Dutka, B.J. (eds.), Bacterial Indicators/Health Hazards Associated with Waters. Am. Soc. Test. Materials, Philadelphia: 131–138.

Levin, M.A., Fischer, J.R. & Cabelli, V.J., 1975. Membrane filter technique for enumeration of enterococci in marine waters. Appl. Microbiol. 30:66–71.

Lin, S.D., 1976. Membrane-filter method for recovery of fecal coliforms in chlorinated sewage effluents. Appl. envir. Microbiol. 32:547–552.

Lorenz, R.C., Hsu, J.C. & Tuorinen, O.H., 1982. Performance variability, ranking, and selection analysis of membrane filters for enumerating coliform bacteria in river water. J. am.

Wat. Wks. Ass. 74:429–437.

Luechtefeld, N.W., Blaser, M.J., Reller, L.B. & Wang, W–L., 1980. Isolation of *Campylobacter-fetus* spp. *jejuni* from migratory waterfowl. J. Clin. Microbiol. 12:406–408.

Luechtefeld, N.W., Wang, W–L.L., Blaser, M.J. & Reller, L.B., 1981. *Campylobacter fetus* subsp. *jejuni*: bacground and laboratory diagnosis. Lab. Med. 12:481–487.

Mathewson, J.J., Keswick, B.H. & DuPont, H.L., 1983. Evaluation of filters for recovery of *Campylobacter jejuni* from water. Appl. envir. Microbiol. 46:985–987.

Matson, E.A., Hornor, S.G. & Buck, J.D., 1978. Pollution indicators and other micro-organisms in river sediment. J. Wat. Pollut. Cont. Fed. 50:13–19.

Maul, A. & Block, J.C., 1983. Microplate fecal coliform method to monitor stream water pollution. Appl. envir. Microbiol. 46:1032–1037.

McDonald, A.T. & Kay, D., 1981. Enteric bacterial concentrations in reservoir feeder streams: baseflow characteristics and response to hydrograph events. Wat. Res. 15:961–968.

McDonald, A.T., Kay, D. & Jenkins, A., 1982. Generation of fecal and total coliforn surges by streamflow manipulation in the absence of normal hydrometerological stimuli. Appl. envir. Microbiol. 44:292–300.

McFeters, G.A., Bissonnette, G.K., Jezeski, J.J., Thomson, C.A. & Stuart. D.G., 1974. Comparative survival of indicator bacteria and enteric pathogens in well water. Appl. Microbiol. 27:823–829.

McFeters, G.A. & Camper, A.K., 1983. Enumeration of indicator bacteria exposed to chlorine. Adv. appl. Microbiol. 29:177–193.

McFeters, G.A., Schillinger, J.E. & Stuart, D.G., 1978. Alternative indicators of water contamination and some physiological characteristics of heterotrophic bacteria in water. In Hendricks, C.W. (ed.), Evaluation of the Microbiology Standards for Drinking Water. US Environmental Protection Agency, Washington, DC EPA–570/9–78–00C. 37–48.

McFeters, G.A. & Stuart, D.G., 1972. Survival of coliform bacteria in natural waters: field and laboratory studies with membrane-filter chambers. Appl. Microbiol. 24:805–811.

McFeters, G.A., Cameron, S.C. & LeChevallier, M.W., 1982. Influence of diluents, media, and membrane filters on detection of injured waterborne coliform bacteria. Appl. envir. Microbiol. 43:97–103.

McFeters, G.A., Stuart, S.A. & Olson, S.B., 1978. Growth of heterotrophic bacteria and algal extracellular products in oligotrophic waters. Appl. envir. Microbiol. 335:383–391.

McKee, J.E. & Wolf, H.W., 1963. Water quality criteria. St. Wat. Qual. Cont. Bd., Sacramento, California.

Mechalas, B.J., Hekimian, K.K., Schinazi, L.A. & Dudley, R.H., 1972. Water quality criteria data book, volume 4– an investigation into recreational water quality. US Environmental Protection Agency, Washington, DC EPA–18040–DAZ–04/72.

Melnick, J.L. & Gerba, C.P., 1980. The ecology of enteroviruses in natural waters. Crit. Rev. envir. Cont. 10:65–93.

Metcalf, T.G. & Melnick, J.L., 1983. Simple apparatus for collecting estuarine sediments and suspended solids to detect solids-associated virus. Appl. envir. Microbiol. 45:323–327.

Miescier, J.J. & Cabelli, V.J., 1982. Enterococci and other municipal indicators in municipal wastewater effluents. J. Wat. Pollut. Cont. Fed. 54:1599–1606.

Mitchell, R.S., Yankofsky, S. & Jannasch, H.S., 1967. Lysis of *Escherichia coli* by marine organisms. Nature (London) 215:891–893.

Moore, B., 1959. Sewage contamination of coastal bathing waters in England and Wales: a bacteriological and epidemiological study. J. Hyg. 57:435.

Mundt, J.O., 1962a. Occurrence of enterococci in animals in a wild environment. Appl. Microbiol. 11:136–140.

Mundt, J.O., 1962b. Occurrence of enterococci on plants in a wild environment. Appl. Microbiol. 11:141–144.

Nemedi, L. & Layni, B., 1971. Incidence and hygienic importance of *Pseudomonas aeruginosa* in water. Acta Microbiol. Acad. Sci. Hung. 18:319–325.

Nix, J.F., Meyer, R.L., Schmitz, E.H., Bragg, J.D. & Brown, R., 1975. Collection of environmental data on DeGray Reservoir and the watershed of the Caddo River, Arkansas. Arkansas Wat. Resurces Res. Center and Ouachita Baptist Univ., US Army Engineers Waterways Experiment Station, Vicksburg, MS.

O'Dell, W.D., 1977. Detection of a potential health hazard in recreational and other surface waters. Nebraska Wat. Resources Res. Inst., Lincoln. NTIS, PB 275 769:19 pp.

O'Dell, W.D., 1979. Isolation, enumeration, and identification of amoebae from a Nebraska, USA, lake. J. Protozool. 26:265–269.

O'Malley, M.L., Lear, D.W., Adams, W.N., Gaines, J., Sawyer, T.K. & Lewis, E.J., 1982. Microbial contamination of continental shelf sediments by wastewater. J. Wat. Pollut. Cont. Fed. 54:1311–1317.

Oragai, J.I. & Mara, D.D., 1983. Investigation of the survival characteristics of *Rhodococcus coprophilus* and certain fecal indicator bacteria. Appl. envir. Microbiol. 46:356–360.

Pagel, J.E., Qureshi, A.A., Young, D.M. & Vlassoff, L.T., 1982. Comparison of four membrane filter methods for fecal coliform enumeration. Appl. envir. Microbiol. 43:787–793.

Pai, C.H., Mors, V. & Oman, E., 1979. Pathogenesis of *Yersinia enterocolitica* gastroenteritis in experimental infection. Abstr. ann. Meet. am. Soc. Microbiol., Washington, D.C.:B21, 19.

Payment, P., Lemieux, M. & Trudel, M., 1982. Bacteriological and virological analysis of water from four fresh water beaches. Wat. Res. 16:939–943.

Pellet, S., Bigley, D.V. & Grimes, D.J., 1983. Distribution of *Pseudomonas aeruginosa* in a riverine ecosystem. Appl. envir. Microbiol. 45:328–332.

Pipes, W.O. (ed.), 1978. Water Quality and Health Significance of Bacterial Indicators of Pollution, Workshop Proceedings Drexel University. National Science Foundation, Washington, DC: 228 pp.

Presswood, W.G., 1978. Modification of M–FC medium by eliminating rosolic acid. Appl. envir. Microbiol. 36:90–94.

Pretorius, W.A., 1962. Some observations on the role of coliphage in the number of *Escherichia coli* in oxidation ponds. J. Hyg., Camb. 60:279–281.

Rippey, S.R. & Cabelli, V.J., 1980. Occurrence of *Aeromonas hydrophila* in limnetic environments: relationship of organism to trophic state. Microb. Ecol. 6:45–54.

Rowbotham, T.J., 1980. Preliminary report on the pathogenicity of *Legionella pneumonophila* for freshwater and soil amoebae. J. clin. Pathol. 33:1179.

Rudolphs, W., Folk, L.L. & Ragotzkie, R.A.L., 1950. Literature review on the occurrence and survival of enteric, pathogenic, and related organisms in soil, water, sewage, and sludges and on vegetation. I. Bacterial and viral diseases. Sew. ind. wastes 22:1261–1281.

Sands, M., Sommers, H.M. & Loewenstein, R., 1981. Is *Campylobacter fetus* a significant enteric pathogen in Chicago? Abstr. ann. Meet. am. Soc. Microbiol., Washington, DC C204, p. 296.

Sayler, G.S., Nelson, J.D., Jr., Justice, A. & Colwell, R.R., 1975. Distribution and significance of fecal indicator organisms in the upper Chesapeake Bay. Appl. Microbiol. 30:625–638.

Scaglia, M., Strosselli, M., Grazioli, V., Gatti, S., Bernuzzi, A.M. & De Jonckeere, J.F., 1983. Isolation and identification of pathogenic *Naegleria australiensis (Amoebida, Vahlkampfiidae)* from a spa in northern Italy. Appl. envir. Microbiol. 46:1282–1285.

Schaiberger, G.E., Edmond, T.D. & Gerba, C.P., 1982. Distribution of enteroviruses in sediments contiguous with a deep sewage outfall. Wat. Res. 16:1425–1428.

Schiemann, D.A., 1978. Isolation of *Yersinia enterocolitica* from surface and well waters in Ontario, Canada. Can. J. Microbiol. 24:1048–1052.

Schiemann, D.A., Brodsky, M.H. & Ciebin, B.W., 1978. *Salmonella* and bacterial indicators in ozonated and chlorine dioxide-disinfected effluent. J. Wat. Pollut. Cont. Fed. 50:158–162.

Schillinger, J.W. & McFeters, G.A., 1978. Survival of *Escherichia coli* and *Yersinia enterocolitica* in stream and tap waters. Abstr. annu. Meet. am. Soc. Microbiol., Washington, DC: N79, p. 175.

Standridge, J.H. & Delfino, J.J., 1981. A–1 medium: alternative technique for fecal coliform organism enumeration in chlorinated wastewaters. Appl. envir. Microbiol. 42:918–920.

Steveson, A.H., 1953. Studies of bathing water quality and health. Am. J. publ. Health 43:529–538.

Sykora, J.L., Keleti, G. & Martinez, A.J., 1983. Occurrence and pathogenicity of *Naegleria fowleri* in artificially heated waters. Appl. envir. Microbiol. 45:974–979.

Taylor, D.N., Brown, M. & McDermott, K.T., 1982. Waterborne transmission of *Campylobacter enteritis* Microb. Ecol. 8:347–354.

Thornton, K.W., Nix, J.F. & Bragg, J.D., 1980. Coliforms and water quality: use of data in project design and operation. Wat. Res. Bull. 16:86–92.

Tison, D.L., Pope, D.H., Cherry, W.B. & Fliermans, C.B., 1980. Growth of *Legionella pneumophila* in association with blue-green algae (*Cyanobacteria*). Appl. envir. Microbiol. 39:456–459.

Tsernoglou, D. & Anthony, E.H., 1971. Particle size, water-stable aggregates, and bacterial populations in sediments. Can. J. Microbiol. 17:217–227.

US Environmental Protection Agency, 1978. Human viruses in the aquatic environment: a status report with emphasis on the EPA research program. Report to Congress. Washington, DC EPA-570/9–78–006. 37 pp.

Van Donsel, D.J. & Geldreich, E.E., 1971. Relationships of *Salmonellae* to fecal coliforms in bottom sediments. Wat. Res. 5:1079–1087.

Van Donsel, D.J., Geldreich, E.E. & Clarke, N.A., 1967. Seasonal variations in survival on indicator bacteria in soil and their contribution to storm-water pollution. Appl. Microbiol. 15:1362–1370.

Velz, C.J., 1970. Applied Stream Sanitation. Wiley Inter-Science, NY: 619 pp.

Vlassoff, L.T., 1977. *Klebsiella*. In Hoadley, A.W. & Dutka, B.J. (eds.), Bacterial indicators/health hazards associated with water. Am. Soc. Test. Mat., Philadelphia: 275–288.

Walker, R.W., Wun, C.K. & Litsky, W., 1982. Coprostanol as an indicator of fecal pollution. Crit. Rev. envir. Cont. 12:91–112.

Weiss, C.M., 1951. Absorption of *Escherichia coli* on river and estuarine silts. Sew. ind. Wastes 23:227–237.

Wellings, F.M., Amuso, P.T., Chang, S.L. & Lewis, A.L., 1977. Isolation and identification of pathogenic *Naegleria* from Florida lakes. Appl. envir. Microbiol. 34:661–667.

Wellings, F.M., Amuso, P.T., Lewis, A.L., Farmelo, M.J.,

Moody, D.J. & Osikowicz, C.L., 1979. Pathogenic *Naegleria*: distribution in nature. US Environmental Protection Agency, Cincinnati, Ohio. EPA–600/1–79–018.

Wentsel, R.S., O'Neill, P. & Kitchens, J.F., 1982. Evaluation of coliphage detection as a rapid indicator of water quality. Appl. envir. Microbiol. 43:430–434.

Wetzel, R.G., 1975. Limnology. W.B. Saunders Co., Philadelphia: 743 pp.

Wilson, R., Anderson, L.J., Holman, R.C., Gary, G.W. & Greenberg, H.B., 1982. Waterborne gastroenteritis due to the Norwalk agent: clinical and epidemiologic investigation. Am J. publ. Health 72:72–74.

Winslow, S.A., 1976. The relationship of bottom sediments to bacterial water quality in a recreational swimming area. M.S. thesis. University of Arizon, Tuscon: 60 pp.

Yea, M., De Wult, E., De Mayer-Cleempoel, S. & Quaghebeur, D., 1982. Coprostanol and bacterial indicators of faecal pollution in the Scheldt Estuary. Bull. envir. Contam. Toxicol. 28:129–134.

Author's address:
G. Allen Burton, Jr.
Environmental Science Program
University of Texas at Dallas
Richardson, Texas, USA
Present address:
Cooperative Institute for Research
in Environmental Sciences
Campus Box 449
University of Colorado
Boulder, CO 80309, USA

Interactions of reservoir microbiota: eutrophication – related environmental problems

GUY R. LANZA and J.K.G. SILVEY

Abstract. Reservoir ecosystems provide diverse and variable habitats for microbial assemblages. Fossil evidence of microbial assemblages date back 3400–3500 million years and highlight the important historical role of microbial interactions in the evolution of aquatic communities. Simple mathematical representations help to conceptualize the interaction of microbial species; however, they are too simplified and assumption-based to accurately describe interactions in reservoirs. Experimental observations of dominant reservoir species remain the best tool for deciphering microbial interactions.

General microbial interactions occur within and between the autotrophic and heterotrophic components of microbial assemblages. Community metabolism, physical and chemical factors, and sedimentation/ sediment systems regulate the structure and function of the reservoir microbial habitat.

Reservoir microbial interactions change with the natural ageing process known as eutrophication. One general characteristic of the advanced stages of eutrophication is the increased frequency of population imbalances or blooms with large numbers of cyanobacteria. Microbial interactions during bloom sequences often result in undesirable water quality characteristics. Recent studies of bloom interactions point to bloom products which add new environmental health significance to traditional eutrophication-related water quality problems. In addition to the taste and odor problems commonly associated with bloom events, episodes of toxicity, mutagenicity and teratogenicity have emerged. Autotrophic/heterotrophic imbalances can also occur with potential public health problems involving pathogens and/or resistance factors (R factor plasmids).

Introduction

The vast majority of streams and rivers in the world exist as modified ecosystems. Regulated lotic ecosystems frequently function as modified riverine habitats supplying water to impounded reservoirs. Increased construction of impoundments is likely to occur with population expansion, and the accompanying urban and industrial growth.

Reservoirs throughout the world differ greatly in climate, size, flow dynamics, geomorphometry, and their physical, chemical, and biological charac-teristics. The arrangement of these characteristics set the stage for organism habitat and species interactions in a specific reservoir. Two general types of reservoir physical habitat exist (Hannon, 1979). Offstream, or closed, reservoirs lack natural inflow or outflow and usually receive pumped water from a nearby lotic system. Since they are more influenced by evaporation than open systems, they are typically more saline. Onstream, or open, reservoirs are located on lotic systems allowing discharge of the stream through the impoundment. Open reservoirs are often classified as mainstream,

Gunnison, D. (ed.) Microbial Processes in Reservoirs.

transitional, and deep-storage impoundments. Most of the impounded water in main-stream reservoirs is restricted to the old river channel, while water in deep-storage reservoirs is extended well beyond the original river bed, forming numerous coves and inlets. Transitional reservoirs provide habitat that is intermediate between main-stream and deep-storage types. All three types may or may not exhibit thermal stratification (Lockett, 1976). Generally, permanent summertime thermal stratification is exhibited only in deep-storage reservoirs (Hannon, 1979).

In the USA, impoundments have been constructed to furnish water power for mills and other small industries, to hold public water supplies and irrigation reserves, for hydroelectric power, flood control, navigation purposes, fish and wildlife protection, and recreation (Ward & Stanford, 1979). Some reservoirs are designed to trap large quantities of silt and/or water. Water releases for various ancillary purposes, e.g. dilution of sewage, improvement of water quality, are provided by this type of reservoir. As dam building techniques improved, engineers utilized our largest lotic systems, building huge impoundments, each with a number of assigned functions. In the United States (excluding Alaska) only 51 rivers greater than 100 km in length remain freeflowing from headwaters to major confluence. Nearly every feasible high-head dam site has been utilized or is proposed for dam construction. (Ward & Stanford, 1979). One major result of the extensive habitat alteration accompanying the conversion of lotic to lentic ecosystems is reflected in changes influencing the interaction of impoundment microbiota.

Structure and function of the reservoir microbial habitat

General microbial interactions

Evolutionary theories describing the development of early planktonic micro-organisms suggest strong interaction between species groups, and highlight the important historical role of microbial interactions. Fossilized mats or filamentous assemblages of procaryotes (i.e. stromatolites) have been found in deposits estimated to be 3400–3500 million years old (Lowe, 1981; Walter et al., 1980). Fossil evidence of the planktonic mode of life in eucaryotes date back at least 650 million years (Vidal, 1984). Eucaryotic microbes may have emerged either as self-generated organisms, or as symbiotic ones. If the latter theory is true, then the first eucaryotic cells were derived from preexisting procaryotes, and represented some of the earliest microbial interactions of major significance. In any event, both theories of microbial evolution require microbial interaction with procaryote aerobic metabolism providing an oxygenated habitat as a prerequisite to eucaryote survival.

In the simplest sense, the interaction of one species in contact with another produces one of three effects on the second species population growth; these are no effect, increased growth or decreased growth. A simple mathematical representation of species interaction can be incorporated into a population growth equation (e.g. the Pearl-Verhulst equation) as:

$$\frac{dN}{dt} = rN - \frac{r}{K}N^2 + CN_2N$$

[growth rate] = [unlimited growth rate] − [self-crowding effects] + [interaction effects with other species]

Where N = size of one species population, N_2 = size of the other species in the interaction, r = intrinsic growth rate, K = carrying capacity of the habitat, t = time, C = constant describing a negative or positive result depending on the intensity of the interaction (Rolan, 1973). Odum (1971) provided an excellent descriptive analysis of the equation constant C with nomenclature for each of nine major types of two species interaction events. The C value in this scheme could be neutral, positive or negative $(0, +, -)$ with regard to interaction effect, and can be simply described by the interaction matrix in Fig. 1. In this matrix, interactions, between two microbes j and i are described as the equilibrium effects of one organism on the other, and vice-versa. The matrix elements A_{ij} and A_{ji} are descriptors of the effects of each microbial population on the other (Bull & Slater, 1982). For exam-

		Effects of population j on i (a_{ij})		
		+	O	−
Effects of population i on j (a_{ji})	+	+ +	+ O	+ −
	O	O +	O O	O −
	−	− +	− O	− −

Bull & Slater, 1982

Fig. 1. Matrix of interactions between two microbial populations, i and j.

ple, if population i represented a species of aquatic-microalgae producing organic carbon compounds, and population j a species of aquatic heterotrophic bacteria in close association with i, the effects of population i on j (A_{ji}) could be positive (+) with species j using the carbon as a source for growth and metabolism. If population i represented a species of cyanobacteria producing antibiotic metabolites (e.g. inhibiting glucose uptake), and population j a species of aquatic heterotrophic bacteria in close association with i, the effects of population i on j (A_{ji}) could be negative (−).

While these simple mathematical representations help to conceptualize the interaction of two species, they are far too simplified and assumption-based to accurately describe real interactions in reservoir ecosystems. In order to begin understanding the multi-species interactions occurring in reservoir microhabitats we still must rely heavily on experimental observations of dominant reservoir species. Although qualitative descriptions of the structure and function of microbial assemblages exist (Silvey & Wyatt, 1971; Cairns & Lanza, 1972; Bull & Slater, 1982), detailed quantitative descriptions are generally lacking.

Microbial assemblages vary greatly in their basic structure from loose associations of mixed species in near proximity (e.g. planktonic microalgae and bacteria in the water column), to close associations involving aggregates or bioflocs (e.g. heterotrophic bacteria epiphytic on algae or several heterotrophic bacterial species attached to a common clay or detrital particle). The specific structure and the trophic status (i.e. sources of ATP energy and carbon) of an assemblage will set the stage for the trophic interaction within and between microbial species. Fig. 2 provides a generalized scheme of typical trophic interaction pathways occurring in common microbial assemblages. As outlined in

Fig. 2, the microbial assemblage can involve a continuum of autotrophic-heterotrophic interaction with a balance of autotrophically-produced carbon compounds supplying energy substrate (electron donors) to microbial heterotrophs. For example, during the aerobic heteretrophic decomposition of substrates formed during microbial autotrophic production, energy is obtained from a stepwise transfer of electrons during oxidative phosphorylation. Dehydrogenase enzymes oxidize or remove hydrogen atoms during electron transport. The autotrophic-heterotrophic continuum plays a significant role for the energy flux in the interacting assemblages. Many of the potential interaction effects outlined in Fig. 1 are metabolically supported by the trophic interaction pathways provided in Fig. 2.

Physical and chemical structure

Limnologists and aquatic ecologists have examined the abiotic and biotic conditions in natural lakes for more than a century. Thermal stratification patterns were described in the USA and Great Britain in the 1800's (Hutchinson, 1957). However, designers of early impoundments largely ignored available lake data, and made little or no allowance for controlling thermal stratification and reservoir circulation in spite of their importance in regulating sediment-water interface phenomena (Ridley & Symons, 1972). Thus, maximum water depth of 40 feet for reservoirs in temperate climates became accepted as the means of limiting the extent of thermal stratification and deoxygenation of the lowest layers, on the assumption that midsummer wind forces would still be adequate for maintaining near-isothermal conditions. During drought summers the combination of excessive solar heating, and the reduction of flow from tributaries resulted in thermal gradients that persisted throughout summer until fall overturn. The need to conserve land resources later led to construction of impoundments where the water depth was considerably more than 40 feet, producing longer thermal stagnation periods often extending into early winter (Ridley & Symons, 1972).

In understanding microbial interactions, it is im-

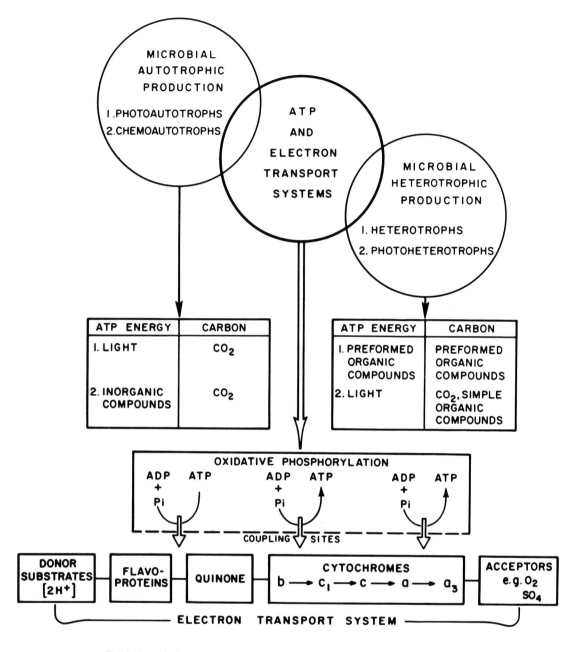

Fig. 2. Trophic interaction pathways occurring in common microbial assemblages.

portant to note that impoundment microhabitats vary in a number of physical details from natural lakes that are interposed in lotic systems. Glaciated lakes usualy have inlets and outlets near the surface, but water may leave an impoundment at one of several depths, or from two or three levels simultaneously. The deepest site in a natural lake may occur anywhere in its basin, unlike an impound-ment where it is always near the dam. Areas where a natural lake is included in the impoundment area are exceptions. Impoundments typically have a regular sloping bottom established by the pre-decessor lotic system prior to damming. A similar slope is common to natural lakes formed by earth-quakes. Basins of glaciated lakes were scooped out below river level, and non-uniformity of bottom slope is to be expected (Neel, 1966).

Chemical habitat factors in natural lakes and impoundments are largely determined by their basin origins. The normal sequence of trophic change occurring over time in a reservoir reflects the physical, chemical, and biological eutrophication process. Glaciated lakes usually begin as low-nutrient oligotrophic bodies of water resting in clay, sand, gravel, and shingle basins. Typically, increases in productivity occur gradually following contributions of allochthonous organic material from terrestrial biota in their drainage basins. Development to a more eutrophic condition usually requires at least several decades.

Impoundments often inundate rich bottom lands and fertile topsoils on river slopes. They normally begin their cycle with high productivity potential fueled by mineral nutrients and organic materials leached directly from these soils. Impoundments may develop productivity declines with passage of time, and some do not regain their initial productivity level. Others undergo gradual increases in productivity when supplied with excessive cultural nutrient additions, e.g. partially treated sewage or agricultural runoff. Stratification, density currents, hydraulic residency time, dominant withdrawal level, and prevalent operation practices all influence productivity rates.

The history of the inflowing water and the path of the flow in the reservoir can directly determine the quality and quantity of conservative chemical ions at the outlet if surface evaporation, groundwater exchange, and sedimentation are low. The longitudinal and vertical conditions of the nonconservative ions in the reservoir are, however, altered by community metabolism. The biological influence on chemical conditions within the reservoir is often greater than the influence of other factors combined (Hannan, 1979).

Sedimentation and sediment systems

The importance of sediments as a reservoir of lake nutrients has been recognized for many years. (Twenhofel & Broughton, 1939; Mortimer, 1941). Lake sediments have been used in attempts to determine the trophic state of a lake. Hansen (1961) pointed out the importance that Naumann and Thienemann placed on bottom deposits when they were preparing their classifications of lakes. Hendricks & Silvey (1973) pointed out the possible roles that the sediments play in affecting lake eutrophication rates. Sediments may also act as a trap for the more refractive organic compounds and immobile inorganic species, thereby effectively removing them from the water (Wetzel, 1983).

Development of anoxic conditions within reservoirs as a result of respiration in the water column and oxygen demand by sediments can have potentially severe consequences. This is particularly true if reduced chemical species are released from the sediment to overlying waters (Nix and Ingols, 1981; Gunnison et al., 1983). In new reservoirs, decomposition of surface layer organic matter flooded during filling acts in concert with microbial degradation of labile materials in the flooded soil to deplete dissolved oxygen and release nutrients and metals to the overlying water column (Sylvester and Seabloom, 1965; Gunnison et al., 1980a,b). Thus, a large portion of the concern in developing a new reservoir is directed towards the oxygen demands of the newly flooded soil which becomes reservoir sediment.

Biological respiration at the sediment-water interface may be strongly dependent upon reduced substances used by micro-organisms in the aerobic-anaerobic microzone layer near the sediment surface (Hargrave, 1972). Brewer et al. (1977) studied oxygen consumption by sediments and suggested that oxygen removal by sediments was predominantly a biological process. Partitioning experiments run on lake sediment cores provided further evidence for the biological nature of oxygen uptake, with bacterial respiration predominating (Belanger, 1981). However, Bowman and Delfino (1980) noted that sediment oxygen uptake methods varied widely between investigators, and disparities in technique were not normally accounted for when results were compared. Moreover, few of these techniques have been applied to the newly flooded soil to determine which kind of oxygen demand, if any, prevails (Gunnison et al., 1983).

Observations of water quality and biological parameters in the field, point toward a relationship between the settling dynamics of particles and increased concentrations of natural oxygen demanding materials in the thermocline region. Microbial interactions occurring in natural particle assemblages common to reservoir water columns are poorly understood. Particle assemblages contain an array of free-floating microbiota (e.g. cells of bacteria, fungi, algae), and aggregations of microbiota attached/associated with inert inorganic and organic particles. The settling dynamics of particulate assemblages are affected by the rapidly changing temperature gradient which significantly changes density and viscosity. The terminal settling velocities of particles accordingly decrease as described by Stokes' Law. Since particles settle into a layer faster than they exit, increased concentrations of oxygen demanding materials result, often producing a metalimnetic dissolved oxygen depletion phenomenon (Gordon & Skelton, 1977).

Drury & Gearheart (1975) described the impact of microbial interactions on reservoir metalimnetic oxygen pulses. Spring diatom blooms and early Summer algal activities produced organic substrates for metalimnetic bacterial communities. At first, bacterial densities increased along with the development of a metalimnetic oxygen minimum. Under anaerobic conditions, bacterial densities decreased, but a bacterial maximum reestablished immediately above the anaerobic layer.

Biologically mediated physical and chemical changes occur within the three different vertical habitats of a thermally stratified reservoir. Photosynthetically-induced decalcification and nutrient depletion can occur within the epilimnion. Both super-saturation and depletion of dissolved oxygen and associated chemical changes occur in the metalimnion. These changes are caused by community metabolism associated with organisms trapped within the mid-depth zone, or by respiration associated with an interflow from upreservoir (Gordon, 1978; Segura, 1978). Hypolimnetic chemical changes are associated with community respiration and related effects on the sediment-water interface.

The spatial and temporal occurrence of these biologically induced chemical conditions within the reservoir are controlled by natural inflow, regulated outflow, thermal stratification, density currents, and overturn. Interactions of these physiochemical and biological factors vary considerably horizontally and vertically within a reservoir and between reservoirs.

Often the trend throughout the year is for the greatest amount of metabolic activity and sedimentation to occur upreservoir, resulting in a progressive downreservoir decrease in concentration of the different chemicals. These changes in longitudinal and vertical conditions are not always associated with the often-used temperature isotherms. Dissolved oxygen depletion is highly variable horizontally, vertically, and seasonally (Straskraba et al., 1973). Commonly, an established hypolimnion will become anoxic in the upper reservoir area. As the summer progresses, hypolimnetic anoxia develops in a downreservoir progression until it reaches the dam, or until overturn occurs. By the time the lower end of the reservoir becomes anoxic, the upreservoir region can become oxygenated. This oxygenation of the upreservoir region is caused by inflow and by the deepening of the epilimnion as a result of hypolimnetic withdrawal (Hannan, 1979).

Downreservoir development of the anoxic hypolimnion results from: (a) an interaction of progressively larger amounts of hypolimnetic dissolved oxygen per unit of surface area downreservoir, (b) reservoir-wide autochthonous plant production in the photogenic zone, (c) benthic oxygen demand (especially in the sediment zone), and (d) downreservoir movement of anoxic water resulting from drawdown at the outlet. Downward vertical migration of water also occurs, and transports photosynthetically-produced organic matter to the anoxic zone as summer progresses. (Hannan, 1979).

Consumption of oxygen at the metalimnetic aerobic-anaerobic interface is due to a combination of biological and inorganic oxidation reactions. Bacteria in this zone use the proximal availability of oxidants and reductants to metabolic advantage by catalyzing their combination, and extracting the resulting energy change. Numerous attempts have been made to discriminate between biological and chemical oxygen demand by poisoning the biological reactions and measuring the resultant chemical oxygen demand (e.g. Brewer *et al.*, 1977; Wang, 1981). However, some oxidation steps can occur with or without biological mediation; poisoning, therefore, can only provide potential fractionations and will not necessarily represent the actual *in situ* partitioning of biological and chemical oxygen demand.

Eutrophication-related environmental health problems

Eutrophication is a term describing the natural ageing processes of lotic and lentic ecosystems. The ageing process is essentially a natural shift in ecosystem trophic status occurring over time. The rate of eutrophication is a complex phenomenon reflecting reservoir basin origin and geomorphology, watershed structure, land use patterns, climate, and successional changes in the physical, chemical, and biological components of the ecosystem. Natural eutrophication rates can be variable within and between reservoirs, and can be dramatically altered (e.g. accelerated) by cultural activities in the reservoir area.

General bloom interactions

Episodes of cultural and/or natural eutrophication are frequently accompanied by periodic microbial population inbalances with large numbers of one or more species of cyanobacteria. These sporadic assamblages are refferred to as blooms, and evidence from the fossil record indicates their occurrence early in the evolution of the cyanobacteria (Vidal, 1984). The first systematic report of a toxic bloom of cyanobacteria was published by Francis (1878), and described cattle mortality from bloom products in Australian reservoir water. However, it is important to note that not all blooms produce incidents of toxicity, and predicting when a bloom is likely to be toxic remains an important research topic for the future.

The stimulated organic producion characterizing accelerated eutrophication often reflects stresses on the structure and function of a microbial as-semblage, and microbial blooms in reservoirs are frequently site specific with regard to the kinds of organisms involved. Taylor *et al* (1981) studied 117 species of algae and cyanobacteria from 250 lakes and reported that certain cyanobacteria are repeatedly associated with water quality problems. Typical microbial bloom sequences often involve population interactions between algae, bacteria, and actinomycetes. The interaction of other microbes, e.g. fungi, protozoa, and viruses, in bloom sequences is not well understood, and poorly documented in reservoirs.

Cairns & Lanza, (1972) noted that waste-induced stress may contribute to microbial community change by a combination of interacting factors including: a) reduction in the number of species present, (b) an increase in the range of numbers of individuals per species, (c) a reduction in colonization rates by creating environmental conditions unfavorable to potential colonizing species before and after their arrival to the community, (d) changes in selective predator or parasite pressure resulting in a shift in balance within the community, and (e) a shift in dominance within the community favoring some species over others. Periodic imbalances favoring cyanobacteria in reservoirs have been explained by numerous hypotheses. Major among these are the role of inorganic carbon availability (King, 1970; Shapiro, 1973), interactions involving cyanobacterial inhibition of metabolism and growth in other microbes (Murphy *et al.*, 1976; Keating, 1976; Juttner, 1978; Juttner, 1979; Reichardt, 1981; Sakevich, 1973), zooplankton predation immunity of cyanobacteria (Porter, 1973), and nutrient ratios favoring cyanobacteria (Schindler, 1977; Soltero & Nichols, 1981; Smith, 1983).

Products resulting from bloom interactions add new environmental health significance to traditional eutrophication-related water quality problems. Recent research on the structural nature of cyanobacterial bloom assemblages indicates great variability in the species composition and numbers of individuals (i.e. density) at various reservoir stations (Carmichael & Gorham, 1981). Thus, microbial bloom interaction products would be expected to vary with the mosaic nature of the microbial assemblages in a reservoir.

A great deal of information concerning seasonal microbiotic cycles has been generated in Southwestern USA water supply reservoirs. Fig. 3 outlines a typical microbial interaction sequence in a Southwestern USA impoundment. During March the microbial community is dominated by cyanobacteria whereas in May, diatoms become the dominant forms. However, neither of these groups appear to reach high concentrations in the spring. This condition, in part, may be due to the fact that reservoirs in the Southwest do not experience an adequate period of winter stratification to assure the formation of an anaerobic zone with subsequent release of nutrients from the reduced sediments.

In July the cyanobacteria come into prominence (Fig. 3). They obain a maximum planktonic population during the latter part of August and may maintain a relatively dense population into the fall.

Two ecological types of cyanobacteria can occur in the open waters of a reservoir, those that are planktonic and those that are benthic. Both forms appear to develop on or near the bottom of Southwestern reservoirs. If the reservoir has adequate depth, there will be a period of thermal stratification in June or July. This situation is not persistent in shallow lakes during exposure to winds of high velocity, yet it is usually of sufficient duration to exist during a period each day, or for periods of time up to several weeks.

During stratification, the microaerobic environment at the sediment-water interface accelerates the liberation of nutrients, particularly phosphorous, ammonia, carbon dioxide, and iron, from the sediments. Thus, while the green algae are maintaining a meager existence in the nutrient-poor surface water, the cyanobacteria are growing luxuriantly on or near the bottom. As these organisms proliferate on the bottom and reach maturity, they develop gas vacuoles. As the vacuoles increase in size and number within the cells, they become bouyant and the cyanobacteria rise toward the surface. Sporulation occurs in the mature filaments, and as the vegetative cells die and lyse, the dense spores slowly settle to the bottom. These spores eventually settle into the sediment to await the stimuli for regeneration.

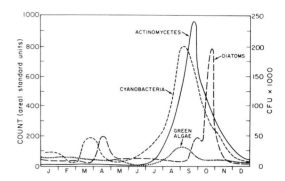

Fig. 3. Typical microbial interaction sequence in a Southwestern USA impoundment.

The phenomena of cyanobacterial blooms in reservoirs is often accompanied by taste and odor problems. Major offenders in this regard are the heterocystous cyanobacteria *Anabaena circinalis* and *Aphanizomenon flos-aquae*. Information collected for a number of years from southwestern reservoirs indicates that most noxious planktonic blooms may begin on or near the bottom of the reservoir during periods of anaerobiosis. Many cyanobacterial blooms develop at the bottoms of lakes under anaerobic conditions and it is the physicochemical and biological parameters of the bottom waters of a reservoir that should be examined for information regarding blooms (Silvey *et al.*, 1972).

Selected factors influencing bloom interactions

Temperature, oxygen, light. Temperature, oxygen and light are the most studied non-nutrient factors influencing microbial growth, accelerated eutrophication, and reservoir water quality. These factors are poorly understood and often misinterpreted when laboratory findings from one microbe are misconstrued as true for an entire group. Problems also arise when data obtained using a single variable are applied to field conditions where a multitude of interactions persist.

Temperature is the factor most constant in its effect on various microbial species and species assemblages. Although wide tolerances and varying optima are shown by various microbes, generalizations may be noted within groups. Lanza & Cairns

(1972) reported laboratory studies identifying 23 °C as the upper critical temperature beyond which added temperature stress produced physiological damage to several species of common diatoms. Similar trends were noted in subsequent field studies of diatom assemblages exposed to thermal discharges (Hein & Koppen, 1979). Many species of cyanobacteria fail to do well at lower temperatures, showing restrained metabolic activity in cold water (Silvey *et al.*, 1972). Hammer (1964) noted that cyanobacteria were seldom found in Canadian lakes until the water temperature reached about 15 °C. Blooms seldom occurred until temperatures reached 23–26 °C. He also reported that temperature influenced the composition and sequence of cyanobacterial blooms. Silvey *et al.*, (1972) reported an almost linear relationship between acetylene reduction by nitrogen-fixing cyanobacteria and temperature, within natural tolerance ranges. Discrepancies did occur at the extremes, e.g. at 2 °C, *A. cylindrica* ceased all acetylene reduction whereas *A. flos-aquae* maintained moderate amounts. At higher range, *A. cylindrica* continued its linear increase much longer than *A. flos-aquae*. Both ceased nitrogen fixation at 47 °C.

The effects of dissolved oxygen on the cyanobacteria are poorly understood. This appears particularly true when considering the nitrogen-fixers. Dissolved oxygen has been considered one necessary requirement for the development of large microbial blooms. However, some evidence indicates that in certain instances, the reverse may be true. Some types of cyanobacteria are known to grow well and fix nitrogen in reduced dissolved oxygen concentrations with high hydrogen sulfide levels (Stewart, 1969). Keating (1976) reported a bloom of *Oscillatoria rubescens* in a pond with high hypolimnetic hydrogen sulfide levels and relatively low light conditions. Stewart & Pearson (1970) showed *A. flos-aquae* and *Nostoc muscorum* to reduce acetylene faster at partial oxygen levels below 0.2 atmosphere, with an optimum of 0.1 atmosphere. Silvey *et al.* (1972) reported experiments with fast-growing (i.e. blooming) cultures of nitrogen-fixing cyanobacteria at or near oxygen saturation producing continued growth and good rates of nitrogen fixation. These examples suggest that the influence of dissolved oxygen on bloom formation is indirect, perhaps by influencing nutrient flux through the regulation of redox potential.

Light is another major factor producing variable effects on different microbial species. Lazaroff (1966) found that the wavelength of light influenced the morphology and life cycle of *N. muscorum*. Optimal light intensity of cyanobacteria in general, i.e. approximately 200 ft-candles, is usually less than that required for the green algae. However, cyanobacteria have been reported to flourish at unmeasurable light intensities. Some investigators have reported heterotrophic growth of several cyanobacteria in complete darkness with a carbohydrate source (Allison *et al.*, 1937; Fay, 1965; Watanabe & Yamamoto, 1967). Most investigators, however, have had little or no success at heterotrophic growth of cyanobacteria (Kratz & Meyer, 1955; Holm-Hansen, 1966). Wyatt (1970) was unable to demonstrate heterotrophic growth of cyanobacteria in numerous trials using different carbohydrate sources.

High intensity light has been shown to inhibit gas vacuole formation in some cyanobacteria (Walsby, 1969). By influencing cell buoyancy through control of vacuole formation, light may help regulate cyanobacterial diurnal vertical migration in the water column. This could result in more efficient use of available nutrients, and support microbial bloom formation where nutrients would otherwise by limiting in one layer (Fogg, 1969). Scott *et al.* (1981) found that *M. aeruginosa* can adapt to varying light intensities over a wide range providing advantages over many species of algae. He noted that the cyanobacterium adapted physiologically to different light intensities by changing it's gas vacuole and pigment composition to permit growth even in turbid reservoirs. Sherer & Boger (1982) studied nine species of lab-cultured cyanobacteria and noted the effect of light on respiration. They demonstrated that carbon dioxide liberated by cyanobacteria was endogenously refixed in the light.

Nutrients. Considerable research on the causes of microbial blooms has focused on limiting nutrients and/or nutrient ratios. Fossil evidence indicates that nutrient shifts controlled early cyanobacterial blooms occurring approximately 650 million years ago (Vidal, 1984). Silvey *et al.* (1972) and Cairns & Lanza (1972) briefly reviewed selected examples of nutrient effects studies in recent microbial assemblages. In attempting to explain causal factors in microbial blooms, a great deal of attention has centered around the principle of the limiting nutrient. Although some work has been done with trace elements, most investigations concerning nutrient influences in the aquatic system have dealt with carbon, nitrogen, and phosphorous. These three are some of the more abundant elements in protoplasm and, if one of them is available in limited quantities, the growth of the organism generally is proportional to the availability of that nutrient. Since algal protoplasm has atom proportions of carbon, nitrogen, and phosphorus as 106:16:1, phosphorus is most likely to be limiting (Sverdrup *et al.*, 1942).

Carbon provides the molecular skeletons for the biochemical constituents of living material and is required in greater quantities than either nitrogen or phosphorous. The possibility of carbon being a limiting factor in the growth of natural phytoplankton populations has been suggested (Kuentzel, 1969; Kerr *et al.*, 1970; King, 1970). In regard to cyanobacterial blooms initiated in micro-aerobic hypolimnetic waters, however, it would appear that available carbon would be far from limiting. The environment under such a situation contains significant amounts of free carbon dioxide generated from microbial degradation of sedimenting organics, and large amounts of bound carbon dioxide as carbonates and bicarbonates. Organic carbon may also make a significant contribution to cyanobacterial growth at low light intensities via photoassimilation (Carr & Pearce, 1966; Hoare *et al.*, 1967; Fogg, 1969).

Nitrogen is seldom limiting for heterocystous cyanobacteria as most possess the ability to fix molecular nitrogen and incorporate it as ammonia in cellular syntheses. It is unlikely that nitrogen becomes a limiting factor in nitrogen fixation due to its solubility, and to denitrification processes in reservoir sediments. Nitrogen fixation appears to be of secondary importance in bloom formations on or near the bottom of reservoirs. An ecologically sound hypothesis would be that ammonia supports the early phase of the bloom followed by nitrogen fixation as the ammonia becomes exhausted.

According to laboratory information, the low light intensities encountered near the bottom of a reservoir are probably limiting to algal nitrogen fixation. It may well be that as the algal bloom rises from the bottom, a secondary bloom or bloom enhancement results from nitrogen fixation at higher light intensities and utilization of previously accumulated phosphorus.

Of the major algal nutrients, phosphorus is without question the most significant factor in cyanobacterial blooms (Shapiro, 1970; Silvey *et al.*, 1972). Smith (1983) suggested that the relative proportion of cyanobacteria in epilimnetic phytoplankton increases when the nitrogen to phosphorus ratio falls below 29 to 1. In aerobic waters phosphorus compounds have a limited solubility, and the phosphate ion has an affinity for sorption onto colloidal particles. Phosphorus under such conditions is locked in the sediments except for the small amounts released from the mud-water interface by bacterial action on sedimenting organic materials. However, during stratification and subsequent formation of a reducing environment, phosphorus is liberated in significant quantities from the sediment (McDonnell, 1975; Bengtsson, 1975). This observation correlates nicely with bloom formation of cyanobacteria at or near the bottom of a reservoir following periods of thermal stratification. May (1981) reported that cyanobacterial blooms only developed when phosphorus levels reached 0.5 mg l^{-1}. She also noted the role of heterotrophic bacterial interaction in phosphorus availability from sediments to the bloom cyanobacteria.

Trace inorganic nutrients, such as sodium, molybdenum, magnesium, calcium, and cobalt, may also play important roles in cyanobacterial blooms, but there has been little work concerning their influences and concentrations in the natural environment.

Cultural activities producing excessive nutrient

loadings and/or nutrient shifts (e.g. imbalances) often accelerate the eutrophication rate of a water supply. Excess phosphorus and/or nitrogen have been implicated in many studies of cultural eutrophication (Beeton & Edmondson, 1972).

Several studies have characterized significant empirical relationships between phosphorus concentration and selected biological and chemical indicators of reservoir trophic status (Walker, 1983). Algal and bacterial growth increases typical of reservoir eutrophication processes have been characterized using several indicators of mixed microbial assemblages including chlorophyll *a*, transparency, hypolimnetic oxygen depletion rate and primary productivity (Vollenweider, 1968, Vollenweider & Dillon, 1974; Dillon & Rigler, 1974; Carlson, 1977; Walker, 1979; Smith, 1979; Vollenweider & Kerekes, 1980; Walker, 1982). Models and indices based on these indicators (e.g. Vollenweider indices, Carlson indices) provide a powerful reservoir management tool for monitoring water quality changes accompanying eutrophication events.

Taste and odor events

Kahn (1920) stressed the sensitivity of a community to taste and odor problems from 'microscopical trouble-makers' in drinking and household water supplies. More recently, Zoetman *et al.* (1978) reported a correlation coefficient (r) of 0.92 (a = .001) when correlating consumer acceptability of drinking water and panel test ratings based on taste and odor. Clearly, taste and/or odor problems in reservoir water continue to be a major and obvious problem in water supply management.

Fig. 4 summarizes field data on the annual cycle of cyanobacteria and actinomycetes in a southwestern reservoir. The actinomycetes and cyanobacterial interactions summarized in Fig. 4 often impart noxious taste and odor substances to the water. Explaining the ecological aspects of this problem has proved to be just as difficult as finding effective methods of control. Both actinomycetes and cyanobacteria have been shown individually to be sources of odor problems in various water supplies. Adams (1929) apparently was the first to ascribe the problem of earthy odors in water to

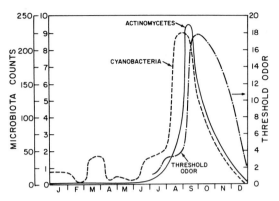

Fig. 4. Average annual cycle of cyanobacteria and actinomycetes with threshold odor production. Scale 1 provides actinomycetes counts × 10^3; scale 2 provides cyanobacterial counts (standard areal units) × 10^2; threshold odor is provided by the scale at the right.

actinomycetes. Other researchers later came to similar conclusions (Ferramola, 1949; Silvey *et al.*, 1950; Morris, 1962).

Gerber and LeChevalier (1965) published studies in which they isolated and identified a strong earthy-odor compound produced by the actinomycete *Streptomyces griseus*. The compound, which they named geosmin, was found to be a neutral oil with a boiling point of 27° C. High-resolution mass spectrometer studies indicated an empirical formula of $C_{12}H_{22}O$ for geosmin (Gerber, 1967). In further work, (Gerber, 1969) determined the structural formula of the compound, and Safferman *et al.*, 1967 published the first evidence that geosmin was the earthy compound produced by organisms other than actinomycetes. They isolated the compound from metabolites of the cyanobacterium *Symploca muscorum*. Henley (1970) isolated geosmin from a common bloom-forming cyanobacterium in southwestern reservoirs, *Anabaena circinalis*. He also isolated geosmin from material collected during a natural cyanobacterial bloom in a Texas reservoir, demonstrating its occurrence in nature.

Geosmin is not the only actinomycete or microbial metabolite that imparts an odor to water. Both the chrysophyte alga *Synura petersenii* (Collins & Kalmis, 1965) and the cyanobacterium *Microcystis flos-aquae* (Jenkins *et al.*, 1967) have been shown to produce odor compounds other than geosmin. Numerous other odors have been attributed to a number of different microbes at bloom concentrations (Palmer, 1962).

The odorous compound 2-methylisoborneol (MIB) has been reported from several actinomycetes (Berber, 1969; Medsker *et al.*, 1969; Rosen *et al.*, 1970) and strains of cyanobacteria of the Oscillatoriaceae (Tabachek & Yurikowski, 1976; Yano & Nakahara, 1976; Tsuchiya *et al.*, 1981; Izaquirre *et al.*, 1982; 1983).

Little is known concerning the ecological effects that odor compounds have in the aquatic environment. Geosmin has been shown to inhibit heterocyst formation initially in *Anabaena circinalis*, but to greatly enhance it later in the life cycle (Henley, 1970). This suggests a possible role of geosmin as an ecological autocontrol agent.

Recent investigations of factors influencing odor production by actinomycetes isolated from hypereutrophic natural waters, water plants, and sediments demonstrated variable odor production (Sivonen, 1982). These studies indicated that general odor production varied with microbial strain and carbon source. Total numbers of actinomycetes did not correlate with odor problems in fish tissue, and odor production probably resulted from ecological factors other than mere microbial density. Juttner (1984) studied the dynamics of low molecular weight (volatile) organic substances associated with cyanobacteria and algae in a shallow eutrophic lake. He found strict correlations of certain microbial species with the occurrence of volatile compounds, but noted that the majority of compounds could not be correlated with definite species. Thus, minor populations in an assemblage (i.e. non-blooming microbes) may be capable of producing significant quantities of volatile organics during certain reservoir interactions.

Toxicity and mutagenicity

General toxicity
Several cyanobacterial species are known to produce potent exotoxins in reservoir water. Carmichael (1981) reported that *Anabaena flos-aquae*, *Microcystis aeruginosa* and *Aphanizomenon flos-aquae* produce about twelve different toxins, only one of which has been identified, synthesized and toxicologically characterized. When bloom events occur in reservoirs, ingestion of cyanobacterial cells with toxin and/or toxins released to the water cause illness and death to domestic and wild animals including waterfowl. Cyanobacterial release of toxins to water is poorly understood in terms of the specific mechanisms involved. Cell metabolic release, cell autolysis, and cell decomposition are probably major routes of toxin release to reservoirs.

Acute oral toxicity from cyanobacterial sources has yet to be confirmed in humans, but reports of contact dermatitis and gastroenteritis in municipal water users are on record (Keleti *et al.*, 1979; Moikeha & Chu, 1971a & b; Carmichael, 1981). Although the ability of some strains of cyanobacteria to produce substances toxic to man is well documented, this relationship has not been widely recognized by official public health agencies (Billings, 1981). Human allergenic and gastrointestinal problems associated with bloom toxics are often reported as 'summer flu' by local physicians.

Toxic strains of cyanobacteria produce alkaloids, polypeptides and pteridines most of which have been designated as anatoxins because they are produced by strains of *Anabaena flos-aquae* (Carmichael & Gorman, 1978). The one *Anabaena* toxin identified to date is an extremely potent alkaloid designated as a very fast death factor (VFDF). *Anabaena* VFDF is probably responsible for many of the very dramatic wildlife poisonings associated with cyanobacterial blooms (Edmonds, 1978). *Anabaena flos-aquae* and *Schizothrix calcicola* also produce a lipopolysaccharide endotoxin which has been implicated in outbreaks of waterborne gastroenteritis (Carmichael, 1981).

Two distinct toxins described from *Microcystis aeruginosa* have been designated as a fast death factor (FDF) and a slow death factor (SDF). The FDF is a cyclic polypeptide produced early in the growth cycle.

Aphanizomenon flos-aquae apparently produces a toxin different from those described for *Anabaena* and *Microcystis*, but similar to saxitoxin produced by the dinoflagellate *Gonyaulax catenella* and responsible for human paralytic shellfish poison (Edmonds, 1978). Although not fully characterized, the *Aphanizomenon* toxin has been responsible for livestock mortality, and has also caused numerous fish kills.

Microbial interactions associated with toxins
Interactions involving cyanobacterial inhibition of other microbes during the establishment of bloom sequences have been reported (see General bloom interactions). Other studies have shown that cyanobacteria produce toxins that inhibit or kill zooplankton. Ranson *et al.*, (1978) described acute toxicity of three species of cyanobacteria (i.e. *Fischerella epiphytica, Gleotrichia echinulata, Nostoc linckia*) to the protozoan *Paramecium caudatum* in lab culture. Snell (1980) noted the role of cyanobacteria in the selection processes in rotifer populations. *Anabaena flos-aquae* and *Lyngbya* sp were seen to suppress reproduction in *Asplanchna girodi*. Eloff (1981) studied toxic strains of *M. aeruginosa* and recorded the influence of interacting heterotrophic bacteria on photoinhibition in the cyanobacterium. He concluded that the interaction produced a protective and/or stimulatory influence on *Microcystis* light responses and growth. Paerl (1980) described microbial interactions producing chemotaxis of aquatic bacteria to and away from the heterocysts of *Anabaena* and *Aphanizomenon* in lakes. Specific bacterial genera had positive chemotaxis to growth-enhancing substrates and/or negative chemotaxis to inhibitory or toxic substances. Ecker *et al.* (1981) studied laboratory and field populations of *Aphanizomenon flos-aquae*, and reported differences in the microbial interactions occurring with toxic and non-toxic filaments. They noted that toxic *Aphanizomenon* heterocysts were heavily populated with rod-shaped bacteria, with few bacteria on vegetative cells. Non-toxic filaments commonly had large ciliate protozoans grazing on epiphytic bacteria thereby keeping the cyanobacterial cells relatively free of microbes. Sykora *et al.* (1980) and Sykora & Keleti (1981) studied the relationship between endotoxins in finished drinking water and algae, heterotrophic bacteria, and cyanobacteria. Based on statistical analyses of field data and laboratory experiments, they concluded that cyanobacteria were one significant source of endotoxins in drinking water. Heterotrophic bacteria, *Chlorella vulgaris* and other green algae were also correlated with endotoxin in finished reservoir water.

Mutagenicity, carcinogenicity, and teratogenicity
Recent evidence indicates that microbial cells and/or their extracellular products and metabolites (ECP) may contribute significantly to the organic nitrogen content of eutrophic reservoirs, especially during bloom episodes. Although many species of micro-organisms are known to liberate organic compounds in water (Fogg, 1966; Fogg & Westlake, 1955), cyanobacteria have been noted as prolific producers of organic nitrogen compounds (Whitton, 1973). Many microbial ECP's are precursors to suspect mutagens, carcinogens, and teratogens (Hoehn *et al.*, 1978; Morris & Baum, 1978; Morris *et al.*, 1980; Kirpenko *et al.*, (1981). Notable among these are the trihalomethane (THM) precursors formed prior to chlorine disinfection of reservoir water (National Cancer Institute, 1976; AWWA Disinfection Committee, 1982). A substantial number of naturally occurring nitrogenous compounds react with chlorine, and some of these produce the THM chloroform. Hoehn *et al.* (1978) noted a direct correlation between THM concentration in finished reservoir water and algal growth measured using chlorophyll levels. Morris *et al.* (1980) reported that the ubiquitous microbial pigment chlorophyll has the potential to react with chlorine to produce chloroform in water. Several investigators have studied the role of intact microbial cells and ECP's in THM production. Hoehn *et al.* (1978) noted that microbial extracellular products and living and decomposing vegetative cell biomass serve as THM precursors in reservoirs. Briley *et al.* (1979) studied *Anabaena cylindrica* in lab culture and observed chlorinated samples of intact cells and ECP's. Both precursors produced THM concentrations comparable to yields from humic and fulvic acids. Oliver & Shindler (1980) reported that intact cells and ECP's from several species of algae (i.e. *Scenedesmus quadricauda, S. brasiliensis, Selanastrum capricornutum, Navicula minima, N. pelliculosa*) and cyanobacteria (i.e. *Anabaena oscillarioides, Ana-*

cystis nidulans) can be potent THM precursors. Hoehn *et al.* (1980) studied lab cultures of two algae (i.e. *Chlorella pyrenoidosa, Scenedesmus quadricauda)*and two cyanobacteria (i.e. *Oscillatoria tenius, Anabaena flos-aquae)* and found that their intact cells and ECP's react with chlorine to produce THM's. Briley *et al.,* (1984) also demonstrated THM formation from chlorinated intact cells and ECP's in lab cultures of *Anabaena cylindrica, S. quadricauda,* and *Pediastrum boryanum.*

Collins (1979) and Collins *et al.* (1981) reported a human epidemiological study demonstrating an association between reservoir water and a high birth defect rate. The reservoir water was tested for general mutagenicity (Ames Test), and for toxicity-mutagenicity in rats and fish. Ames test results showed mutagenicity during reservoir microbial blooms dominated by the cyanobacterium *O. subbrevis.* No mutagenicity was noted in the absence of bloom activity. Unibacterial cultures of the dominant bloom component *O. subbrevis* were not mutagenic pointing to the possibility of other microbial sources of mutagens and/or interaction-produced mutagenicity in the reservoir. Rats drinking the reservoir water gave birth to pups developing neurological disorders, and these abnormalities persisted for six generations although only the F_0 and F_1 generations received mutagenic reservoir water. Reservoir water was also toxic to the fish *Gambusia affinis,* but the relationship between mutagenicity and ichthyotoxicity was not determined. Kirpenko *et al.* (1981) reported remote after-effects in rats of toxins produced in reservoir populations of the cyanobacterium *M. aeruginosa..* They examined the effects of purified toxins and whole cell biomass, and noted embryolethal toxicity, teratogenic and gonadotoxic events in rats.

Autotrophic-heterotrophic imbalances

General imbalances in microbial assemblages
Reservoir ecosystems represent balanced food webs with integrated autotrophic-heterotrophic organism assemblages. Imbalances of food web components can select for less favorable microbial interactions producing bloom sequences that lead to potential public health problems. Hall & Hyatt

(1974) stressed the importance of the sediment microbial-detritus food chain as an important link between organic matter cycling and fish populations at the upper levels of the food web. Autotrophically-fixed carbon in phytoplankton and benthic algae was noted as a major energy substrate source for the heterotrophic microbial community. Gunnison & Alexander (1975) reported differing susceptibilities of algae and cyanobacteria to decomposition, with cell wall composition playing a major role in the heterotrophic breakdown process. Federle & Vestal (1982) described the development and succession of interacting autotrophic and heterotrophic microbes (heterotrophic bacterial rods and filaments with pennate diatoms), and noted that decomposition of allochthonous organics was dependent on microbial interaction.

The specific interactions producing autotrophic-heterotrophic imbalance in reservoir microbial assemblages are poorly understood. Some of the mechanisms that may contribute to overall imbalances in reservoirs have been noted in laboratory studies. Examples are provided by Juttner (1978; 1979) who demonstrated inhibition of algal growth by norcarotenoid production of cyanobacteria, and Reichardt (1981) who reported inhibition of glucose uptake by bacteria due to cyanobacterial metabolites. Either or both interactions may play an important role in setting the stage for bloom sequences and community imbalances.

Cultural acceleration of natural reservoir eutrophication processes often produce nutrient enrichment, excess microbial biomass, and major reservoir management/treatment problems. One major effect of this imbalance is excess autotrophic biomass producing heterotrophically-induced anoxia in the reservoir water column. Verhoff & Depinto (1977) reported field and laboratory experiments along with modelling studies indicating a significant oxygen consumption and nutrient regeneration in the thermocline region. They also noted that autotrophic bloom sequences are dependent upon nutrient release from microbial heterotrophic decomposition. More than fifty percent of the nutrients supporting the bloom assemblage originated from the heterotrophic nutrient regeneration of autotrophic biomass.

Selected public health problems associated with imbalances

Prolonged anoxia in the reservoir hypolimnion results in a breakdown of the oxidized microzone barrier at the sediment-water interface. Nutrients, noxious gasses (e.g. H_2S, CH_4) and toxic metals/metalloids (e.g. arsenic) sequestered in the sediment are released to the water column producing lowered water quality and increased risk of toxicity. Stratified eutrophic reservoirs can produce successive crops of algae from early spring to late fall, confronting a reservoir filtration plant with gross algal overloading for several months (Ridley & Symons, 1972). Silverman *et al.* (1983) reported that reservoir algae were associated with increased turbidity and particulate load, possibly enhancing bacterial activity and survival through physical protection against disinfection.

Potential problems with pathogens and R factors

The impact of impounding eutrophic waters is often expressed in profound changes in the biotic components of the reservoir. Frankland in 1896 noted that storage of river water contaminated with sewage produced a decrease in the water content of pathogenic bacteria (Ridley & Symons, 1972). More recently, Kapuscinski & Mitchell (1983) showed that U-V light plays a major role in the demise of transient bacteria entering the water column. However, many pathogenic and non-pathogenic bacteria entering the reservoir survive and are sequestered in the sediments (Von Donsel & Geldreich, 1971; Laliberte & Grimes, 1982).

Aeromonas hydrophila is a native aquatic bacterium with a cosmopolitan distribution (Rippey & Cabelli, 1980). The organism is an opportunistic human pathogen associated with aquatic injuries, and is also pathogenic to fish and amphibians. Reported human pathogenicity includes meningitis, gastroenteritis, septicemia, and peritonitis (Davis *et al.*, 1978). *A. hydrophila* density varies with the trophic state of the water and shows a strong correlation with a proposed relative trophic index for freshwater (Rippey & Cabelli, 1980). *A. hydrophila* density increases correlated well with autotrophic parameters (i.e. chlorophyll a, total phosphorus, and Secchi depth) in the oligotrophic to mesotrophic range, leveling off in eutrophic waters. Whether or not the increased density of the pathogen accompanying change in the autotrophic-heterotrophic balance is of public health significance remains to be demonstrated.

Legionella pneumophila is the etiological agent of Legionnaires disease producing acute respiratory illness and death in humans. Paradoxically, the bacterial pathogen is fastidious in lab growth requirements, but widely distributed in natural aquatic habitats (Fliermans *et al.*, 1979). *L. pneumophila* survives in water distribution systems where infections occur associated with inhaling aerosols during secondary water use (Dufour & Jakubowski, 1982). Tison *et al.* (1980) reported the growth of *L. pneumophila* in association with autotrophic mats of cyanobacteria. They explained the wide distribution and survival of the pathogen as the result of heterotrophic growth on natural substrates produced by the cyanobacterium, *Fischerella* sp. Recently, Fields *et al.* (1984) reported interactions of *L. pneumophila* and the protozoan *Tetrahymena pyriformis*. The bacterium was seen to proliferate as an intracellular parasite in the protozoan cultured in sterile tap water lab cultures.

Coliform bacteria are currently monitored as a bacteriological water quality parameter reflecting relative health risk from associated pathogens. Although traditionally viewed as harmless indicator bacteria (Grabow, 1970; Wolf, 1972), the coliform group has been implicated in the transfer of resistance factors (R plasmids) to and from pathogens (Grabow *et al.*, 1974). The R-factors convey resistance to antibacterial drugs used to treat disease, viral phages, heavy metals (e.g. mercury and nickel) and UV light. Grabow *et al.*, (1975) report a decrease in the ratio of resistant to total coliforms between sewage treatment plant influent and the dam in a reservoir. However, the ratio increased between the influent and effluent of the dam with a higher ratio in dam effluent than in water upstream of the sewage treatment plant. They noted R factor transfer in dialysis bags with *Escherichia coli* strains immersed in water at bacterial input densities of 4×10^5 ml^{-1} (recipient and donor).

Actual *in situ* densities of recipient to donor bacteria in reservoir water and sediment are highly variable, and rates of transfer of resistance are poorly understood in nature. Bell *et al.* (1980) reported more than fifty percent of fecal coliforms from water and sediment with multiple resistance to antibiotics could transfer R factors to recipient *Salmonella typhimurium,* while more than forty percent could transfer to *Escherichia coli.* They calculated fecal coliform (FC) populations with R factors as high as 1400/100 ml., and noted that accidental intake of a few milliliters of the water could produce transient or permanent colonization of the digestive tract.

More information is needed on the influence of common variables in the microhabitat in order to predict the magnitude of public health risk from non-pathogenic to pathogenic transfer of R-factors. Singleton (1983) reported that colloidal montmorillonite clay strongly inhibited conjugal transfer of a R factors between strains of *E. coli* at cell densities higher than those found in many water columns (i.e. donor bacteria at 2×10^8 with recipients at 10^8). These results from a single parameter study point to the need for more information on microbial interaction phenomena as they influence public health risk.

Summary

Reservoirs throughout the world differ greatly in climate, size, flow dynamics, geomorphometry, and their individual physical, chemical and biological structure. As a result, reservoir microbial habitats, and their corresponding microbial interactions, are diverse and highly variable within and between sites.

General microbial interactions occur within and between the autotrophic and heterotrophic components of reservoir microbial assemblages. Community metabolism, physical and chemical factors, and sedimentation/sediment systems regulate the structure and function of the reservoir microbial habitat.

Eutrophication is the natural ageing process occurring in lotic and lentic ecosystems. The process is essentially a complex shift in ecosystem trophic status occurring over time. One general characteristic of more advanced stages of eutrophication

is the increased frequency of population imbalances or blooms, with large numbers of cyanobacteria. Microbial interactions occurring in bloom sequences often result in undesirable water quality characteristics. Recent studies of bloom interactions point to bloom products which add new environmental health significance to traditional eutrophication-related water quality problems.

Acknowledgements

The authors thank Ms. Mary Saathoff for typing the manuscript.

References

Adams, B.A., 1929. The *Cladothrix dichotoma* and allied organisms as a cause of 'indeterminate' taste in chlorinated water. Wat. and Wat. Eng. 31:309–321.

Allison, F.E., Hoover, S.R. & Morris, H.J., 1937. Physiological studies with the nitrogen fixing alga, *Nostoc muscorum.* Bot. Gaz. 98:433–445.

American Water Works Association, 1982. Disinfection. Committee Report. J. am. Wat. Wks. Ass. 74:376–380.

Beeton, A.M. & Edmondson, W.T., 1972. The eutrophication problem. J. Fish. Res. Bd. Can. 29:673–682.

Belanger, T.V., 1981. Benthic oxygen demand in Lake Apopka, Florida. Wat. Res. 15:267–274.

Bell, J.B., Macrae, W.R. & Elliott, G.E., 1980. Incidence of R factors in coliform, fecal coliform, and *Salmonella* populations of the Red River in Canada. Appl. envir. Microbiol. 40:486–491.

Bengtsson, L., 1975. Phosphorus release from a highly eutrophic lake sediment. Verh. int. Ver. Limnol. 19:1107–1116.

Billings, W.H., 1981. Water-associated human illness in northeast Pennsylvania and its suspected association with blue-green algae blooms. In Carmichael, W.W., (ed.), The water environment: algal toxins and health: 243–255.

Bowman, G.T. & Delfino, J.J., 1980. Sediment oxygen demand techniques: a review and comparison of laboratory and *in situ* systems. Wat. Res. 14:491–499.

Brewer, W.S., Abernathy, A.R. & Paynter, M.J.B., 1977. Oxygen consumption by freshwater sediments. Wat. Res. 11:471–473.

Briley, K.F., Williams, R.F., Longley, K.E. & Sorber, C.A., 1979. Trihalomethane production from algal precursors. In Jolley, R.L., *et al.,* (eds.), Water Chlorination: Environmental Impact and Health Effects 3, Ann Arbor Press, Ann Arbor: 117–129.

Briley, K.F., Williams, R.F. & Sorber, C.A., 1984. Alternative water disinfection schemes for reduced trihalomethane formation 2. Algae as precursors for trihalomethanes in chlorinated drinking water. US Environmental Protection Agency (EPA–600/S2–84–005), Washington, DC.

Bull, A.T. & Slater, J.H., 1982. Microbial interactions and communities. Academic Press, London. 545 pp.

Cairns, J. & Lanza, G.R., 1972. Pollution controlled changes in algal and protozoan communities. In Mitchell, R., (ed.), Water Pollution Microbiology. Wiley-Interscience, New York: 245–272.

Carlson, R.E., 1977. A trophic state index for lakes. Limnol. Oceanogr. 22:361–369.

Carmichael, W.W., 1981. Freshwater blue-green algae (cyanobacteria) toxins – a review. In Carmichael, W.W., (ed.), The Water Environment: Algal Toxins and Health. Plenum Press, New York: 1–13.

Carmichael, W.W. & Gorham, P.R., 1978. Anatoxins from clones of Anabaena flos-aquae isolated from lakes of western Canada. Mitt. int. Ver. Limnol. 21:285–295.

Carmichael, W.W. & Gorham, P.R., 1981. The mosaic nature of toxic blooms of cyanobacteria. In Carmichael, W.W., (ed.), The Water Environment: Algal Toxins and Health. Plenum Press, New York: 161–172.

Carr, N.G. & Pearce, J., 1966. Photoheterotrophism in blue-green algae. Biochem. J. 99:28P – 29P.

Collins, M., 1979. Possible link between water supply and high birth defect rate. Proc. am. Wat. Wks. Ass. 1978 Annual Conf. Part II (Atlantic City, New Yersey), Paper 24–5.

Collins, M., Gowans, C.S., Garro, F., Estervig, D. & Swanson, T., 1981. Temporal association between an algal bloom and mutagenicity in a water reservoir. In Carmichael, W.W., (ed.), The Water Environment: Algal Toxins and Health: 271–284.

Collins, R.P. & Kalmis, K., 1965. Volatile constituents of Synura petersenii Part I. The carbonyl fraction. Lloydia 28:48–52.

Davis, W.A., Kane, J.G. & Garagusi, V.F., 1978. Human Aeromonas infections: a review of the literature and a case report of endocarditis. Medicine (Baltimore) 57:267–277.

Dillon, D.J. & Rigler, F.H., 1974. The phosphorus-chlorophyll relationship in lakes. Limnol. Oceanogr. 19:767–773.

Drury, D.D. & Gearheart, R.A., 1975. Bacterial-population dynamics and dissolved oxygen minimum. J. am. Wat. Wks. Ass. 67:154–158.

Dufour, A.P. & Jakubowski, W., 1982. Drinking water and legionnaires disease. J. am. Wat. Wks. Ass. 74:631–637.

Ecker, M.M., Foxall, T.L. & Sasser, J.J., 1981. Morphology of toxic versus non-toxic strains of Aphanizomenon flos-aquae. In Carmichael, W.W., (ed.), The Water Environment: Algal Toxins and Health. Plenum Press, New York: 113–126.

Edmonds, P., 1978. Microbiology: an environmental perspective. Macmillan Publishing Co., Inc. New York: 293–313.

Eloff, J.N., 1981. Autecology of Microcystis. In Carmichael, W.W., (ed.), The Water Environment: Algal Toxins and Health. Plenum Press, New York: 71–96.

Fay, P., 1965. Heterotrophy and nitrogen fixation in Chlorogloea fritschii. j. gen. Microbiol. 39:11–20.

Federle, T.W. & Vestal, J.R., 1982. Evidence of microbiol

succession on decaying leaf litter in an arctic lake. Can. J. Microbial. 28: 686–695.

Ferramola, R., 1949. Summary: earthy odor produced by Streptomyces in water. Assoc. Interam. ingenieria sanitaria 2:371–380.

Fields, B.S., Shotts, E.B., Feeley, J.C., Gorman, G.W. & Martin, W.T., 1984. Proliferation of Legionella pneumophila as an intracellular parasite of the ciliated protozoan Tetrahymena pyriformis. Appl. envir. Microbiol. 47:467–471.

Fliermans, C.B., Cherry, W.B., Orrison, L.H. & Thacker, L.. 1979. Isolation of Legionella pneumophila from non-epidemic related aquatic habitats. Appl. envir. Microbiol. 37:1239–1242.

Fogg, G.E., 1966. Algal Cultures and Phytoplankton Ecology. The University of Wisconsin Press, Madison: 59–81.

Fogg, G.E., 1969. The physiology of an algal nuisance. Proc. r. Soc. Lon., Ser. B., 173:175–189.

Fogg, G.E. & Westlake, D.F., 1955. The importance of extracellular products of algae in freshwater. Verh. int. Ver. Limnol. 12:219–232.

Francis, G., 1878. Poisonous Australian lake. Nature 18:11–12.

Gerber, N.N., 1967. Geosmin, an earthy-smelling substance isolated from actinomycetes. Biotechnol. Bioeng. 9:321–327.

Gerber, N.N., 1969. A volatile metabolite of actinomycetes, 2–methylisoborneol. J. Antibiot. (Tokyo) 22: 508–509.

Gerber, N.N., 1979. Volatile substances from actinomycetes: their role in the odor pollution of water. CRC Crit. Rev. Envir. Cont. 9:191–214.

Gerber, N.N. & Lechevalier, H.A., 1965. Geosmin an earthy-smelling substance isolated from actinomycetes. Appl. Microbiol. 13:935–938.

Gordon, J.A., 1978. Definition of dissolved oxygen depletion mechanisms operating in the metalimnia of deep impoundments. Water Resources Research Center Res. Rep. No. 63, University of Tennessee. 110 pp.

Gordon, J.A. & Skelton, B.A., 1977. Reservoir metalimnion oxygen demands. J. env. eng. Div. 103(EE6):1001–1011.

Grabow, W.O.K., 1970. Literature survey: the use of bacteria as indicators of fecal pollution in water. South African Council for Scientific and Industrial Research (Special Report o/Wat. 1), Pretoria.

Grabow, W.O.K., Prozesky, O.W. & Smith, L.S., 1974. Review Paper. Drug resistant coliforms call for review of water quality standards. Wat. Res. 8:1–9.

Grabow, W.O.K., Prozesky, O.W. & Burger, J.S., 1975. Behaviour in a river and dam of coliform bacteria with transferable or non-transferable drug resistance. Wat. Res. 9:777–782.

Gunnison, D. & Alexander, M., 1975. Resistance and susceptibility of algae to decomposition by natural microbial communities. Limnol. Oceanogr. 20:64–70.

Gunnison, D., Brannon, J.M., Smith, I. & Burton, G.A., 1980a. Changes in respiration and anaerobic nutrient regeneration during the transition phase of reservoir development. In Barcia, J. & Mur, L.R., (eds.). Hypertrophic ecosystems, SIL Workshop on hypertrophic ecosystems. W. Junk, The Hague.

116

Gunnison, D., Brannon, J.M., Smith, I. & Preston, K.M., 1980b. A reaction chamber for study of interactions between sediments and water under conditions of static or continuous flow. Wat. Res. 14:1529–1532.

Gunnison, D., Chen, R.L. & Brannon, J.M., 1983. Relationship of materials in flooded soils and sediments to the water quality of reservoirs-I oxygen consumption rates. Wat. Res. 17:1609–1617.

Hall, K.J. & Hyatt, K.D., 1974. Marion Lake (IBP)-from bacteria to fish. J. Fish. Res. Bd. Can. 31:893–911.

Hammer, U.T., 1964. The succession of bloom species of blue-green algae and some causal factors. Verh. int. Ver. Limnol. 15:829–843.

Hannon, H.H., 1979. Chemical modifications in reservoir-regulated streams. In Ward, J.V. & Stanford, J.A., (eds.), The Ecology of Regulated Streams. Plenum Press, New York: 75–91.

Hansen, K., 1961. Lake types and lake sediments. Verh. int. Ver. Limnol. 14:285–303.

Hargrove, B.T., 1972. Oxidation-reduction potentials, oxygen concentration and oxygen uptake of profundal sediments in a eutrophic lake. Oikos 23:167–177.

Hein, M.K. & Koppen, J.D., 1979. Effects of thermally elevated discharges on the structure and composition of estuarine periphyton diatom assemblages. Estuar. coast. mar. Sci. 9:385–401.

Hendricks, A.C. & Silvey, J.K.G., 1973. Nutrient ratio variation in reservoir sediments. J. Wat. Pollut. Cont. Fed. 45:490–497.

Henley, D., 1970. Odorous metabolites and other selected studies of cyanophyta. Ph.D. Dissertation, North Texas State University, Denton, Texas.

Hoare, D.S., Hoare, S.L. & Moore, R.B., 1967. The photoassimilation of organic compounds by autotrophic blue-green algae. J. gen. Microbiol. 49–351–370.

Hoehn, R., Randall, C.W., Goode, R.P. & Shaffer, T.B., 1978. Chlorination and water treatment for minimizing trihalomethanes in drinking water. In R.L. Jolley et al. (eds.), Water Chlorination: Environmental Impact and Health Effects 2, Ann Arbor Press, Ann Arbor: 519–535.

Hoehn, R.C., Barnes, D.B., Thompson, B.C., Randall, C.W., Grizzard, T.J. & Shaffer, P.T.B., 1980. Algae as sources of trihalomethane precursors. J. am. Wat. Wks, Ass. 72:344–349.

Holm-Hansen, O., 1966. Recent advances in the physiology of blue-green algae. Proc. Symp. Fed. Wat. Pollut. Cont. Ass., Environmental requirements of blue-green algae.

Hutchinson, G.E., 1957. A treatise on limnology, 1. J. Wiley & Sons, New York: 426.

Izaquirre, G., Hwang, C.J., Krasner, S.W. & McGuire, M.J., 1982. Geosmin and 2-methyl-isoborneol from cyanobacteria in three water supply systems. Appl. envir. Microbiol. 43:708–714.

Izaquirre, G., Hwang, C.J., Krasner, S.W. & McGuire, M.J., 1983. Production of 2-methyl-isoborneol by two benthic cyanophyta. Wat. Sci. Tech. (Finland) 15:211–220.

Jenkins, D., Medsker, L.L. & Thomas, J.F., 1967. Odorous compounds in natural waters. Some sulfur compounds associated with blue-green algae. Envir. Sci. Technol. 1:731–735.

Juttner, F., 1978. Nor-carotenoids as the major volatile excretion products of Cyanidium. Z. Naturforsch 34:186–191.

Juttner, F., 1979. The algal excretion product, Geranylacetone: a potent inhibitor of carotene biosynthesis in Synechococcus. Z. Naturforsch 34:957–960.

Juttner, F., 1984. Dynamics of volatile organic substances associated with cyanobacteria and algae in a eutrophic shallow lake. Appl. envir. Microbiol. 47:814–820.

Kahn, M.C., 1920. Microscopical trouble-makers in the water supply. Nat. Hist. 20:83–90.

Kapuscinski, R.B. & Mitchell, R., 1983. Sunlight-induced mortality of viruses and Escherichia coli in coastal seawater. Envir. Sci. Technol. 17:1–6.

Keating, K.I., 1976. Algal metabolite influence on bloom sequence in eutrophied freshwater ponds. US Environmental Protection Agency (EPA–600/3–76–081), Washington, DC 148 pp.

Keleti, G., Sykora, J.L., Lippy, E.C. & Shapiro, M.A., 1979. Composition and biological properties of lipopolysaccharides isolated from Schizothrix calcicola (Ag.) Gomont (Cyanobacteria). Appl. envir. Microbiol. 38:471–477.

Kerr, P.C., Paris, D.F. & Brockway, D.L., 1970. The interrelation of carbon and phosphorus in regulating heterotrophic and autotrophic populations in aquatic ecosystems. US Department of the Interior, Federal Water Quality Administration, Water Pollution Control Research Series (16050 FGS), Washington, DC.

King, D.L., 1970. The role of carbon in eutrophication. J. Wat. Pollut. Cont. Fed. 42: 2035–2051.

Kirpenko, Yu. A., Sirenko, L.A. & Kirpenko, N.I., 1981. Some aspects concerning remote after-effects of blue-green algae toxins: impact on warm-blooded animals. In Carmichael, W.W. (ed.), The Water Environment: Algal Toxins and Health: 257–269.

Kratz, W.A. & Meyers, J., 1955. Nutrition and growth of several blue-green algae. Am. J. Bot. 42:282–287.

Kuentzel, L.E., 1969. Bacteria, carbon dioxide, and algal blooms. J. Wat. Pollut. Cont. Fed. 41:1737–1747.

LaLiberte, P. & Grimes, D.J., 1982. Survival of Escherichia coli in lake bottom sediment. Appl. envir. Microbiol. 43:623–628.

Lanza, G.R. & Cairns, J., 1972. Physio-morphological effects of abrupt thermal stress on diatoms. Trans. am. Microscop. Soc. 91:276–298.

Lazaroff, N., 1966. Photoinduction and photoreversal of the Nostocacean development cycle. J. Phycol. 2:7–17.

Lockett, C.L., 1976. Classification of seventeen central Texas reservoirs. M.S. Thesis, Southwest Texas State University, San Marcos, Texas.

Lowe, D.R., 1980. Stromatolites, 3,400 Myr old from the Archean of Western Australia. Nature 284:441–443.

May, 1981. The occurrence of toxic cyanophyte blooms in Australia. In Carmichael, W.W., (ed.), The Water Environment: Algal Toxins and Health, Plenum Press, New York: 127–141.

McDonnell, J.C., 1975. *In situ* phosphorus release rates from anaerobic lake sediments M.S. Thesis, University of Washington, Seattle, Washington.

Medsker, L.L., Jenkins, D., Thomas, J.F. & Koch, C., 1969. Odorous compounds in natural waters: 2-*exo*-hydroxy-2-methylbornane, the major odorous compound produced by several actinomycetes. Envir. Sci. Technol. 3:476–477.

Moikeha, S.N., Chu, G.W. & Berger, L.R., 1971a. Dermatitis-producing alga *Lyngbya majuscula* Gomont in Hawaii I. isolation and chemical characterization of the toxic factor. J. Phycol. 7:4–8.

Moikeha, S.N. & Chu, G.W., 1971b. Dermatitis-producing alga *Lyngbya majuscula* Gomont in Hawaii II. biological properties of the toxic factor. J. Phycol. 7:8–13.

Morris, R.L., 1962. Actinomycetes studied as taste and odor cause. Wat. and Sew. Wks. 109–76–77.

Morris, J.C. & Baum, B., 1978. Precursors and mechanisms of haloform formation in the chlorination of water supplies. In Jolley, R.L., *et al.* (eds.), Water Chlorination: Environmental Impact and Health Effects 2, Ann Arbor Press, Ann Arbor: 29–48.

Morris, J.C., Ram, N., Baum, B. & Wajon, E., 1980. Formation and significance of N-chloro compounds in water supplies. US Environmental Protection Agency (EPA–600/2–80–031), Washington, DC 353 pp.

Mortimer, C.H., 1941. The exchange of dissolved substances between mud and water in lakes. J. Ecol. 29:280–329.

Murphy, T.P., Lean, D.R.S. & Nalewajko, C., 1976. Blue-green algae: their excretion of iron-selective chelators enables them to dominate other algae. Science 192:900–902.

National Cancer Institute, 1976. Report on carcinogenesis bioassay of chloroform. National Cancer Institute Carcinogenesis Program, Bethesda, Maryland.

Neel, J.K., 1966. Impact of reservoirs. In Frey, D.G. (ed.). Limnology in North America. University of Wisconsin Press, Madison: 575–593.

Nix, J. & Ingols, R., 1981. Oxidized manganese from hypolimnetic water as a possible cause of trout mortality in hatcheries. Progess. Fish. Cult. 43:32–36.

Odum, E.P., 1971. Fundamentals of Ecology, 3rd ed. W.B. Saunders Co., Philadelphia. 574 pp.

Oliver, B.G. & Shindler, D.B., 1980. Trihalomethanes from the chlorination of aquatic algae. Envir. Sci. Technol. 14:1502–1505.

Paerl, H.W., 1980. Attachment of micro-organisms to living and detrital surfaces in freshwater systems. In Bitton, G. & Marshall, K.C., (eds.) Adsorption of Micro-organisms to Surfaces. John Wiley & Sons, New York: 375–402.

Palmer, C.M., 1962. Algae in water supplies. US Department of Health, Education, and Welfare (Public Health Service Publication No. 657), Washington, DC.

Porter, K.G., 1973. Selective effects of grazing by zooplankton on the phytoplankton of Fuller Pond, Kent, Connecticut. Ph.D. Dissertation, Yale University, Hartford, Connecticut.

Ransom, R.E., Nerad, T.A. & Meier, P.G., 1978. Acute toxicity of some blue-green algae to the protozoan *Paramecium caudatum*. J. Phycol. 14:114–116.

Reichardt, W., 1981. Influence of methylheptenone and related phytoplankton norcarotenoids on heterotrophic aquatic bacteria. Can. J. Microbiol. 27:144–147.

Ridley, J.E. & Symons, J.M., 1972. New approaches to water quality control in impoundments. In R. Mitchell (ed.). Water Pollution Microbiology. Wiley-Interscience, New York: 389–412.

Rippey, S.R. & Cabelli, V.J., 1980. Occurrence of *Aeromonas hydrophila* in limnetic environments: relationship of the organism to trophic state. Microb. Ecol. 6:45–54.

Rolan, R., 1973. Laboratory and field investigations in general ecology. The Macmillan Co., New York: 39–41.

Rosen, A.A., Mashni, C.I. & Safferman, R.S., 1970. Recent developments in the chemistry of odour in water: the cause of earthy/musty odor. Wat. Treat. Exam. 19:106–119.

Safferman, R.S., Rosen, A., Mashni, C.I. & Morris, M.E., 1967. Earthy-smelling substance from a blue-green alga. Envir. Sci. Technol. 1:429–430.

Sakevich, A., 1973. Volatile growth-inhibiting metabolites of blue-green algae. Gidrobiologicheskii Zhurnal 9:25–29.

Scherer, S. & Boger, P., 1982. Respiration of blue-green algae in the light. Arch. Mikrobiol. 132:329–332.

Schindler, D.W., 1977. Evolution of phosphorus limitation in lakes. Science 195:260–262.

Scott, W.E., Barlow, D.J. & Hauman, J.H., 1981. In Carmichael, W.W. (ed.), The Water Environment: Algal Toxins and Health. Plenum Press, New York: 49–69.

Segura, R.D., 1978. Non-conservative cation dynamics in a deep-storage reservoir in Central Texas. M.S. Thesis, Southwest Texas State University, San Marcos, Texas.

Shapiro, J., 1970. A statement on phosphorus. J. Wat. Pollut. Cont. Fed. 42:772–775.

Shapiro, J., 1973. Blue-green algae: why they become dominant. Science 179:382–384.

Silverman, G.S., Nagy, L.A. & Olson, B.H., 1983. Variations in particulate matter, algae, and bacteria in an uncovered, finished drinking-water reservoir. J. am. Wat. Wks, Ass. 75:191–195.

Silvey, J.K.G., Russell, J.C., Redden, D.R. & McCormick, W.C., 1950. Actinomycetes and common tastes and odors. J. am. Wat. Wks. Ass. 42:1018–1026.

Silvey, J.K.G. & Wyatt, J.T., 1971. The interrelationship between freshwater bacteria, algae, and actinomycetes in southwestern reservoirs. In Cairns, J. (ed.), The Structure and Function of Fresh-water Microbial Communities. Research Division Monograph 3, Virginia Polytechnic Institute and State University, Blacksburg, Virginia: 250–275.

Silvey, J.K.G., Henley, D.E. & Wyatt, J.T., 1972. Planktonic blue-green algae: growth and odor-production studies. J. am. Wat. Wks. Ass. 64:35–39.

Singleton, P., 1983. Colloidal clay inhibits conjual transfer of R-plasmid R1 drd-19 in *Escherichia coli*. Appl. envir. Microbiol. 46:756–757.

Sivonen, K., 1982. Factors influencing odour production by actinomycetes. Hydrobiol. 86:165–170.

Smith, V.H., 1983. Low nitrogen to phosphorus ratios favor dominance by blue-green algae in lake phytoplankton. Science 221:669–671.

Smith, V.H., 1979. Nutrient dependence of primary productivity in lakes. Limnol. Oceanogr. 24:1051–1064.

Snell, T.W., 1980. Blue-green algae and selection in rotifer populations. Oecologia (Berl.) 46:343–346.

Soltero, R.A. & Nichols, D.G., 1981. In Carmichael, W.W. (ed.), The Water Environment: Algal Toxins and Health. Plenum Press, New York: 143–159.

Stewart, W.D.P., 1969. Biological and ecological aspects of nitrogen fixation by free-living micro-organisms. Proc. r. Soc. Lon., Ser. B., 172:367–388.

Stewart, W.D.P. & Pearson, H.W., 1970. Effects of aerobic and anaerobic conditions on growth and metabolism of blue-green algae. Proc. r. Soc. Lon., Ser. B., 175:293–311.

Straskraba, M., Hrbacek, J. & Javornicky, P., 1973. Effects of an upstream reservoir on the stratification conditions in Slapy Reservoir. Hydrobiol. Stud. 2:7–82.

Svedrup, H.V., Johnson, M.W. & Fleming, R.H., 1942. The oceans, Their Physics, Chemistry, and General Biology. Prentice Hall, Englewood Cliffs, New Jersey. 1087 pp.

Sykora, J.L., Keleti, G., Roche, R., Volk, D.R., Kay, G.P., Burgess, R.A., Shapiro, M.A. & Lippy, E.C., 1980. Endotoxins, algae, and *Limulus* amoebocyte lysate test in drinking water. Wat. Res. 14:829–839.

Sykora, J.L. & Keleti, G., 1981. Cyanobacteria and endotoxins in drinking water supplies. In Carmichael, W.W. (ed.), The Water Environment: Algal Toxins and Health. Plenum Press, New York: 285–301.

Sylvester, R.O. & Seabloom, R.W., 1965. Influence of site characteristics on the quality of impounded water. J. amer. Wat. Wks. Ass. 57:1528–1546.

Tabachek, J.L. & Yurkowski, M., 1976. Isolation and identification of blue-green algae producing muddy odor metabolites, geosmin, and 2-methylisoborneol, in saline lakes in Manitoba. J. Fish. Res. Bd. Can. 33:25–35.

Taylor, W.D., Williams, L.R., Hern, S.C., Lambou, V.W., Howard, C.L., Morris, F.A. & Morris, M.K., 1981. Phytoplankton water quality relationships in US lakes, Part VII: algae associated with or responsible for water quality problems. US Environmental Protection Agency (EPA-600/S3-80-100), Washington, DC.

Tison, D.L., Pope, D.H., Cherry, W.B. & Fliermans, C.B., 1980. Growth of *Legionella pneumophila* in association with blue-green algae (Cyanobacteria). Appl. envir. Microbiol. 39:456–459.

Tsuchiya, Y., Matsumoto, A. & Okamoto, T., 1981. Identification of volatile metabolites produced by blue-green algae, *Oscillatoria splendida*, *O. amoena*, *O. germinata,* and *Aphanizomenon* sp. Yakagaku Zasshi 101:852–856.

Twenhofel, W.H. & Broughton, W.A., 1939. The sediments of Crystal Lake, and oligotrophic lake in Villas County, Wisconsin. Amer. J. Sci. 237:231–252.

Verhoff, F.H. & DePinto, J.V., 1977. Modeling and experimentation related to bacterial-mediated degradation of algae and its effect on nutrient regeneration in lakes. In Developments in Industrial Microbiology 18:213–229.

Vidal, G., 1984. The oldest eucaryotic cells. Sci. Amer. 250:48–57.

Vollenweider, R.A., 1968. Scientific fundamentals of the eutrophication of lakes and flowing waters with particular reference to nitrogen and phosphorus as factors in eutrophication. Organization Economic Cooperation and development Technical Report (DAS/SCI/68. 27.), Paris. 159 pp.

Vollenweider, R.A. & Dillon, P.J., 1974. The application of the phosphorus loading concept to eutrophication research. National Research Council Canada (NRCC No. 13690), Ottawa, Ontario, 42 pp.

Vollenweider, R.A. & Kerekes, J.J., 1980. Background and summary results of the OECD cooperative program on eutrophication. Restoration of lakes and inland waters. US Environmental Protection Agency (EPA-440/5-81-010), Washington, DC.

Von Donsel, D.J. & Geldreich, E.E., 1971. Relationships of *Salmonellae* to fecal coliforms in bottom sediments. Wat. Res. 5:1079–1087.

Walsby, A.E., 1969. The permeability of blue-green algal gas-vacuole membranes to gas. Proc. r. Soc. Lon., Ser. B., 173:235–255.

Walker, W.W., 1979. Use of hypolimnetic oxygen depletion rate as a trophic state index for lakes. Wat. Res. Res. 15:1463–1470.

Walker, W.W., 1982. Empirical methods for predicting eutrophication in impoundments, phase II model testing. US Army, Office of the Chief of Engineers, Washington, DC.

Walker, W.W., 1983. Significance of eutrophication in water supply reservoirs. J. am. Wat. Wks. Ass. 75:38–42.

Walter, M.R., Buick, R. & Dunlop, J.S.R., 1980. Stromatolites 3,400–3,500 Myr old from the North Pole area, Western Australia. Nature 284:443–445.

Wang, W., 1981. Kinetics of sediment oxygen demand. Wat. Res. 15:475–482.

Ward, J.V. & Stanford, J.A., 1979. Symposium summary and conclusions. In Ward, J.V. & Stanford, J.A., (eds.), The Ecology of Regulated Streams. Plenum Press, New York: 377–385.

Watanabe, A. & Yamamoto, Y., 1967. Heterotrophic nitrogen fixation by the blue-green alga *Anabaenopsis circularis*. Nature 214: 738.

Wetzel, R.G., 1983. Limnology. Saunders College Publishing, Philadelphia: 591–614.

Whitton, B.A., 1973. Interactions with other organisms. In Carr, N.G. & Whitton, B.A., (eds.), Biology of Blue-green Algae. Bot. Monogr. 9, University of California Press, Berkelly: 415–433.

Wolf, H., 1972. The coliform count as a measure of water quality. In Mitchell, R., (ed.) Water Pollution Microbiology. Wiley-Interscience, New York: 333–345.

Wyatt, J.T., 1970. Selected physiological and biochemical studies on blue-green algae. Ph.D. Dissertation, North Texas

State University, Denton, Texas.

Yano, Y. & Nakahara, S., 1976. Research on 2-methyliso-borneol production by *Phormidium tenue*. Japan Kansai District Water Supply Association Conference Proceedings: 24–25.

Zoeteman, B.C.J., Piet, G.J. & Morra, C.F.H., 1978. Sensorily perceptible organic pollutants in drinking water. In Hutzinger, H., *et al.* (eds.), Aquatic pollutants, transformation and biological effects. Pergammon Press, New York: 359–368.

Author's address:

Guy R. Lanza
Environmental Sciences Program
University of Texas at Dallas
Post Office Box 830688
Richardson, Texas 75083 – 0688

J.K.G. Silvey
Institute of Applied Sciences
North Texas State University
Denton, Texas 76203

CHAPTER 7

Sediment-water interactions and mineral cycling in reservoirs

JAMES M. BRANNON, REX L. CHEN, and DOUGLAS GUNNISON

Abstract. This study was conducted to examine the influences of sediment-water interactions on releases of iron (Fe), manganese (Mn), and nutrients to the overlying water in established reservoirs. Emphasis was placed on sediment-water interactions under anaerobic conditions, and on sediment-water interactions with and without an externally supplied source of biologically available organic carbon. These studies were conducted in large, 250 l reactor chambers containing 20 cm of sediment and 210 l of water. Results showed that at the end of an anaerobic incubation, the overlying water more closely resembles the composition of sediment interstitial water rather than aerobic surface waters. Dissolved oxygen depletion was shown to result in remobilization of metals and nutrients from sediment into the overlying water. Addition of organic matter to the overlying water resulted in enhanced release rates of Fe and Mn. It is postulated that Fe and Mn trapped in the sediment surface layer under aerobic conditions were reduced and released to the water column under anaerobic conditions.

Introduction

Reservoir sediments, as is the case for sediments elsewhere, can exert a strong influence upon mineral cycling in the reservoir waters and upon overlying water quality. Lerman & Brunskill (1971) have suggested that some fraction of the total chemical budget of a lake may be supported by the diffusional flux of dissolved constituents from sediment into the water. There has been work reported on soil-water interactions at proposed impoundment sites (Sylvester & Seabloom, 1965; Visco *et al.*, 1979; Gunnison *et al.*, 1983). The transition from terrestrial ecosystems that occurs upon flooding occupies the first 6–10 years of a new reservoir's life, after which most degradable organic matter initially present has been utilized (Chapter 3). At this point, the reservoir settles down and develops the mineral cycling characteristics of an established reservoir ecosystem. Little research has been reported, however, on the impact of sediment-water interactions on mineral cycling and water quality in established reservoirs. While sediment-water processes occurring in established reservoirs would be expected to be similar to those found in adjacent lakes, the processes themselves must be considered within the context of a system that is physically dominated by hydrodynamic properties and fueled largely by detrital inputs. Thus, the inorganic properties of sediments entering into and being deposited in the reservoir ecosystem are probably well-represented by sediments presently within the system. However, the biologically available organic carbon (BAOC) that drives the in-reservoir microbial processes may not be so evident because of the ephemeral nature of this material.

There are a number of major pathways by which nutrients and metals associated with sediment can

Gunnison, D. (ed.) Microbial Processes in Reservoirs.

122

be cycled into the overlying water. One of the most important factors in controlling the release of metals and nutrients from sediments to the overlying water is the oxygen status of the overlying water at the sediment-water interface. Numerous workers have reported that release of Fe^{2+}, Mn^{2+}, NH_4^+-N, and ortho-P is higher under anaerobic compared to aerobic conditions (Mortimer, 1941, 1942; Fillos & Molof, 1972; Theis & McCabe, 1978; Fillos & Swanson, 1975; Visco et al., 1979). Mortimer (1941, 1942) concluded that chemical exchange from sediment to the overlying water may exert a measurable, but quantitatively unimportant, influence upon chemical concentrations of materials in overlying waters as long as oxygen concentrations at the sediment-water interface remained above 1 to $2\,mg\,l^{-1}$. Fillos & Molof (1972) showed that the release of phosphate and ammonia from benthic deposits increased appreciably when dissolved oxygen concentrations in the overlying water fell below $1.5\,mg\,l^{-1}$. They attributed the increase in the release rate of phosphorus to two factors: (a) release from bacteria occurring at low levels of dissolved oxygen, and (b) destruction of the absorptive capacity of the mud surface under reducing conditions, the latter presumably resulting from the reductive destruction of the oxidized status of the surface layer at the mud-water interface. Processes occurring at the sediment-water interface (Fillos & Swanson, 1975; Eaton, 1979; Klump & Martens, 1981), benthic remineralization and diffusion (Theis & McCabe, 1978; Eaton, 1979; Klump & Martens, 1981), and microbial activity (Bates & Neafus, 1980) are also instrumental in determining the rates and magnitude of sediment releases.

This chapter considers the influence sediment-water interactions exert on releases of Fe, Mn, N, and P to the overlying water in established reservoirs. Emphasis will be placed on sediment-water interactions under anaerobic conditions, and on the interactions of reservoir sediments and water with and without an externally-supplied source of BOAC. The information presented here is derived from recent studies of established reservoirs. The results of microbial processes, rather than the activities of individual micro-organisms or groups of organisms will be stressed.

Material and methods

Sediments from existing Corps of Engineers (CE) reservoirs were sampled for study of anaerobic sediment-water interactions. Sediment samples from the following reservoirs were used:

CE Reservoir	Location (State)
Eau Galle Reservoir	Wisconsin
East Lynn Reservoir	West Virginia
Browns Lake	Mississippi
Beech Fork Reservoir	West Virginia
Eagle Lake	Mississippi
DeGray Reservoir	Arkansas
Lavon Lake	Texas
Red Rock Reservoir	Iowa
Greers Ferry Reservoir	Arkansas

Sediment samples from existing reservoirs were taken with a clamshell dredge or similar devices. After sampling, sediments were placed in 208-l steel drums with polyethylene liners, sealed with airtight lids, and transported to the Waterways Experiment Station (WES). Upon arrival at WES, sediments in each drum were stored at 20°C until used shortly thereafter in this study.

Sediment-water interaction studies

At WES, sediment samples from each reservoir were mechanically mixed, then placed in large scale reactor units to a depth of 15 to 20 cm for experimental analysis. The general design of the 250-l reactor units is shown in Figure 1. Construction and operation of the reactor units have been presented in detail elsewhere (Gunnison et al., 1980). Reactor chambers were maintained in a constant temperature (e.g., 20°C) environmental chamber throughout each experiment.

Prior to initiation of an experiment, 210-l of distilled water were added to each reactor unit, and the sediment-water contents of each unit were permitted to equilibrate for at least 30 days with constant aeration and mixing. After equilibration, an

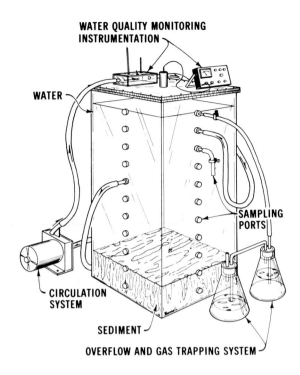

WATER QUALITY MONITORING INSTRUMENTATION

WATER

SAMPLING PORTS

CIRCULATION SYSTEM

SEDIMENT

OVERFLOW AND GAS TRAPPING SYSTEM

Fig. 1. General design of large sediment-water reactor units.

initial water sample was taken to provide baseline data under aerobic conditions. Then aeration was discontinued, and the reaction columns were sealed off from the atmosphere. In some cases, 150 mg l^{-1} of cellulose was added to reactor units as a source of BOAC. The circulation pump, which was used for mixing, achieved a complete turnover of reaction column water once every 2 min; this was used to ensure complete mixing of inflows with the water column and to enable samples to be representative of the entire water column.

Reaction columns were run steadily for at least 100 days and sampled for various physical and chemical parameters except dissolved oxygen (DO) at O, 10, 15, 25, 50, 75, and 100 days. The DO content was measured daily from the initiation of the experiment up to the point where it was no longer detectable or for a period of 30 days, whichever occurred first. In the latter case, DO was subsequently measured at 10-day intervals.

Methods of water sample collection, preservation, and analysis

Dissolved oxygen was measured on samples collected by permitting water to flow gently from a reaction column sampling port into a standard BOD bottle. Dissolved oxygen was determined with the azide modification of the Winkler method as described in *Standard Methods* (APHA, 1980) or estimated by using an oxygen meter (Yellow Springs Instrument, model 54) equipped with an oxygen/temperature probe.

If a reaction chamber became anoxic, all procedures were conducted under a nitrogen atmosphere to maintain the anaerobic integrity of the samples; otherwise, tests were done under air. Samples to be analyzed for soluble nutrients were cleared of particulate matter by passage through a 0.45-μm membrane filter. Samples for total or soluble nutrients were preserved by acidification with concentrated HCl to pH 2 and immediate freezing and storage at $-40°$ C. Samples for dissolved iron and manganese analysis were passed through 0.10-μm membrane filters (Kennedy *et al.*, 1974) then preserved by acidification with concentrated HCl. Samples for total sulfide were taken and preserved simultaneously using zinc acetate; analysis was conducted immediately using the methylene blue method described in *Standard Methods* (APHA, 1980).

Total Kjeldahl nitrogen and total phosphorus were converted to ammonium and inorganic phosphate, respectively, by digesting water samples on a semi-automatic digestion block (Ballinger, 1979). Nutrient concentrations in water samples were determined using a Technicon Autoanalyzer II, in accordance with procedures recommended by the US Environmental Protection Agency (Ballinger, 1979). Metal concentrations were determined using direct flame aspiration with an atomic absorption spectrophotometer.

Sediment characterization

Total iron (Fe) and manganese (Mn)
A 2.0-g subsample of oven-dried sediment was weighed into a Teflon beaker containing 25 ml of

$8\,N$ HNO_3. The mixture was digested for 1 hr at approximately 82° C on a hot plate (Carmody *et al.*, 1973). The digest was then filtered through a Whatman No. 5 filter, brought to a final volume of 50 ml with distilled water, and stored prior to analysis.

Total Kjeldahl nitrogen (TKN) and total phosphorus (TP)

A subsample of water or sediment was weighed into a Technicon digestion tube containing the salt/acid/catalyst digestion mixture (Technicon Method No. 376-75 W/B). The mixture was digested for 4 hrs at 380° C after the digest had cleared. The digest was allowed to cool, diluted with distilled water, and filtered through Whatman No. 5 filter paper when necessary. After cooling to room temperature, the digest was then analyzed using methods 351.2 and 365.4 (Ballinger, 1979), respectively, for TKN and TP concentrations.

Carbon

Total sediment organic carbon was estimated after assessing total organic matter by weight loss following heating 10 g (oven-dry weight) of sediment for 5 hr in a muffle furnace at 550° C (Davies, 1974). The resulting value was then multiplied by 58 percent to determine TOC (Allison, 1965).

Sediment sampling following anaerobic incubation

At the conclusion of anaerobic incubation, the sediment-water reactors were opened and the overlying water removed. Sediment cores, 7.5 cm in diameter were then taken. Cores were immediately transferred to a glove box containing a nitrogen atmosphere. The sediment core was extruded from its liner into a flat plastic container in the glove box. The core was subsequently divided into sections 2.5 cm in thickness. Multiple cores were normally taken from each chamber, and similar depth segments were composited to obtain sufficient sample for analyses. A portion of each composited depth segment was placed into an oxygen-free, 250-ml polycarbonate centrifuge bottle in the glove box, followed by centrifugation at 9000 rpm (13,000 × g) for 5 min. Following centrifugation, the interstitial water was filtered under nitrogen through a 0.10-μm pore-size membrane filter and immediately

acidified to a pH of 1 with concentrated HCl. The interstitial water was stored in polyethylene bottles and frozen until analyzed for NH_4-N, orthophosphate-P, Fe, and Mn using methods previously described.

A subsample of each depth segment was placed into a preweighed container of known volume, weighed, then dried for 24 hr at 103° C. The sample was then reweighed and sediment porosity (cm^3 of interstitial water/cm^3 of sediment) determined.

Results and discussion

Physical and chemical characteristics of the reservoir sediments used in this study are presented in Table 1. These sediments exhibited a wide range of physical and chemical characteristics. For example, clay content ranged from 13.7 to 60 percent, total Kjeldahl nitrogen ranged from 0.22 to 2.22 percent, and total iron ranged from 613 to 20,142 $\mu g\,g^{-1}$.

Dissolved oxygen depletion

In established reservoirs, the transition period that occurs in the 5 to 7 years following filling has ended and much of the biologically available organic carbon (BAOC) has been utilized (Gunnison *et al.*, 1983). Established reservoir sediments generally exhibit high chemical oxygen demands, ranging from 62.7 to 89.9 percent of the total oxygen demand, and have proportionately lower biological oxygen demands than newly flooded soils (Gunnison *et al.*, 1983).

Presented in Figure 2 are dissolved oxygen depletion results for water overlying sediments from established reservoirs obtained after sealing the large reactor units. In this and all other figures, data points each represent the mean of three replicates. Without the addition of cellulose as an external carbon source, sediments required from 30 to 70 days to lower dissolved oxygen concentration in the overlying water to approximately 1.5 mg l^{-1}. Releases of reduced products from the sediment into the overlying water can be expected to begin when dissolved oxygen concentrations have

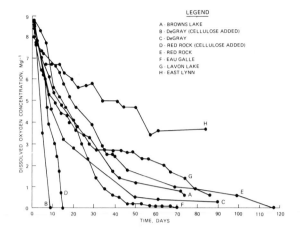

Fig. 2. Dissolved oxygen concentration in waters overlying reservoir sediments.

reached this level (Mortimer, 1941, 1942; Fillos & Molof, 1972). However, as illustrated for DeGray and Red Rock sediments, addition of cellulose as an energy source for micro-organisms considerably hastened the rate of oxygen depletion. It can therefore be expected that if established reservoirs are subject to large loadings of BAOC, dissolved oxygen depletion and subsequent releases of reduced products would occur earlier than could be predicted, if sediment-water interactions alone were considered.

Effects of dissolved oxygen depletion on metal and nutrient releases

In this section we will examine the effects that depletion of dissolved oxygen in the overlying water has on releases of nitrogen (N), phosphorus (P), iron (Fe), and manganese (Mn). This is of vital interest because large amounts of reduced substances and nutrients can be released from the sediment that will adversely affect water quality both within the reservoir and in the downstream area. In this study, sediments were the sole source of chemical constituents, allowing a detailed examination of sediment-water interactions. In the field, reduction of constituents associated with particulate matter in the water column may also serve to increase water column chemical concentrations (Davison *et al.*, 1982).

Nitrogen

In aerobic waters of established reservoirs, NO_3^-–N and Total Kjeldahl Nitrogen (TKN) are the major forms of dissolved N. If sediment oxygen demand is low or if mixing of the overlying water is high, dissolved oxygen concentrations in the overlying water may not fall below $2 \, mg \, l^{-1}$ and no changes will be observed in the N species in the overlying water. This is illustrated by the data presented in Fig. 3 for East Lynn reservoir sediments. Overlying water dissolved oxygen concentrations in reactor

Table 1. Characteristics of selected sedimens used in the laboratory studies.

Sediment	Particle size distribution, percent			pH	Total carbon, percent	Total Kjeldahl nitrogen, percent	Total phosphorus $\mu g \, g^{-1}$	Total iron $\mu g \, g^{-1}$	Total manganese $\mu g \, g^{-1}$
	Sand	Silt	Clay						
Beech Fork Reservoir	10.0	51.3	38.7	6.3	1.36	1.74	567.5	2,260	587
Browns Lake	13.8	57.5	28.7	7.0	2.77	2.22	478.1	19,467	794
Eagle Lake	17.5	27.5	55.0	7.3	1.94	2.12	782.9	20,142	644
Eau Galle Reservoir	70.0	16.3	13.7	8.3	3.87	1.81	480.2	6,585	462
Red Rock Reservoir	37.5	45.0	17.5	6.7	1.45	1.12	423.3	613	1182
Greers Ferry Reservoir	46.3	26.3	27.4	5.4	1.32	1.05	225.0	14,392	404
Lavon Lake	27.5	12.5	60.0	–	3.63	0.13	483.8	16,180	672
East Lynne Lake	0.0	52.5	47.5	5.4	1.74	0.71	37.8	6,167	283
DeGray Lake	9.0	61.0	30.0	7.0	2.05	1.60	80.9	2,365	401

126

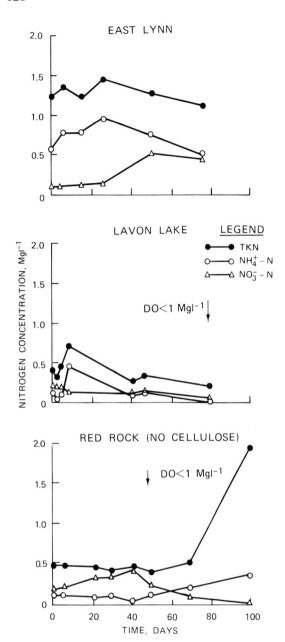

Fig. 3. Dissolved nitrogen concentrations in the waters overlying reservoir sediments.

systems containing East Lynn sediment did not fall below 4.0 mg l^{-1} during the incubation (Fig. 2). Consequently, there was little change in either TKN, NH$_4^+$-N, or NO$_3^-$-N concentrations other than a small decrease in NH$_4^+$-N and a corresponding increase in NO$_3^-$-N concentrations. These same general patterns were repeated in the reactor units

containing Lavon Lake sediments, where dissolved oxygen concentrations reached 1.0 mg l^{-1} at the end of the incubation.

In Red Rock sediment-water systems, dissolved oxygen concentrations reached 1.0 mg l^{-1} after 48 days of incubation (Fig. 2). At this time, NH$_4^+$-N and TKN began to increase and NO$_3^-$-N concentrations to decrease (Fig. 3). At the end of 100 days, NH$_4^+$-N concentrations had risen to 0.51 mg l^{-1} from a low of 0.01 mg l^{-1} at day 40. This was accompanied by a steep rise in TKN concentration to 1.96 mg l^{-1} and a complete disappearance of NO$_3^-$-N. Figure 4 depicts the differences in the various forms of nitrogen in reactor units containing Red Rock sediments with and without cellulose added. Cellulose addition caused an immediate drop in dissolved oxygen levels, and the reactor units were anoxic by 15 days. This was accompanied by a precipitous decline in NO$_3^-$-N levels in the reactor units containing cellulose amendments, while the non-amended units, which were still aerobic, exhibited increasing levels of NO$_3^-$-N (Fig. 4). By contrast, the NH$_4^+$-N levels in both cellulose amended and non-amended units followed similar patterns until dissolved oxygen concentrations in the non-amended units fell below 1 mg l^{-1}. After this, NH$_4^+$-N levels in non-amended units began a linear increase while levels in amended units remained nearly constant. (Fig. 4).

The data indicate that the presence or absence of dissolved oxygen in the overlying water was a critical factor affecting the amount and species of nitrogen released from a sediment. This parallels the findings of Austin & Lee (1973) for lake sediments. This is not to say that NH$_4^+$-N is not being released from sediments when the overlying water is aerobic. Unless the bottom waters are completely anoxic, however, nitrification should proceed just above the water-sediment interface (Lee, 1970). However, in the reactor units, the water columns were completely mixed, permitting an instantaneous distribution of released NH$_4^+$-N throughout the water column and promoting the opportunity for NH$_4^+$-N oxidation by nitrifying micro-organisms. In fact, processes such as nitrification, denitrification, and NO$_3^-$-N immobilization that can reduce the concentration of both

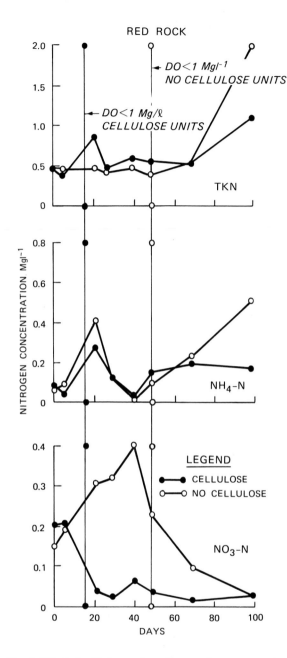

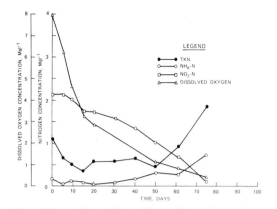

Fig. 5. Nitrogen and oxygen concentrations in the water overlying DeGray reservoir sediments.

Fig. 4. Effect of added organic matter on nitrogen concentrations in the waters overlying reservoir sediments.

NO_3^--N and NH_4^+-N in the water column, may proceed simultaneously in an aerated hypolimnion (Keeney, 1972).

Changes in concentration of the nitrogen species monitored, especially NO_3^--N, begin as the overlying water is undergoing transition to anaerobic conditions. This is illustrated in Fig. 5, which presents results of a second, consecutive, incubation of De-Gray sediments. These sediment-water systems were sealed and incubated for 103 days, aerated for 30 days, then sealed again. During the aeration period, NO_3^--N concentrations increased from <0.01 mg l^{-1} at the conclusion of the anaerobic incubation to >2.0 mg l^{-1}. These high NO_3^--N concentrations began to decrease within 9 days when dissolved oxygen concentration was >4.0 mg l^{-1} then decreased steadily. Concentrations of NH_4^+-N rose steadily during the transition to anaerobic conditions, but increased sharply, as did TKN, when dissolved oxygen concentrations fell below 1 mg l^{-1}.

Phosphorus

The enhanced release of sediment P under anaerobic conditions is well documented (Golterman, 1977; Visco et al., 1979; Klump & Martens, 1981); increasing temperature or lowering the O_2 concentration in the overlying water increase PO_4^{3-}–P release rates (Holdren & Armstrong, 1980). The effect of oxygen concentration on overlying water total-P and PO_4^{3-}–P concentrations was clearly evident in established reservoir sediments. As illustrated in Fig. 6, there was no release of either total-P or PO_4^{3-}–P from East Lynn sediment-water reactors that did not go anaerobic. Release of both total-P and ortho-P were noted from Red Rock sediments when dissolved oxygen concentrations reached <2.0 mg l^{-1} well before the end of the

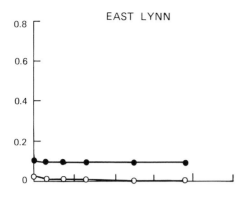

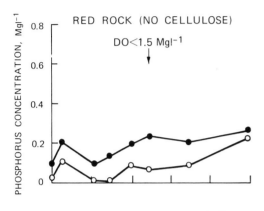

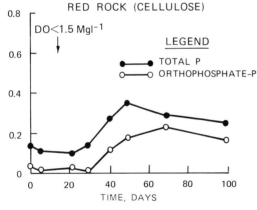

Fig. 6. Phosphorus concentrations in waters overlying reservoir sediments.

incubation. Reactor units containing Red Rock sediments and cellulose exhibited a more pronounced release of these constituents with increased levels becoming evident soon after the deplention of dissolved oxygen in the water column was completed (15 days) (Fig. 6).

Manganese and iron

In soils and sediments where reducing conditions are present, interstitial water Mn and Fe will consist primarily of Mn^{2+} and Fe^{2+}. A discussion of the reduction processes that result in these reduced elements is given in Chapter 3. When dissolved oxygen is present in the overlying water, surficial sediments develop an aerobic surface layer that acts as a barrier to the transfer of reduced substance to the overlying water (Eaton, 1979; Duchart *et al.*, 1973; Gorham & Swaine, 1965). During the aeration period prior to sealing the reactor units, a pronounced aerobic zone formed on the surface all reservoir sedements studied.

The existence of an oxidized layer at the sediment surface does not, however, stop migration of chemical constituents from occuring within the sediments, nor does it stop sediment-water exchanges (Duchart *et al.*, 1973; Bischoff & Ku, 1971; Li *et al.*, 1969). The presence of the sediment oxidized layer may, however, severely decrease the flux of chemical constituents into the overlying water. Results for nitrogen and phosphorus in this study support these findings, as do the results for Mn and Fe. When overlying water dissolved oxygen concentrations fell below 1.5. mg l^{-1}, substantial increases in overlying water Mn and Fe concentrations were noted (Fig. 7). Dissolved oxygen concentrations in the overlying water of 1 to 2 mg l^{-1} have been previously indentified as the critical level for suppressing release of anaerobic products such as Mn^{+2} and Fe^{+2} from sediments (Mortimer, 1941, 1942; Fillos & Molof, 1972). This is illustrated more dramatically by the data for reactor units containing Red Rock sediments with and without cellulose amendments depicted in Fig. 8. Here, the cellulose amendment hastened the initiation of iron and manganese appearance in the water column and also caused an increase in the amounts of these constituents released over the 100 day incubation period. However, it is also apparent that once the dissolved oxygen levels in the unamended reactor units dropped below 1 mg l^{-1} the rates of iron and manganese release nearly paralleled those in the units containing cellulose. Fillos & Molof (1972) attributed these increased release rates of ana-

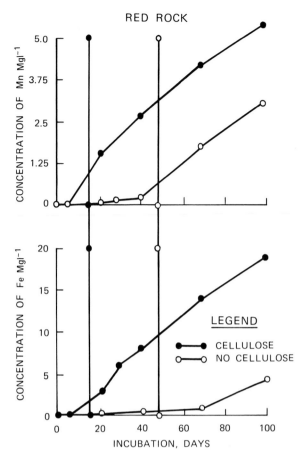

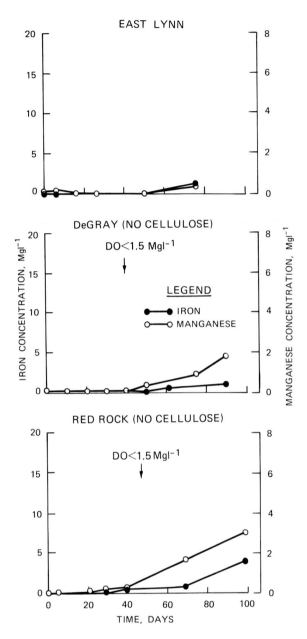

Fig. 7. Iron and manganese concentrations in waters overlying reservoir sediments.

Fig. 8. Iron and manganese concentrations in water overlying Red Rock reservoir sediment in the presence and absence of added organic matter.

erobic products at low dissolved oxygen levels to destruction of the adsorptive capacity of the sediment surface under reducing conditions.

General observations

In the previous discussions on N, P, Fe, and Mn,

we have seen how the transition from aerobic to anaerobic conditions enhances releases to the overlying water. The transition from aerobic to anaerobic conditions in water overlying Red Rock sediment is summarized in Fig. 9. It is noteworthy to contrast water quality at the beginning and end of the incubation. Dissolved oxygen and NO_3^--N were present in higher concentrations than other soluble constituents initially measured. At the end of the incubation, however, dissolved oxygen and NO_3^--N had almost disappeared from the water column, and much higher concentrations of Fe, Mn, NH_4^+-N, and PO_4^{3-}-P were present. These higher concentrations in the water column at the end of the incubation were much more comparable to interstitial water Fe $(22.5 \, \text{mg} \, l^{-1})$, Mn $(26.2 \, \text{mg} \, l^{-1})$, NH_4^+-N $(8.9 \, \text{mg} \, l^{-1}$, and PO_4^{3-}-P $(0.3 \, \text{mg} \, l^{-1})$ con-

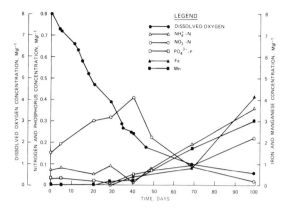

Fig. 9. Concentrations of selected chemical constituents in the water overlying Red Rock sediment.

centrations than at the start of the incubation. This was also the case with other reactor units where the overlying water became anaerobic or reached low dissolved oxygen concentrations (Table 2). This supports the contention of Ponnamperuma (1972) that the overlying hypolimnetic water essentially becomes an extension of the sediment interstitial water following the onset of anaerobic conditions. Redox potential and chemical concentrations in the water directly overlying the sediment may not be identical to sediment interstitial water conditions, but the opportunity exists for reduced substances to accumulate in the water column without being subjected to aerobic oxidation.

The data indicate that during and following the transition to anaerobic conditions, the composition of water overlying established reservoir sediments

more closely resembles that of anaerobic interstitial water than aerobic surface water. During the transition, dissolved oxygen and NO_3^--N concentrations decrease after the supply of air to the system is curtailed. When dissolved oxygen concentrations in the overlying water reach sufficiently low levels ($1.5–2.0 \, mg \, l^{-1}$) releases of Fe, Mn, NH_4^+-N, ortho-P, total-P, and TKN to the water column are observed.

Stratification and subsequent development of anaerobic conditions in reservoir hypolimnetic waters will result in a large transfer of nutrients and metals to the overlying water. During overturn, oxidation and adsorption processes will reduce the soluble concentrations of some constituents such as Fe, ortho-P, and to a lesser extent Mn (Chen *et al.*, 1983). Ammonium-N will be subject to nitrification, but will remain in the water column as a soluble, readily available nutrient. Anaerobic conditions will therefore result in the cycling of large amounts of metals and nutrients from sediment to the overlying water.

Addition of BAOC to the sediment surfaces in the reservoir ecosystem can occur autochthonously through algal blooms and growth of aquatic plants or allochthonously through organic matter carried into the system by reservoir inflows and the wind. In either case, addition of BAOC will considerably hasten the onset of anaerobic conditions and intensify the release of nutrients and metals from the sediments to the overlying water column by supplying the reservoir microflora with fuel to drive natu-

Table 2. Maximum concentrations ($mg \, l^{-1}$) of Fe, Mn, NH_4^+-N and ortho-P observed in reactor unit water following 100 days of incubation and interstitial water concentrations.

Reservoir	Fe, $mg \, l^{-1}$		Mn, $mg \, l^{-1}$		NH_4^+, $mg \, l^{-1}$		ortho-P, $mg \, l^{-1}$	
	IW*	OW+	IW	OW	IW	OW	IW	OW
DeGray	46.6	0.8	6.4	0.6	–	1.1	–	0.01
Browns Lake	3.4	2.1	7.2	1.9	15.2	0.5	1.2	0.17
Lavon Lake	9.5	1.1	3.0	0.2	2.7	0.5	–	0.07
Eau Galle	21.9	3.2	8.5	1.8	–	0.6	–	0.02
Greers Ferry	55.7	13.8	18.9	5.0	2.0	0.1	1.1	0.03
Eagle Lake	19.1	8.7	6.5	2.3	1.7	1.8	1.2	0.30

* Interstitial water
+ Overlying water

ral cycles. The impact of these cycles is magnified by the manner in which the reservoir is operated. Bottom withdrawal will tend to draw anaerobic releases along with hypolimnetic waters and discharge these materials downstream. Pump-back storage may mix nutrient-rich bottom waters with surface waters in the immediate vicinity of the dam, creating an ideal environment for growth of nuisance algae. Alternatively, the project may be operated in a manner that precludes adverse effects.

Sediment interface processes

In theory, fluxes of anaerobic constituents from sediment to overlying waters should be predictable. Flux of dissolved constituents across the sediment-water interface is driven by advective transport of the species in pore water solution and attached to sediment particles as well as by molecular diffusion (Lerman, 1979). However, in the reactor units in this study, there was no sedimentation, and fluxes of Fe, Mn, ortho-P, and NH_4^+-N were measured when the sediment column and overlying water were anaerobic. Therefore, in the absence of oxygen, there should be little chance for precipitation reactions to reduce the flux of these dissolved species. Under these conditions, advective transport and precipitation will be minimal and can be ignored, and the appropriate flux equation reduces to (Berner, 1971)

$$J_s = -\emptyset_o D_s \left.\frac{\partial c}{\partial z}\right|_{pw} \tag{1}$$

where

J_s = flux (mg cm$_s^{-2}$ sec^{-1}, where s denotes bulk wet sediment)

$\emptyset_o$ = porosity at the interface (cm$_{pw}^3$ cm$_s^{-3}$, where pw denotes pore water)

D_s = bulk sediment molecular diffusion coefficient (cm$_s^2$ sec^{-1})

$\left.\dfrac{\partial c}{\partial z}\right|_{pw}$ = pore water concentration gradient at the sediment-water interface (mg cm$_{pw}^{-3}$ cm$_s^{-1}$)

Porosities and interstitial water concentrations of Fe, Mn, NH_4^+-N, and ortho-P measured in the various sediments are presented in Table 3. Bulk sediment diffusion coefficients were estimated using the value of the tracer diffusion coefficients (D_o) at 18° C of Li & Gregory (1974) and the relationship $D_s = D_o \emptyset^2$ to obtain an estimate of the bulk sediment diffusion coefficient (Lerman, 1979). Values of D_o for Fe, NH_4^+-N, Mn, and ortho-P used in calculating predicted fluxes were 5.82×10^{-6}, 16.8×10^{-6}, 5.75×10^{-6}, and 7.15×10^{-6} cm sec^{-1}, respectively. Predicted fluxes calculated using equation (1) must be multiplied by the factor 8.64×10^8 to convert units to mg m$_s^{-2}$ day^{-1}.

Fluxes of NH_4^+-N, Fe, Mn, and ortho-P com-

Table 3. Porosities and interstitial water concentrations in reservoir sediments.

Sediment	Porosity $\emptyset$	Concentration, mg l^{-1}			
		Fe	Mn	NH_4^+-N	Ortho-P
Beech Fork	0.71	25.8	8.8	17.6	1.7
Browns Lake	0.73	3.4	7.2	15.2	1.2
East Lynn	0.68	71.9	6.0	13.0	0.3
Lavon Lake	0.82	9.5	3.0	2.7	ND
Red Rock	0.53	22.5	26.6	8.9	0.3
DeGray	0.73	46.6	6.4	ND	ND
Eagle Lake	0.74	19.1	6.5	1.7	1.2
Eau Galle	0.55	21.9	8.5	ND	ND
Greers Ferry	0.64	55.7	18.9	2.0	1.1

ND = not determined.

puted using equation (1), and appropriate sediment and interstitial water data are presented in Table 4. Comparison of these predicted fluxes with observed fluxes (Chapter 9) showed that observed fluxes, except for that of NH_4^+-N, consistently exceeded those predicted by equation 1. If net sediment-water interface processes are unimportant, observed and predicted fluxes should agree (Klump & Martens, 1981). Their failure to do so for Fe, Mn, and ortho-P indicates that processes at the sediment-water interface in addition to simple molecular diffusion were influencing releases, a phenomenon also noted by other workers (Eaton, 1979; Klump, 1980; Klump & Martens, 1981: Cook, 1984).

The most likely sediment interface effect was remobilization during anaerobic conditions of iron, manganese, and ortho-P from compounds and complexes formed in the aerobic surface layer of sediments when the overlying water was aerobic (Cook, 1984; Eaton, 1979, Duchart et al., 1973; Gorham and Swaine, 1965) in addition to concentration gradient supported diffusion. Prior to being sealed, sediment-water reactors were allowed to incubate aerobically for at least 30 days. Coupled with the time required for the water columns to go anaerobic, sufficient opportunity was provided for an oxidized zone enriched in Fe, Mn, and ortho-P to form.

If enhanced releases of Fe, Mn and ortho-P under reduced conditions are indeed due to sediment-water interface processes, an increase in microbial activity at the sediment-water interface should result in higher release rates. Cellulose was added to the overlying water of DeGray and Red Rock sediment to accelerate the rate of reduction at the interface and test this hypothesis. Addition of cellulose to the overlying water as an energy source did increase the rate of microbial activity. This is indicated by the greatly enhanced rates of oxygen depletion (Fig. 2) in sediment-water systems so amended.

Addition of cellulose to the overlying water of Red Rock and DeGray sediments resulted in enhanced release rates of Fe and Mn, especially Fe (Table 5). Because these release rates were accel-

Table 4. Fluxes (mg m^{-2}day^{-1}) calculated from equation 1.

Reservoir	Parameter			
	NH_4^+-N	Fe	Mn	Ortho-P
Red Rock	1.6	2.6	2.1	0.2
Lavon Lake	2.1	2.6	0.6	0.8
East Lynn	4.7	ND	1.1	0.1
Browns Lake	10.8	0.9	1.4	ND
Beech Fork	9.1	4.6	1.6	0.2
Greers Ferry	0.8	7.7	2.7	ND
Eagle Lake	ND	16.4	1.3	0.3
DeGray	ND	11.5	1.2	ND

ND = not determined

Table 5. Release rates of Fe and Mn from sediment to water.*

Parameter	Red rock		DeGray	
	Cellulose	No cellulose	Cellulose	No cellulose
Fe	181.7 ± 12.1	66.3 ± 14.7	246.2 ± 45.3	34.1 ± 9.6
Mn	48.3 ± 1.6	39.1 ± 4.6	20.7 ± 1.7	9.7 ± 4.1

* Release rates units are mg m^{-2} day^{-1} ± 1 standard deviation

erated with no discerbible change in sediment interstitial water Fe and Mn profiles, it is felt that remobilization of Fe and Mn in the surface layer of sediment was occurring.

In our studies, ortho-P was released from most sediment during anaerobic conditions. Specific release rates, in agreement with results of Theis & McCabe (1978), could not be predicted from the gradient of interstitial water ortho-P to overlying water ortho-P. This indicates that interface processes probably played an important role in ortho-P release. However, very little or no ortho-P was released from either unamended or cellulose amended Red Rock or DeGray sediment-water systems. This complicated evaluation of the effects of added organic matter on ortho-P sediment-water interface processes in our studies.

These interface effects greatly complicate the predictability of sediment releases under anaerobic conditions. Releases of Fe and Mn under anaerobic conditions consistently exceeded releases predicted using interstitial water profiles (Chen et al., 1983) even when no cellulose was added. Predicted releases of Fe were correlated with measured releases of Fe (Chapter 9) but predicted and actual releases of Mn were not. This makes it difficult to predict releases of Mn from reservoir sediments under anaerobic conditions.

Conclusion

During and following the transition from aerobic to anaerobic conditions sediment releases of nitrogen, phosphorus, iron, and manganese to the overlying water are enhanced. During the transition phase, dissolved oxygen and NO_3^--N concentrations decrease. When dissolved oxygen concentrations reach levels of approximately $1.5\ mg\,l^{-1}$, releases of NH_4^+-N, ortho-P, Fe, and Mn commence. At the end of the anaerobic incubation, the overlying water more closely resembles the composition of sediment interstitial water rather than that of aerobic surface waters.

Reservoir stratification and subsequent oxygen depletion will therefore result in remobilization of metals and nutrients from sediment into the overlying water. During overturn, oxidation and absorption processes will reduce the soluble concentrations of Fe, ortho-P, and to a lesser extent Mn (Chen et al., 1983). Over the short term, nitrogen concentrations in the water column will be enhanced, even though its form may change. The overall impact of anaerobic conditions is to return many sediment bound metals and nutrients to the water column.

Processes operating at the sediment-water interface were shown to be affecting anaerobic releases of Fe and Mn. Releases of these metals consistently exceeded those predicted using interstitial water concentration profiles, a trait shared by ortho-P. In addition, increasing the microbial activity by addition of organic matter to the overlying water resulted in enhanced release rates of Fe and Mn. It is postulated that Fe and Mn trapped in the sediment surface layer under aerobic conditions were reduced and released to the water column under anaerobic conditions. This process apparently occurred at an accelerated rate in the presence of increased microbial activity.

Acknowledgements

This research was supported by the Environmental and Water Quality Operational Studies Program of the US Army Corps of Engineers. Permission was granted by the Chief of Engineers to publish this information.

References

Allison, L.E., 1965. Organic carbon. In Black, C.A., (ed.), Methods of Soil Analysis, Part 2, Chemical and Microbiological Properties. American Society of Agronomy, pp. 1367–1378.

American Public Health Association, 1980. Standard methods for the examination of water and wastewater, 15th Ed. Washington, DC, 1134 pp.

Austin, E.R. & Lee, G.F., 1973. Nitrogen release from lake sediments. J. Wat. Pollut, Contr. Fed. 45:870–879.

Ballinger, D.C., 1979. Methods for chemical analysis of water and wastes. EPA 600/4-79-020, EPA, Cincinnati, OH.

Bates, M.H. & Neafus, N.J.E., 1980. Phosphorus release from sediments from Lake Carl Blackwell, Oklahoma. Wat. Res. 14:1477–1481.

Berner, R.A., 1971. Principles of chemical sedimentology, McGraw-Hill Book Co., New York, 240 pp.

Bischoff, J.L. & Ku, T.L., 1971. Pore fluids of recent marine sediments; II., anoxic sediments of 35 to 45°N Gibralter to Mid Atlantic Ridge. J. Sed. Pet. 41:1008.

Carmody, D.J., Pearce, J.B. & Yasso, W.E., 1973. New York Bight. Mar. Pollut, Bull. 4:132–135.

Chen, R.L., Brannon, J.M. & Gunnison, D., 1984. Anaerobic and aerobic rate coefficients for use in CE–QUAL–R1. Miscellaneous Paper E-84-5. US Army Engineer Waterways Experiment Station, CE, Vicksburg, Miss.

Chen, R.L., Keeney, D.R. & McIntosh, T.H., 1983. The role of sediments in the nitrogen budget of Lower Green Bay, Lake Michigan. Jour. Great Lakes Res. 9:23–31.

Cook, R.B., 1984. Distribution of ferrous iron and sulfide in an anoxic hypolimnion. Con. J. Fish. Aquat. Sci. 41:286–293.

Davies, B.E., 1974. Loss-on-ignition as an estimate of soil organic matter. Proc. soil Sci. Soc. am. 38:150–151.

Davison, W., Woof, C. & Rigg, E., 1982. The dynamics of iron and manganese in a seasonally anoxic lake; direct measurement of fluxes using sediment traps. Limnol. Oceanogr, 27:987–1003.

Duchart, P., Calvert, S.E. & Price, N.B., 1973. Distribution of trace metals in the pore waters of shallow water marine sediments. Limnol. Oceanog. 18:605–610.

Eaton, A., 1979. The impact of anoxia on Mn fluxes in the Chesapeake Bay. Geochim. Cosmochim. Acta 43:429–432.

Fillos, J. & Molof, A.H., 1972. The effect of benthal deposits on oxygen and nutrient economy of flowing waters. J.Wat. Pollut. Contr. Fed. 44:644–662.

Fillos, J. & Swanson, W.R., 1975. The release rate of nutrients from river and lake sediments. J.Wat. Pollut. Contr. Fed. 47:1032–1041.

Golterman, H.L. (ed.), 1977. Interactions between sediments and fresh waters. Proc. Intl. Symp. Amsterdam, Holland.

Gorham, E. & Swaine, D.J., 1965. The influence of oxidizing the reducing conditions upon the distribution of some elements in lake sediments. Limnol. Oceanogr. 10:268–279.

Grizzard, T.J., Randall, C.W. & Jennelle, E.M., 1982. The influence of sediment-water interactions in an impoundment on downstream water quality. Wat. Sci. Tech. 14:227–244.

Gunnison, D., Brannon, J.M., Smith Jr., I., Burton Jr., G.A. & Preston, K.M., 1980. A reaction chamber for study of interactions between sediments and water under conditions of static or continuous flow. Wat. Res. 14:1529–1532.

Gunnison, D., Chen, R.L. & Brannon, J.M., 1983. Relationship of materials in flooded soils and sediments to the water quality of reservoirs. I. Oxygen consumption rates. Wat. Res. 17:1609–1617.

Holdren, G.C., Jr. & Armstrong, D.E., 1980. Factors affecting phosphorus release from intact lake sediment cores. Environ. Sci. Technol. 14:79–87.

Keeney, D.R., 1979. Prediction of the quality of water in a proposed impoundment in southwestern, Wisconsin, USA. Environ. Geol. 2:341–349.

Kennedy, V.C., Zellweger, G.W. & Jones, B.F., 1974. Filter pore size effects on the analysis of Al, Fe, Mn, and Ti in water. Wat. Resources Res. 10:785–791.

Klump, J.V., 1980. Benthic nutrient regeneration and the mechanisms of chemical sediment-water exchange in an organic-rich coastal marine sediment, Ph.D. Thesis, University of North Carolina at Chapel Hill, 160 pp.

Klump, J. Val. & Martens, C.S., 1981. Biogeochemical cycling in an organic rich coastal marine basin – II. Nutrient sediment-water exchange processes. Geochim. Cosmochim. Acta 45:101–121.

Lee, G.F., 1970. Factors affecting the transfer of materials between water and sediments. Lit. Rev. Eutrophication Info. Program, Univ. of Wisconsin, Madison.

Lerman, A. & Brunskill, G.J., 1971. Migration of major constituents from lake sediments into lake water and its bearing on lake water composition. Limnol. Oceanog. 16:880–890.

Li, Y. & Gregory, S., 1974. Diffusion of ions in sea water and in deep sea sediments. Geochim. Cosmochim. Acta 38:703–714.

Li, Y., Bischoff, J.L. & Mathieu, G., 1969. The migration of manganese in the artic basin sediment. Earth & Plant. Sci. Let. 7:265.

Mortimer, C.H., 1941. The exchange of dissolved substances between mud and water in lakes, Parts I and II. J. Ecol. 29:280–329.

Mortimer, C.H., 1942. The exchange of dissolved substances between mud and water in lakes, Parts III and IV. J. Ecol. 30:147–201.

Ponnamperuma, F.N., 1972. The chemistry of submerged soils. Adv. Agron. 24:29–96.

Theis, T.L. & McCabe, P.J., 1978. Phosphorus dynamics in hypereutrophic lake sediment. Wat. Res. 12:677–685.

Visco, S., Coler, R. & Zajicek, O.T., 1979. An evaluation of benthic nutrient regeneration in a projected flood control impoundment. J. Environ. Sci. Health A14:399–414.

Sylvester, R.O. & Seabloom, R.W., 1965. Influence of site characteristics on quality of impounded water. J. am. Wat. Wks. Ass. 57:1528–1546.

Author's address:
James M. Brannon
Rex L. Chen
Douglas Gunnison
US Army Engineer Waterways Experiment Station
P.O. Box 631
Vicksburg, Mississippi 39180 USA

CHAPTER 8

Pathogenic microorganisms in thermally altered reservoirs and other waters

R.L. TYNDALL

Abstract. A variety of naturally and artificially heated waters were studied for the presence of the causative agent of Legionnaires' Disease, i.e. *Legionella* and the encephalitic free-living amoebae *Naegleria fowleri.* In addition to temperature effects, a variety of other physical and chemical parameters were also analyzed for possible correlation with the presence of *Legionella* and *N. fowleri.* Results of these various analyses showed a statistically significant correlation between thermal additions and the presence of *Legionella* and *N. fowleri.* To date, two species of *Legionella*, i.e. *L. oakridgensis* and *L. cherrii* have only been isolated from thermally altered waters. No correlation has yet been consistently found between the presence of *Legionella* and *N. fowleri* and various other physical/chemical parameters.

In addition to thermal effects on the presence of *Legionella* and *N. fowleri* in reservoirs and other natural waters, thermal additions may also affect and amplify the presence these microbial pathogens in potable water. Ensuing public health consequences of such amplification and dispersion have been documented.

Introduction

Some aquatic microorganisms normally present in reservoirs have recently been recognized as pathogenic for man. Among these are Legionnaires' Disease Bacteria (*Legionella*), and free-living amoebae of the genera *Naegleria* and *Acanthamoeba,* the causative agents of various human infections.

Tests for the common water-born enteric pathogens are well established, and factors germane to their presence in aquatic environs are known and generally controllable. Factors affecting the distribution of *Legionella* and pathogenic free-living amoebae, however, are not well understood. Simple, rapid tests for their detection and procedures for their control are not yet developed.

The clinical significance of *Legionella* was dramatized by the namesake outbreak in 1976 at the Amercian Legion Convention in Philadephia (McDade *et al.*, 1977). At this convention over one hundred people became ill and thirty-four died. After an intensive effort, laboratory isolations were made of the causitive agent, i.e. *Legionella pneumophila.* Since 1977, eight serogroups of *L. pneumophila* and twenty-two species of *Legionella* have been discovered (Brenner, 1984). *Legionella* are gram-negative rods approximately $0.5 \times 2.4\ \mu m$ in size. Infection generally occurs by inhalation of the aerosolized bacteria. Two disease syndromes can be manifested by infection with *Legionella*. Legionnaires' Disease (LD) is a pneumonia with associated cough, fever and malaise (Beaty, 1984). The disease can be fatal, although Erythromycin is effective in treating LD. Conversely, Legionellosis may also be expressed as Pontiac Fever, a non-pneumonic, flu-like illness which also responds to Erythromycin therapy (Beaty, 1984).

Gunnison, D. (ed.) Microbial Processes in Reservoirs.
© 1985, Dr W. Junk Publishers, Dordrecht, Boston, Lancaster. ISBN 90 6193 751 5.

Estimates of the number of cases of Legionellosis range from 25,000 to 200,000 per year (Wilkinson, 1982). Some of the known Legionellosis outbreaks were traceable to the aerosolization of water-borne *Legionella* by cooling towers and evaporative condensers (Dondero *et al.*, 1980; Miller, 1979). The cooling devices are presumably seeded with *Legionella* from their potable and natural water supplies. That *Legionella* are in fact normal components of the aquatic flora was first demonstrated by Fliermans *et al.* (1979) and has been confirmed by subsequent studies (Tyndall, 1981; Christensen *et al.*, 1983). In view of this ubiquity in reservoirs and other natural surface waters, it is not surprising that water in cooling towers and evaporative condensers contains *Legionella*. These devices can then amplify *Legionella* concentrations and disperse the pathogen via aerosolization.

In contrast to *Legionella*, the presence of *Naegleria* in water and soil was known before their clinical significance was recognized. Butt (1966) and Carter (1970) described the first cases of *Naegleria* infection in Floridian and Australian children who were infected by swimming or bathing in *Naegleria*-infested waters. *Naegleria* are small amoebae capable of using dissolved organic material or gram-negative bacteria as a food source (Page, 1976). They are eukaryotic cells generally with a single nucleus and a centrally located nucleolus (Maitra *et al.*, 1974). Locomotion is by means of eruptive pseudopodia. Four species of *Naegleria* have been isolated. *N. gruberi* and *N. jadini* (Willaert, 1974) have not shown any pathogenic potential in experimental animals or in man. *N. australiensis* (De Jonckheere, 1981) is pathogenic for mice but as yet has not been implicated in human diseases. *N. fowleri* is pathogenic for man and mice (Willaert, 1974).

On entry into the nasal passages of a susceptible individual, *N. fowleri* will penetrate the nasal mucosa and migrate along the olfactory nerve through the cribriform plate to the cerebrum. The ensuing infection results in a rapidly fatal meningoencephalitis (Duma *et al.*, 1969). Antibiotic therapy is generally ineffectual. Fortunately, primates in general are resistant to infection with *Naegleria*. This has been demonstrated in laboratory studies with chimpanzees and in epidemiologic studies at sites where fatal cases of primary amoebic meningoencephalitis (PAME) occurred (Wellings *et al.*, 1979; Wong *et al.*, 1975). In such cases, hundreds of individuals were exposed and yet only a single case of PAME occurred. Reasons for the susceptibility of the occasional individual are unknown. Subsequent to the reports of fatal cases of PAME in Australia and Florida, other cases of PAME were reported. Sources of infections included heated swimming pools (Cerva, 1971), thermal springs (Hecht *et al.*, 1972) and a variety of naturally or artificially heated surface waters (De Jonckheere *et al.*, 1975; Willaert, 1974). One of the largest clusters of PAME occurred in Virginia where fifteen cases were reported over a nine year period (Duma *et al.*, 1971). Unlike the tens-of-thousands of cases of Legionellosis per year in the United States alone, only one to two hundred cases of PAME have been reported to date world-wide. As with *Legionella*, simple rapid assays for detecting and quantitating pathogenic free-living amoebae in aquatic environments are not generally available.

The possible influence of thermal additions on the emergence and/or propagation of *Legionella* and *Naegleria* was suggested by the association of Legionellosis outbreaks with thermally altered waters in cooling towers and evaporative condensers, and by the association of PAME cases with naturally and artificially heated waters. In view of these associations, various studies on the effect of thermal additions on *Legionella* and *Naegleria* were initiated. The results and ramifications of these studies are the subjects of this review.

Legionella

As previously mentioned, Fliermans *et al.* (1979) first reported the presence of *Legionella* in natural surface waters. The bacteria were initially detected by the direct fluorescent antibody (DFA) test of 500–fold concentrates. Known volumes of concentrates were placed in 6 mm diameter wells on glass slides. The samples were treated with fluorescein-tagged antibodies specific for various serogroups of

Legionella. Concentrations of *L. pneumophila* in undiluted samples were then determined by enumerating the average number of cells per well (Table 1).

Some of these *Legionella* populations were also shown infectious for guinea pigs (Fliermans *et al.*, 1979). The animals were injected intraperitoneally with concentrates and were subsequently observed for signs of illness. When overtly ill, guinea pigs were sacrificed and their tissues were plated on charcoal yeast extract (CYE) agar for isolation of *Legionella.* Isolates from the tissues of inoculated guinea pigs in this preliminary study were serogroup 1 of *L. pneumophila.*

Interestingly, analysis of warm and hot spring waters also revealed the presence of *Legionella* by DFA and guinea pig inoculations (Fliermans, 1985). Because of the apparent fastidiousness of *Legionella* growth under laboratory conditions, it was surprising that *L. pneumophila* was isolated through guinea pig injections from waters with temperatures as high as 60° C. The temperature of waters in which *Legionella* were detected by DFA ranged from 8.5 to 64.5° C (Table 2).

Subsequently, extensive studies by Fliermans *et al.* (1981) in the central and eastern United States confirmed and expanded these initial observations. These studies also included artificially heated waters at the Savannah River Plant (SRP) in South Carolina.

Water samples were collected from both thermally altered and ambient waters at the Savannah River Plant. Many samples were collected from the Par Pond system. This system consists of a 1092–ha cooling reservoir which is used as a source for 90 percent of the water required to cool the heat exchangers of a nuclear production reactor. The resulting thermal effluent (approximately 200,000 gal/min) is discharged to the environment through an 11–Km long series of canals with associated troughs and cooling ponds before being discharged again into Par Pond.

Legionella were detected by DFA in all of the 67 sampling sites and in >95 percent of the 793 water samples tested. As in the case of naturally heated waters, *Legionella* were detected in and isolated from waters as high as 70° C and 63° C, respectively. In these studies, water parameters other than temperature were measured at the time of sampling. These included dissolved oxygen, pH, conductivity, chlorophyll, pheophytin, and sechii disk as a measure of water clarity. When comparing the frequency of isolation of *Legionella* from inoculated guinea pigs with the various ecological parameters, only high water temperature correlated with isolation of *Legionella* (Table 3).

Consequently, studies of temperature variations in the Par Pond system relative to concentrations of *Legionella* (determined by the DFA test) were carried out (Fliermans, 1985). These data showed that at extreme temperatures, i.e. 60–70° C, the concentrations of *Legionella* were low but increased as the temperature dropped such that the highest *Legionella* concentrations were found in a temperature range of 45–55° C (Figs. 1–3). These results contrasted sharply with data regarding laboratory cultures of *Legionella*, which will not grow at 45–55° C. Optimal temperature for growth of

Table 1. Habitat characteristics for isolates of *L. pneumophila* from nonepidemic sources (Fliermans, et al., 1979).

Isolate and sample no.	Depth (m)	Temp (°C)	Conductivity (μmho/cm)	pH	Dissolved O$_2$ (mg l^{-1})	No. of cells staining/liter in serogroup:[a]			Serogroup of guinea pig isolates
						Knoxville 1	Togus 1	Control	
SRP–2 (85–13)	2	17.2	62	6.4	5.85	9.6×10^6	BD[b]	Neg.[c]	1
SRP–3 (64–14)	7	28.0	57	6.7	3.80	2.7×10^5	1.8×10^4	Neg.	1
SRP–4 (85–1)	Surf	18.4	24	6.8	9.80	3.0×10^5	BD	Neg.	1
SRP–5 (86–1)	3	17.0	52	6.6	6.90	5.0×10^6	BD	Neg.	1

[a] All specimens were screened with the working dilutions of conjugates for *L. pneumophila* of serogroups 3 (Bloomington 2) and 4 (Los Angeles 1), but none were detected.
[b] BD = Below detectability of 9 *L. pneumophila*/ml.
[c] Neg. = Negative.

laboratory cultures is 35–40° C. These observations dramatically demonstrate the tenuousness of ecological conclusions drawn from laboratory experiments alone.

In addition to the extensive studies of thermal impact on *Legionella* distribution and isolatability at SRP, a major study of thermal additions relative to effects on *Legionella* populations was also carried out at electric power plants (Tyndall, 1981).

The preliminary results of studies of eleven thermally altered waters from electric power plants compared with ambient source waters confirmed Fliermans' observation of the ubiquity of *Legionella* in both natural and thermally altered waters. One sample of plant-affected and ambient source water was taken during each of three successive years (1980–1982). The geographically disparate sites included Oregon, California, Colorado, Vermont, Michigan, Pennsylvania, Minnesota, Arkansas and Iowa. In spite of this geographic disparity, the concentrations of *L. pneumophila* (serogroups 1–4) as determined by DFA analysis

Table 2. Densities of *L. pneumophila* in selected natural warm and hot springs in the United States (Fliermans, 1985).

Habitat	Location	Temperature °C	*L. pneumophila*/liter				Isolation
			Knoxville	Togus	Bloomington	Los Angelos	
Cyanidium Creek	YNP	42	BD	BD	BD	BD	NT
Cyanidium Creek	YNP	45	BD	BD	BD	BD	NT
Cyanidium Creek	YNP	50	BD	BD	BD	BD	NT
Cyanidium Creek	YNP	55	2.3×10^8	BD	BD	BD	K
Bath Lake	YNP	36	1.5×10^8	BD	BD	4.1×10^7	K, LA
Octopus Lake	YNP	45	BD	BD	BD	2.7×10^7	Neg
Octopus Lake	YNP	50	BD	BD	BD	BD	Neg
Octopus Lake	YNP	55	6.8×10^7	BD	BD	BD	K
Octopus Lake	YNP	60	9.6×10^7	BD	BD	BD	K
Mineral Hot Springs	CO	54	5.8×10^5	BD	BD	4.1×10^4	Neg
Madison River	YNP	13	6.5×10^6	BD	1.1×10^5	9.1×10^3	Neg
Terrace Spring	YNP	48.5	1.1×10^5	BD	BD	6.6×10^8	LA
Old Bath	YNP	36.5	BD	2.8×10^5	1.7×10^4	1.7×10^4	Neg
Obsidian Creek Springs	YNP	35.5	8.4×10^4	7.1×10^4	BD	8.2×10^4	NT
Amphitheatre Springs	YNP	31.3	BD	BD	BD	BD	NT
Tantalus Creek	YNP	28	2.0×10^4	9.1×10^3	BD	9.1×10^3	NT
Gray Lake	YNP	64.5	3.8×10^5	BD	BD	1.1×10^5	Neg
Pool A	YNP	65	BD	BD	BD	BD	NT
Pool A Outfall	YNP	45	5.7×10^5	4.0×10^4	BD	8.1×10^4	Neg
White Creek	YNP	51	3.0×10^4	BD	BD	7.1×10^7	Neg
White Creek	YNP	47	BD	BD	BD	6.5×10^5	Neg
White Creek	YNP	41	3.1×10^7	1.1×10^5	4.0×10^4	1.1×10^5	Neg
Lifsey Springs	NC	21.3	2.5×10^5	1.6×10^5	BD	1.7×10^4	Neg
Thundering Springs	VA	10.3	2.2×10^5	1.6×10^5	BD	1.7×10^4	Neg
Barker Springs	VA	18.5	BD	BD	BD	BD	NT
Warm Springs	GA	31	BD	8.2×10^4	BD	BD	Neg
Parkman Springs	GA	9.5	4.5×10^5	3.8×10^5	BD	8.2×10^4	Neg
Tom Brown Springs	GA	8.5	1.1×10^6	1.9×10^5	BD	1.4×10^5	Neg

YNP = Yellowstone National Park
NT = Not tested
BD = Below detection ($<9.1 \times 10^3$)
K = Knoxville
LA = Los Angeles

showed little variation, ranging from $<1.3 \times 10^4$ to 9.2×10^5/liter in ambient source water and from 1.5×10^4 to 7.3×10^6/liter in thermally altered waters (Table 4). No marked plant effects were seen in the concentrations of *L. pneumophila*. *Legionella* were recovered from inoculated guinea pigs in seven of eleven plant waters and five of eleven source waters (Tables 5, 6). *Legionella* isolated from guinea pigs included *L. pneumophila* serogroups 1, 2, 4 and 6 as well as a new species, i.e. *Legionella oakridgensis* (Orrison *et al.*, 1983; Tyndall *et al.*, 1983).

A second, more comprehensive study of thermally altered waters associated with electric power plants in the United States was then undertaken (Christensen *et al.*, 1983). Water samples were collected during each of the four seasons (1981–1982) at various plant-affected locations within each of nine power plants (e.g., precondenser, postcondenser, cooling tower basin, discharge canal) and from source waters at each site. Test sites included

four plants from the northern midwest, four from the southeast, and one from the east. Five of the plants used once-through cooling systems and the remaining four were predominantly closed-cycle. Temperature, dissolved oxygen, pH, conductivity, alkalinity, phosphate, nitrate, ammonia, and inorganic and total carbon were measured at the time of sampling.

In addition to determining *Legionella* concentrations by DFA, viability of the *Legionella* was determined by testing for a functional electron transport system using a tetrazolium dye. Infectivity of the sample was determined by guinea pig inoculation. A sample was considered infectious if after inoculation the guinea pigs showed signs of illness, had temperature rises ($>0.6°$C), and if *Legionella* were isolated from the tissues of the test animal.

As was seen in the previous environmental studies, *Legionella* were detectable by DFA in source water and plant-affected water from all nine sites in all seasons (Table 7). *Legionella* were detected by

Table 3. Relationship of high- and low-parameter characteristics to the frequency of *L. pneumophila* isolations[a] (Fliermans, et al., 1981).

Range of characteristic	No. of samples	No. of isolates	% of isolates	χ^2	P value
Temp (°C)					
0–35	233	32	13.9	9.48	0.0021[b]
36–70	38	16	42.1		
Conductivity (μS cm^{-1})					
0–55	120	12	10.0	2.75	0.0971
56–110	185	35	18.9		
pH					
5–6.75	176	29	16.5	0.07	0.7873
6.76–8.5	132	19	14.4		
Dissolved oxygen (ppm)					
0.5	68	14	20.6	0.72	0.2757
6–10	242	32	13.2		
Chlorophyll a (mg m^{-3})					
0–12	98	9	9.2	2.36	0.1244
13–24	40	9	22.5		
Pheophytin (mg m^3)					
0–10	112	16	14.3	0.006	0.9371
11–20	22	4	18.2		
Secchi disk (m)					
0–2	119	21	17.6	0.16	0.6857
2.1–4	24	6	25.0		

[a] High 50% of range and low 50% of range.
[b] Significant at <5% probability level.

Table 4. Combined concentration[a] of serogroups 1–4 of *L. pneumophila* in cooling tower water and ambient control source waters (Tyndall, 1981).

Site	Summer 1980	Spring/summer 1981	Fall 1981
A-test[b]	1.0×10^5	7.6×10^5	6.9×10^5
A-control[c]	2.6×10^4	4.0×10^4	7.6×10^4
B-test	6.7×10^4	3.0×10^5	8.0×10^4
B-control	4.0×10^5	$<1.3 \times 10^4$	7.3×10^4
C-test	8.1×10^5	6.3×10^5	6.8×10^4
C-control	2.7×10^5	1.5×10^5	1.0×10^5
D-test	1.7×10^5	8.0×10^5	1.7×10^5
D-control	1.5×10^4	4.0×10^5	1.3×10^5
E-test	4.4×10^4	4.4×10^4	1.1×10^5
E-control	4.2×10^4	4.2×10^4	6.4×10^4
F-test	1.7×10^6	1.0×10^5	4.8×10^5
F-control	$<1.3 \times 10^4$	4.1×10^5	7.3×10^4
G-test	5.4×10^5	5.7×10^5	1.5×10^4
G-control	6.0×10^5	5.4×10^5	3.9×10^4
H-test	4.0×10^5	1.0×10^5	7.3×10^6
H-control	4.0×10^4	1.7×10^5	9.2×10^5
I-test	1.3×10^5	8.0×10^4	1.6×10^5
I-control	3.8×10^5	2.6×10^5	2.2×10^5
J-test	1.0×10^5	4.8×10^5	6.4×10^5
J-control	1.3×10^4	1.6×10^5	3.6×10^4
K-test	1.3×10^5	NT[d]	8.7×10^5
K-control	2.4×10^4	NT	1.1×10^5

[a] Legionella l^{-1} of water.
[b] Thermally altered water from cooling tower basin.
[c] Ambient source water.
[d] NT = not tested.

Table 5. Presence of infectious *Legionella* in cooling tower waters and ambient control source waters sampled in spring/summer 1981 (Tyndall, 1981).

Site	No. *Legionella* injected	Infectious for guinea pigs	Serogroup isolated
A-test[a]	7.6×10^5	Neg	NA[c]
A-control[b]	7.2×10^4	Neg	NA
B-test	3.0×10^5	Neg	NA
B-control	$<10^4$	Neg	NA
C-test	6.3×10^5	Pos	Knox.[d]
C-Control	1.5×10^5	Pos	Bloom.
D-test	9.2×10^5	Neg	NA
D-control	4.1×10^5	Neg	NA
E-test	4.4×10^4	Pos	Chic., Knox.
E-control	4.2×10^4	Pos	Chic.
F-test	1.7×10^5	Neg	NA
F-control	2.6×10^5	Neg	NA
G-test	5.5×10^4	Pos	OR
G-control	3.0×10^5	Neg	NA
H-test	5.3×10^4	Neg	NA
H-control	2.1×10^5	Neg	NA
I-test	2.4×10^4	Neg	NA
I-control	5.1×10^4	Neg	NA
J-test	3.9×10^5	Pos	Chic., LA
J-control	7.9×10^4	Pos	Knox., LA

[a] Thermally altered water from cooling tower basin.
[b] Ambient source water.
[c] NA = not applicable.

[d] Knox. = Knoxville, Bloom. = Bloomington, Chic. = Chicago, and LA = Los Angeles (serogroups of *L. pneumophila*), OR = *L. oakridgensis*.

Table 6. Presence of infectious *Legionella* in water concentrates from cooling towers and ambient control source waters sampled in fall 1981 (Tyndall, 1981).

Site	Concentration of *Legionella* injected	Infectious for guinea pigs	Serogroup isolated
A-test[a]	6.9×10^5	Pos	Chic., Knox.[c]
A-control[b]	7.6×10^4	Neg	NA
B-test	8.0×10^4	Pos	Knox.
B-control	7.3×10^4	Pos	Knox.
C-test	4.2×10^4	Neg	NA
C-control	4.2×10^4	Neg	NA
D-test	1.7×10^5	Neg	NA
D-control	1.3×10^5	Neg	NA
E-test	1.1×10^5	Neg	NA
E-control	6.4×10^4	Neg	NA
F-test	4.8×10^5	Neg	NA
F-control	7.3×10^4	Neg	NA
G-test	1.5×10^4	Pos	OR, LA
G-control	3.9×10^4	Pos	LA
H-test	7.3×10^6	Neg	NA
H-control	9.2×10^5	Neg	NA
I-test	1.6×10^5	Neg	NA
I-control	2.2×10^5	Neg	NA
J-test	6.4×10^5	Pos	Knox., LA
J-control	3.6×10^4	Neg	NA
K-test[d]	8.7×10^5	Pos	Knox.
K-control	1.1×10^5	Neg	NA

[a] Thermally altered water from cooling tower basin.

[b] Ambient source water.

[c] Chic. = Chicago, Knox. = Knoxville, LA = Los Angeles (serogroups of *L. pneumophila),* and OR = *L. oakridgensis.*

[d] NA = not applicable.

Table 7. Mean cell densities of *Legionella* (number of cells ml^{-1})[a] (Christensen, et al., 1983).

Season	Sample location	Plants operating in once-through mode					Plants operating in closed-cycle mode			
		A	B	C	D	E	F	G	H	I
Spring	Ambient	35	48	22	227	15,733 *	3,967 *	8,767	10,133 *	11,333 *
	Plant-exposed	33	35	28	233	343	508	19,400	300	248
Summer	Ambient	520	140	71	56	300[b]	263[c]	183	650 *	327 *
	Plant-exposed	235	225	105	128	[b]	4,081[c]	368	146	95
Fall	Ambient	63	80	33	21	93	40	45	49	372 *
	Plant-exposed	70	59	45	22	42	36	112	30	55
Winter	Ambient	665	185	320	114	1.205	1,360	957	4,450 *	45
	Plant-exposed	337	352	710	158	1,070	2,050	1,640	315	56

[a] Mean cell densities separated by an asterisk represent a significant ($P<0.05$) decrease with passage through a power plant.

[b] Plant was shut down.

[c] Mixed operating mode.

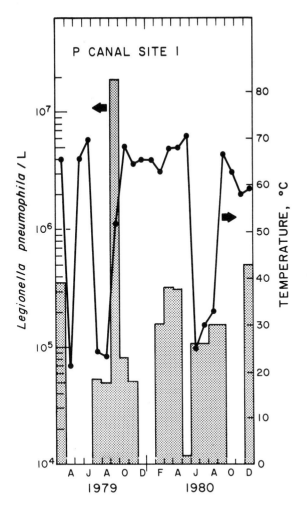

Fig. 1. Density estimates of *L. pneumophila* serogroups 1–4 at site 1 along the thermal canal (Fliermans, 1985).

DFA in 265 of 270 samples. Average concentrations of *Legionella* were higher in the source water in winter and spring than in summer and fall. No effect of passage through a once-through cooling system was found, but retention within a closed-cycle system resulted in decreases in total cell density and the density of viable cells in both spring and summer (Tables 8, 9). Attempts to relate density and viability with water quality parameters in these field samples were inconclusive due to multiple correlations among the parameters.

In spite of lower concentrations and diminished densities of viable *Legionella* in thermally altered water of closed-cycle power plants, the infectivity of the *Legionella* populations were higher than those of source water or from water of once-through systems (Fig. 4, p<0.05). This difference may relate in part to the source water, since increased infectivity was associated with source water at the closed-cycle sites.

Analyses of the *Legionella* from tissues of inoculated guinea pigs showed that serogroup 1 of *L. pneumophila* was most frequently isolated and serogroup 4 was the second most prevalent (Tables 10, 11). *Legionella pneumophila* serogroup 1 was isolated from 14 plant-affected and 5 ambient samples while serogroup 4 was isolated from 8 plant-affected and 6 ambient samples. Of interest was the isolation of two newly discovered *Legionella* species (Brenner, 1984; Orrison *et al.*, 1983; Tyndall *et al.*, 1983). *Legionella oakridgensis* was isolated from three different plant sites and *L. cherrii* was isolated at one plant site. Other *Legionella* isolates included *L. pneumophila* serogroups 2, 3, and 6, *L. gormanii* and *L. bozemanii*. Unlike the *L. pneumophila* isolates, *L. oakridgensis* isolations were more site-specific. While *L. pneumophila* was isolated from 8 of 9 plant-affected waters and from 7 of 9 ambient waters, *L. oakridgensis* was isolated from only 3 of 9 plant-affected waters and none of the ambient waters.

DFA analysis of various sites showed that *L. oakridgensis* was widely distributed and in concentrations similar to that of *L. pneumophila* (Tyndall, 1983). To date, *L. oakridgensis* and *L. cherrii* have been isolated only from thermally altered waters. Thus, thermally altered water can significantly affect *Legionella* populations. These studies have shown that both concentrations and infectivity of the populations may be affected by thermal additions. Thermally altered sites may also serve as habitats for previously undiscovered *Legionella* species.

Naegleria fowleri

The effect of thermal additions on the propagation or isolatability of *N. fowleri* was suggested by the epidemiologic aspects of the original cases of PAME. As previously indicated, children bathing in potable waters warmed by transit across arid

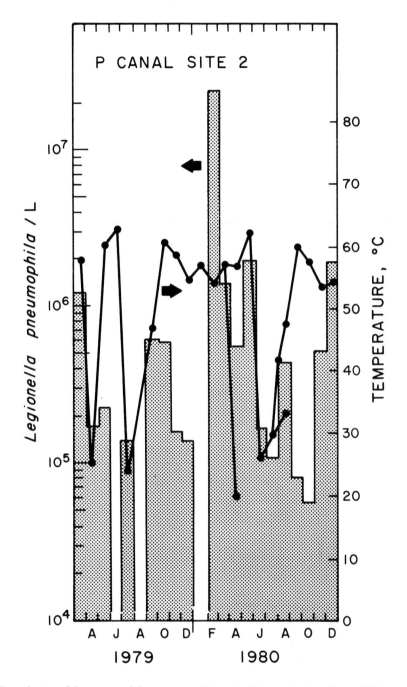

Fig. 2. Density estimates of *L. pneumophila* serogroups 1–4 at site 2 along the thermal canal (Fliermans, 1985).

Australian deserts were some of the first cases of PAME. De Jonckheere then described a fatal case of PAME in a child swimming in a artificially heated waters from an industrial plant (De Jonck-heere *et al.,* 1975). Naturally warmed waters in Florida and other Southern states (Willaert, 1974), as well as thermal springs (Hecht *et al.,* 1972), were sites of other fatal PAME infections. All of these reports indicated that thermal additions, either natural or artificial, contributed to the presence of *N. fowleri.* In addition, laboratory studies by Griffin (1972) showed that the pathogenic free-

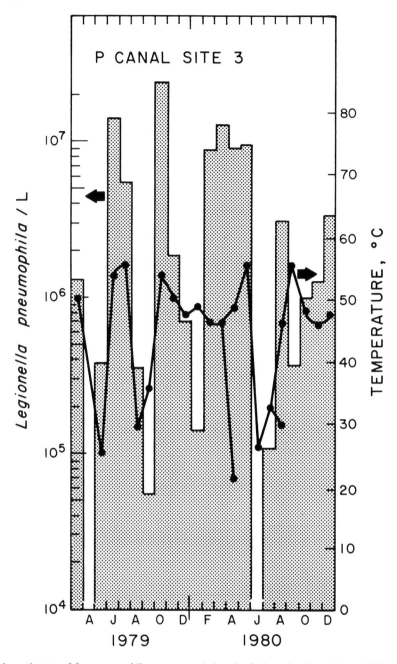

Fig. 3. Density estimates of *L. pneumophila* serogroups 1–4 at site 3 along the thermal canal (Fliermans, 1985).

living amoebae could grow at higher temperatures than the nonpathogenic species.

The influence of natural thermal additions on the presence of detectable *N. fowleri* in a Florida lake was strikingly depicted in a study by Wellings *et al.* (1979). In this study, a lake in central Florida was

sampled monthly for one year. Multiple water and sediment samples were analyzed for *N. fowleri*. During the winter months when water temperatures dropped below 25° C, the pathogen was not isolated from water samples. When the temperature exceeded 25° C, some samples yielded *N.*

fowleri. As the water temperature approached 30° C in the summer months, approximately 50–65 percent of the water samples were positive for *N. fowleri* (Fig. 5). The isolation of *N. fowleri* from sediment samples followed the same general trend except that a percentage of sediment samples (2–30 percent) yielded *N. fowleri* even during the winter months, indicating that sediments harbored the organism during the colder part of the year (Fig. 5).

A study by Stevens *et al.* (1977) confirmed the presence of *N. fowleri* in Florida waters not associated with cases of PAME. In this study, *N. fowleri* was isolated from both naturally warmed ambient waters and from artificially heated waters.

Table 8. Comparison of mean cell densities[a] of *Legionella* (number of cells ml^{-1}) before and after plant passage for the two operating modes (Christensen, et al., 1983).

Season	Operating mode	Location[b]	
		Ambient	Plant-exposed
Spring	Once-through	116	67
		*	*
	Closed-cycle	7795 ------*--------	884
Summer	Once-through	116	118
	Closed-cycle	289 ------*--------	65
Fall	Once-through	44	32
	Closed-cycle	46	35
Winter	Once-through	311	390
	Closed-cycle	549	647

[a] An analysis of variance with season, operating mode and location as main effects and specific plant as a blocking variable was used to derive the mean square error for later use in Duncan's multiple range test.
[b] An asterisk between two numbers indicates that these two means are significantly different from one another ($P<0.05$).

Table 9. Comparison of mean densities of viable *Legionella* cells[a] (number of cells ml^{-1}) before and after plant passage for the two operating modes (Christensen et al., 1983).

Season	Operating mode	Location[b]	
		Ambient	Plant-exposed
Spring	Once-through	21	17
		*	*
	Closed-cycle	645 ------*--------	162
Summer	Once-through	20	32
			*
	Closed-cycle	14 ------*--------	2
Fall	Once-through	7	8
	Closed-cycle	7	6
Winter	Once-through	83	103
	Closed-cycle	243	126

[a] An analysis of variance with season, operating mode and location as main effects and specific plant as a blocking variable was used to derive the mean square error for later use in Duncan's multiple range test.
[b] An asterisk between two numbers indicates that these two means are significantly different from one another ($P <0.05$).

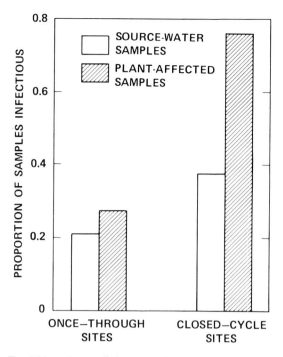

Fig. 4. Mean *Legionella* infectivity (proportion of samples infectious) by type of cooling system and location of sample (Christensen, et al., 1983).

Probably the most extensive studies of the effect of solar heating on the *Naegleria* populations in reservoirs was carried out by Robinson & Lake (1981) and Walters *et al.* (1981). These studies of Australian reservoirs were undertaken because of the implication of potable waters as the source of *N. fowleri* in fatal cases of PAME. One aspect of these investigations compared the temperature tolerance of *Naegleria* isolates with the temperature of the water from which they were isolated. In 1978/1979 the temperature tolerance of *Naegleria* isolates from the Paskeville No. 1 reservoir correlated significantly ($P<0.05$) with the 3-week running mean of the water temperature from which the *Naegleria* were isolated (Fig. 6a, b). In 1979/1980 the temperature tolerance of the *Naegleria* isolates also correlated significantly with water temperatures at the time of sampling. A significant correlation between temperature tolerance of *Naegleria* isolates and the temperature of the water sample was also seen in studies of Nelshaly reservoir. Analysis of the Hope Valley reservoir also showed a significant correlation ($P<0.001$) of temperature

Table 10. Types of pathogens isolated from plant-affected water samples[a] (Christensen, et al., 1983).

Site	Spring	Summer	Fall	Winter
A		K	CH	GO
B				
C		CH	'VIBRIO'	
D		LA		LA, L?
E	K	[b]		LA
F	OR	K	K, BL, LA, OR	K, CH, OR
G	K, OR, BL, LA	CH, OR	K, OR	K, CH, OR
H	[c]	LA	LA, CH	LA
I	K	K, CH	K, TO	

[a] Key to symbols:
Legionella pneumophila serogroups:
BL = Bloomington
CH = Chicago
K = Knoxville
LA = Los Angeles
TO = Togus

Other *Legionella* species:
GO = *Legionella gormanii*
L? = Species of *Legionella* not typeable with antiserum prepared against known species and/or serogroups of *Legionella*
OR = *Legionella oakridgensis*
Other organisms:
'VIBRIO' = Vibrio-like organism, at present unidentifiable by the Centers for Disease Control (CDC).

[b] Plant was not operating; samples were not injected.

[c] One or more samples were injected. Samples were either toxic to the guinea pigs or resulted in contaminated plates upon subsequent plating of tissue; therefore, *Legionella* could not have been isolated if present.

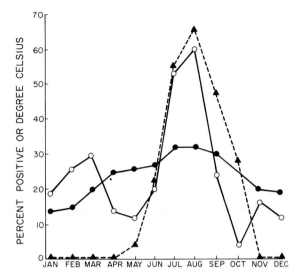

Fig. 5. Percent of positive water and sediment samples as related to the average water temperature by month. (●—●) Average water temperature. (▲---▲) Percent of water samples positive. (○—○) Percent of sediment samples positive (Wellings, et al., 1979).

tolerance of *Naegleria* isolates and seasonal changes in water temperature. Not surprisingly, the *Naegleria* isolates most tolerant of high temperatures were more prevalent in the summer months. In addition, *N. fowleri* was isolated only from summer water samples (Fig. 6b).

A unique example of the effect of thermal additions on *N. fowleri* presence was described by Duma *et al.* (1981). In this study, a reservoir (Lake

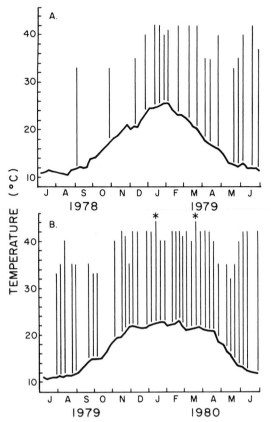

Fig. 6a, b. Seasonal variation in water temperature and temperature tolerance of *Naegleria* isolated from Paskeville No. 1 reservoir. (a) 1978/1979. (b) 1979/1980. Range of temperatures over which growth of *Naegleria* occurred, truncated below at water temperature. Bold line indicates water temperature at time of sampling (3-week running mean). *Naegleria fowleri* isolates indicated by asterisks (Robinson and Lake, 1981).

Table 11. Types of pathogens isolated from ambient water samples[a] (Christensen, et al., 1983).

Site	Spring	Summer	Fall	Winter
A		K	K, LA, GO	
B				
C		K	'VIBRIO'	K
D			LA, CH	
E			OR	
F	[b]	[b]	J19,[c]LA	
G	[b]	K, CH		BOZ[d]
H		[b]	LA	LA
I				LA

[a] Key to symbols: Abreviations given in Table 10, footnote a.
[b] See Table 10, footnote b.
[c] *Legionella* spp not typable with antisera prepared against known species of *Legionella* but reactive with antisera prepared against a previously isolated as yet unnamed *Legionella* spp.
[d] *L. bozemannii.*

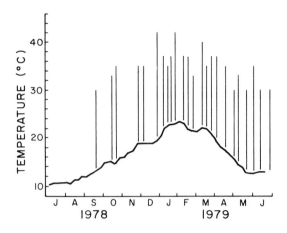

Fig. 7. Seasonal variation in water temperature and temperature tolerance of *Naegleria* isolated from Hope Valley reservoir, 1978/1979. Range of temperatures over which growth of *Naegleria* occurred, truncated below at water temperature. Bold line indicates water temperature at time of sampling (3-week running mean) (Robinson and Lake, 1981).

Anna, Virginia) receiving thermal additions from an electric power plant was studied in detail before and after the power plant began operations.

The 17-mile long fresh water Lake Anna was created by damming the North Anna River. The lake was completely filled by 1972. The three sites, i.e. Lake Anna I, II and III, represented three separate cooling lagoons. The sample site at Lake Anna I was at the outlet of the thermal discharge from the first cooling canal. The Lake Anna II sampling site was approximately 0.65 miles from the first sampling site and along the banks of a cooling canal. The third sampling site was along the shore of the main reservoir. Sampling depth at all three sites was 4 to 6 inches. A variety of area lakes not receiving thermal additions were also sampled for comparison.

As seen in Table 12, *N. fowleri* was not detected in Lake Anna samples prior to thermal addditions. During the same time period, a few samples from nearby smaller lakes were positive for the pathogen.

After initiation of thermal additions in April of 1978, thermophilic amoebaeflagellates were isolated from Lake Anna I in 12 of 16 months, from Lake Anna II in 13 of 16 months, and from Lake Anna III in 4 of 16 months. Coincident with in-

creased isolation of thermophilic amoebaeflagellates from the thermally altered waters of Lake Anna was the detection of *N. fowleri*. The pathogen was isolated from all three sampling sites in 1978 but from only one site in 1979. Water temperatures were several degrees warmer in 1978 than in 1979. During the period of thermal additions to Lake Anna when *N. fowleri* was detectable, fewer isolations were made in the surrounding smaller control lakes, with only one sample yielding *N. fowleri*. Another effect of the thermal additions to Lake Anna was the continuing presence of thermophilic amoebaeflagellates during winter as well as summer months. This was not observed prior to thermal additions.

During the studies of Lake Anna, various parameters of the sampling sites were assayed. These included temperature, pH, salinity and conductivity. Of these various parameters, only temperature was altered after initiation of thermal additions.

In other studies of electric power plants, increased numbers of isolations of *N. fowleri* from thermally altered waters was also observed (Tyndall, 1979; Tyndall *et al.*, 1981). In these studies, heated waters of seventeen predominantly northern power plants were tested and compared with their ambient source waters or nearby control waters. As seen in Tables 13 and 14, there was an increased frequency of thermophlic amoebae, thermophilic amoebaeflagellates, and *N. fowleri* in waters receiving thermal additions compared to the ambient source waters. One hundred and thirty-one samples from ambient control waters and three hundred and eighty-five samples from thermally altered waters were analyzed for thermophilic amoebae, amoebaflagellates and *N. fowleri*. Twenty-nine percent and seven percent of the control samples were positive for thermophilic amoebae and amoebaeflagellates, respectively. *N. fowleri* was not isolated from any of the control samples. In contrast, fifty percent and twenty-two percent of the samples of thermally altered waters were positive for thermophilic amoebae and amoebaeflagellates, respectively. *N. fowleri* was found in four percent of the samples and was isolated from four of the twenty-one sites. The difference in

Table 12. Amoeboflagellates and pathogenic amoebae isolated from lakes in Richmond, Virginia area (Oct. 1976–Sept. 1979)* (Duma, 1981).

Lake	1976 D	1977 J	F	M	A	M	J	J	A	S	O	N	D	1978 J	F	M	A	M	J	J	A	S	O	N	D	1979 J	F	M	A	M	J	J	A
Overhill	–	–	O	–	–	–	–	–	+	–	–	–	–	–	–	–	–	–	–	+	+	–	–	–	–	–	–	–	–	–	–	–	–
Overhill Feeder	O	O	+	–	–	–	+	O	O	+	+	+	+	–	+	+	+	+	+	+	+	+	+	+	+	+	+	–	–	+	+	+	+
Anna I	+	+	–	–	–	+	+	+	+	+	+	+	+	+	+	+	+	–†	–†	–†P	P	P	+	+	+	+	+	+	+	+	+	+	+
Anna II	–	–	–	–	–	–	P	P	+	+	+	+	–	+	+	–	–	+	P	+	+	P	+	+	+	+	+	P	–	+	–	–	+
Anna III	–	–	+	–	–	–	–	+	+	+	+	+	+	–	–	–	–	–	–	–	–†	–†	+	+	–	+	+	–	–	–	–	–	–
Pocahontas	–	–	–	–	P	P	+	+	+	+	–	–	–	–	–	–	P	–	–	–	–†	–†	+	–	–	–	–	–	–	–	–	–	–
Pocahontas Deep	–	–	–	–	–	–	+	+	+	+	+	+	+	–	–	–	–	–	–†	–†	+	+	+	+	+	–	–	–	–	–	–	–	–
Salisbury	+	+	+	–	–	–	+	+	+	+	+	+	+	+	+	+	+	+	+	+	+	+	+	+	+	+	+	–	–	–	–	–	–
Salisbury Deep	–	–	–	–	+	–	+	+	+	+	+	+	+	–	–	–	–	–	–	+	+	+	+	+	–	–	–	–	–	–	–	–	–
Manchester	+	–	–	+	–	–	–	–	–	+	+	–	–	–	–	+	+	+	+	+	+	+	+	+	+	–	–	–	–	–	–	–	–
Falling Creek	–	–	–	–	–	+	–	–	–	+	+	+	+	+	–	–	+	+	–	–†	P	+	+	–	+	+	–	–	–	–	–	–	–
Falling Creek Deep	–	–	–	–	–	–	+	+	+	+	+	+	+	–	–†	–†	–†	–†	–†	+	+	–	+	+	–	+	+	–	–	–	–	–	–
Moore's	O	O	O	O	O	O	O	O	O	O	O	O	O	O	O	O	O	O	O	O	O	O	O	O	O	O	O	O	O	O	O	O	O
Moore's Deep	O	O	O	O	O	O	O	O	+	O	O	O	O	O	O	O	O	O	O	O	O	O	O	O	O	O	O	O	O	O	O	O	O
Pallisades	–	+	–	–	–	P	+	+	+	+	+	–	–	–	+	+	–	–	–	–	+	+	–	–	–	+	–	–	–	–	–	–	–
Swift Creek	–	–	–	–	P	–	–	+	+	+	+	–	–	+	+	+	+	+	+	–	+	+	+	+	–	+	+	–	–	–	–	–	–
Swift Creek Deep	–	–	–	–	–	–	+	+	+	+	+	–	–	+	+	+	+	–	–	+	–†	–	–	–	–	+	+	–	–	–	–	–	–

* (+) = Amoeboflagellates growing at 44°C; (–) = Nonflagellating amoebas growing at 44°C; (P) = Pathogen; (O) = Not tested.
† = Grew well in Chang's liquid media at 37°C, but growth not sustained.

percent of samples positive for thermophilic amoebae, amoebaeflagellates and *N. fowleri* between unheated and heated waters was significant at the 0.02 level.

Perhaps the most statistically dramatic data relative to the effects of thermal additions on the presence of thermophilic amoebae in general and *N. fowleri* in particular is found in comparison of two impoundments in a northern state (Tyndall & Domingue, 1984). Both sites were sampled in August of 1983. One site had been receiving thermal additions from an electric power plant since 1970. Also, solar thermal additions were somewhat higher at this site compared to the unheated site. Water temperatures at the three sampling locations at this site ranged from 34.8°C to 41.5°C. Fourteen water and fifteen substrate samples (i.e. algae, leaves, wood, foam, etc.) were tested for the presence of thermophilic amoebae, amoebaeflagellates, and *N. fowleri*. The percent of samples positive were 100, 76 and 41, respectively (Table 15).

The other test site was an unheated lake approximately one hundred and eighty-five miles from the heated site. Fifty-eight total substrate/sediment samples and twenty-four water samples from a

Table 13. Thermophilic amoebae in unheated control waters.

# sites	# states	Temperature range (°C)	Number of samples			
			Incubated at 45°C	Positive for amoebae at 45°C	Positive for flagellate amoebae at 45°C	Positive for pathogenic isolates
17	12	13–32	131	38	9	0

Table 14. Thermophilic amoebae in power plant cooling systems.

# sites	# states	Temperature range (°C)	Number of samples			
			Incubated at 45°C	Positive for amoebae at 45°C	Positive for flagellate amoebae at 45°C	Positive for pathogenic isolates
17	12	12–41	385	195	85	14

Table 15. Distribution of thermophilic free-living amoebae in water and substrate/sediment samples from cooling lake and control unheated lake, August, 1983.

Sites	Sample type	Percent of samples positive		
		Thermophilic amoebae	Thermophilic *Naegleria*	*N. fowleri*
Cooling lake	Water	100 (14/14)[a]	71 (10/14)	57 (8/14)
	Substrates	100 (15/15)	80 (12/15)	27 (4/15)
Total		100 (29/29)	76 (22/29)	41 (12/29)
Control lake	Water	33 (8/24)	17 (4/24)	0 (0/24)
	Substrates/Sediments	67 (39/58)	31 (18/58)	0 (0/58)
Total		57 (47/82)	27 (22/82)	0 (0/82)

[a] Numbers in parentheses are number of samples positive over total number of samples taken.

total of eight locations at this site were tested. The percentages of samples positive for thermophilic amoebae and amoebaeflagellates were 57 and 27, respectively (Table 15). *N. fowleri* was not isolated at this site. Statistical comparison of the heated site with the unheated site showed differences in the presence of thermophilic amoebae, amoebaflagellates and *N. fowleri* significant at the <0.001 level.

As in other studies, various water quality parameters were also measured at the time of sampling. Temperature was the only measured parameter which differed markedly between the two sites.

As in the case of the *Legionella* studies, the investigations of thermally altered waters for the presence of *N. fowleri* shows a significant correlation between thermal input and the presence of *N. fowleri*. Also, studies of various water quality parameters have yet to reveal physical/chemical characteristics other than temperature that account for the enhanced presence of *N. fowleri*.

Ramifications

In the past three decades the amount of thermal additions to the environment has markedly increased. The production of electricity, for example, has increased from 0.033×10^{15} BTU in 1947 to 78.9×10^{15} in 1979 (Fliermans, 1985). The impact of the increased, associated thermal discharges on promoting the presence of human microbial pathogens in the thermally heated waters has only recently been investigated. As discussed in this review, thermal addition is one parameter known to affect the *Legionella* and *Naegleria* populations in reservoirs and other surface waters. Heated waters harbor known species of both types of pathogens and also have been the source of isolation of new species. At least two species of *Legionella* (i.e., *L. oakridgensis* and *L. cherrii*) have thus far been isolated only from thermally altered habitats. *N. australiensis*, a newly described *Naegleria* species, was also isolated from thermally altered habitats (Scaglia *et al.*, 1983).

The pathogenic potential of these new species has not been investigated, and possible pathologies elicited by these new species are unknown. It is interesting in this respect that some thermophilic *Naegleria* and *Acanthamoeba* isolates from thermally altered sites tested in our laboratory for pathogenicity in mice resulted in syndromes other than rapidly fatal encephalitis. The test mice were injected intranasally and observed for development of PAME. No rapidly fatal encephalitis occurred in the test mice. Rather than discarding the mice, they were observed for a period of one year. Overtly altered behavior was obvious in some of the test mice several months after the original inoculation. Mice were subsequently sequestered in a small room which allowed their observation without their being able to observe the investigators. Test mice were placed in a cage fitted with a sonic detector such that each movement across the beam which bisected the cage was automatically registered on an external counter. Mice were also visually observed in the morning and afternoon for three minute periods and the number of specific types of behavioral occurrences during the period were recorded. Types of behavior tabulated included grooming, rearing, hanging from the cage top, and crossing. Three groups of mice were observed: inoculated mice showing obviously altered behavior; inoculated mice without marked behavioral changes; and control mice that had been inoculated with water devoid of *Naegleria*. As seen in Table 16, some of the inoculated mice had significantly altered behavior compared with control mice.

Autopsy of the severely affected mice and subsequent analysis of brain tissue revealed foci of perivascular inflammation and amoebae localized throughout the tissue. The amoebae apparently entered the brain randomly via the systemic circulation. Host response was manifested by a marked infiltration of white blood cells, inflammation and subsequent sclerotic sequelae. If such lesions occurred in areas of the brain affecting behavioral characteristics, altered behavior would not be unexpected.

While this was a limited study, the results suggest that syndromes other than PAME can occur in laboratory animals exposed to these pathogens. Whether similar syndromes result in humans on infection with *Naegleria* or other free-living amoebae is as yet unknown.

Thus, the available data show that previously unknown species of *Naegleria* and *Legionella* can be found in heated waters. Previously unrecognized sequelae resulting from infection with microbes from thermally altered sites is also a possibility although not yet proven.

These potential problems are further complicated by the detection of *Legionella* and *Naegleria* in potable waters. As previously noted, some of the cases of PAME and presence of *N. fowleri* have been linked to potable water. We have isolated *Naegleria* and other free-living amoebae from such waters (unpublished data). *Legionella* has also been repeatedly demonstrated in potable waters, albeit in very low numbers (Bartlett, 1984; Stout *et al.*, 1982; Wadowsky, 1982). Chlorination dramatically reduces the levels of both *Naegleria* and *Legionella* in contaminated waters (Fliermans *et al.*, 1982; Tyndall *et al.*, 1983); however, the microbes are not totally eliminated. Enough remain such that they can seed potable water systems and can then respond to thermal additions present in hot water systems. The few remaining *Legionella* may be amplified in hot water systems by repeated thermal additions in selected niches such as shower heads. This may pose a problem for immunosuppressed or otherwise compromised individuals when subjected daily to aerosols of *Legionella* via showers. Numerous Legionellosis cases in hospital populations are traceable to *Legionella*-contaminated shower heads (Cordes *et al.*, 1981).

Finally, the survival of both *Legionella* and free-living amoebae in potable water and subsequent exposure to thermal additions may be further complicated by recent evidence documenting their interaction. Laboratory studies have shown that *Naegleria* and *Acanthamoebae* can support the growth of *Legionella* (Rowbotham, 1980; Tyndall & Domingue, 1982). *Legionella* concentrations of 10^8–10^9 cells per ml of culture fluid can be maintained in the presence of amoebae in media which will not support growth of *Legionella* in the absence of amoebae. In addition to free living amoebae, the protozoan *Tetrahymena pyriformis* can also support the growth of *Legionella* (Fields *et al.*, 1984). *Legionella* increased over one-hundred-fold when cultured in the presence of the protozoa without any other nutrients. Even some non-*Legionella* bacteria in potable waters may influence *Legionella* growth. Such bacteria in laboratory experi-

Table 16. Behavioral analysis of mice inoculated[a] with free-living amoebae.

A. Average number of movements in six, three minute observation periods

Control mice	*Naegleria*-inoculated	*Acanthamoebae*-inoculated
5.8 (10)[b]	11.8 (24)[c]	21.2 (17)[c]

B. Time[d] spent in specific behavior in six, three minute observation periods

Test mice	Rearing	Hanging	Grooming	Crossing
Controls (10)	63 [30–97][e]	36 [0–90]	105 [0–138]	41 [31–76]
Acanthamoebae[f]	13	4	688	11
Naegleria[f] - mouse #1	3	50	11	>400
Naegleria[f] - mouse #2	5	0	51	>400

[a] Mice inoculated intranasally with amoebae.

[b] Number in parenthesis indicates number of test animals per group.

[c] Differs significantly from the controls by the Mann-Whitney test.

[d] In seconds.

[e] Number in square brackets indicates range of time in seconds of the individuals in the group.

[f] Individual test mice from the amoebae-inoculated groups shown in Table 16A.

ments can apparently supply cysteine which permits growth of *Legionella* on cysteine-free media (Wadowsky & Yee, 1983). Therefore the presence of amoebae and other non-*Legionella* micro-organisms in potable waters and concomitant thermal additions may be important to the survival and growth of *Legionella* in such systems and many contribute to the resultant Legionellosis associated with potable waters.

Legionella and amoebae populations in potable waters may reflect in part the populations in reservoirs and may be influenced by the thermal history of such waters. Both the natural microbial populations and the subsequent thermal additions of the reservoir may thus determine the degree and microbial spectrum of subsequent human exposure.

Conclusions

Recognition of the potential of thermal additions to influence human pathogens is important in management of reservoir waters. A greater understanding of the factors other than thermal additions which may contribute to the emergence of the pathogens is also needed as is a fuller understanding of the possible pathologies associated with such micro-organisms.

Acknowledgements

The author acknowledges the cooperation of Dr C.B. Fliermans. His advice and counsel are highly valued. The cooperation of Dr G.R. Bratton in the pathological examination of mouse tissues infected with pathogenic amoebae is also acknowledged and greatly appreciated.

References

Bartlett, C.L.R., 1984. Potable water as reservoir and means of transmission. In Thornsberry, C., Balows, A., Feeley, J.C. & Jakubowski, W. (eds.), *Legionella*. Proceedings of the 2nd International Symposium. American Society for Microbiology, Washington, DC: 210–215.

Beaty, H.N., 1984. Clinical features of Legionellosis. In Thornsberry, C., Balows, A., Feeley, J.C. & Jakubowski, W. (eds.), *Legionella*. Proceedings of the 2nd International Symposium. American Society for Microbiology, Washington, DC: 6–10.

Brenner, D.J., 1984. Classification of Legionellae. In Thornsberry, C., Balows, A., Feeley, J.C. & Jakubowski, W. (eds.), *Legionella*. Proceedings of the 2nd International Symposium. American Society for Microbiology, Washington, DC: 55–60.

Butt, C., 1966. Primary amebic meningoencephalitis. New Engl. J. Med. 274:1473–1476.

Carter, R.F., 1970. Description of a *Naegleria* sp. isolated from two cases of primary amoebic meningoencephalitis. J. Pathol. 100:217–244.

Cerva, L., 1971. Studies of *Limax amoebae* in a swimming pool. Hydrobiologia 38:141–161.

Christensen, S.W., Solomon, J.A., Gough, S.B., Tyndall, R.L. & Fliermans, C.B., 1983. Legionnaire's Disease Bacterium in power plant cooling systems: Phase I final report. EPRI EA–3153. Electric Power Research Institute, Palo Alto, CA.

Cordes, L.G., Wiesenthal, A.M., Gorman, G.W., Phair, J.P., Sommers, H.M., Brown, A., Yu, V.L., Magnussen, M.H., Meyer, R.D., Wolf, J.S., Shands, K.N. & Fraser, D.W., 1981. Isolation of *Legionella pneumophila* from hospital shower heads. Ann. Intern. Med 94(2):195–197.

De Jonckheere, J., Van Dijck, P.J. & van de Voorde, H., 1975. The effect of thermal pollution on the distribution of *Naegleria fowleri*. J. Hyg. Comb. 75:7–13.

De Jonckheere, J. F, 1981. *Naegleria australiensis* sp. *Nov.*, another pathogenic *Naegleria* from water. Protistologica 27:423–429.

Dondero, T.J., Rendtorff, R.C., Mallison, G.F., Weeks, R.M., Levy, J.S., Wong, E.W. & Schaffner, W., 1980. An outbreak of Legionnaires' disease associated with a contaminated air-conditioning cooling tower. N. Engl. J. Med. 302:365–370.

Duma, R.J., 1981. Study of pathogenic free-living amebas in fresh-water lakes in Virginia. EPA–600/S1–80–037, February.

Duma, R.J., Ferrell, H.W., Nelson, E.C. & Jones, M.M., 1969. Primary amebic meningoencephalitis. New Engl. J. Med. 281:1315–1323.

Duma, R.J., Schumacker, J.B. & Callicott, J.H., 1971. Primary amoebic meningoencephalitis; a survey in Virginia. Arch. Envir. Health 23:43–47.

Fields, B.S., Shotts, E.B., Feeley, J.C., Gorman, G.W. & Martin, W.T., 1984. Proliferation of *Legionella pneumophila* as an intracellular parasite of the ciliated protozoan, *Tetrahymena pyriformis*. In Thornsberry, C., Balows, A., Feeley, J.C. & Jakubowski, W. (eds.), *Legionella*. Proceedings of the 2nd International Symposium. American Society for Microbiology, Washington, DC: 327–328.

Fliermans, C.B., 1985. Philosophical ecology: *Legionella* in historical perspective. In Thornsberry, C., Balows, A., Feeley, J.C. & Jakubowski, W. (eds.), *Legionella*. Proceedings of the 2nd International Symposium. American Society for Microbiology, Washington, DC: 285–289.

154

Fliermans, C.B., 1985. Ecological niche of *Legionella pneumophila*. Critical Reviews in Microbiology, in press.

Fliermans, C.B., Bettinger, G.E. & Fynsk, A.W., 1982. Treatment of cooling systems containing high levels of *Legionella pneumophila*. Water Research 16:903–909.

Fliermans, C.B., Cherry, W.B., Orrison, L.H. & Thacker, L., 1979. Isolation of *Legionella pneumophila* from nonepidemic-related aquatic habitats. Appl. envir. Microbiol. 37:1239–1242.

Fliermans, C.B., Cherry, W.B., Tison, D.L., Smith, S.J. & Pope, D.H., 1981. Ecological distribution of *Legionella pneumophila*. Appl. envir. Microbiol. 41(1):9–16.

Griffin, J.L., 1972. Temperature tolerance of pathogenic and nonpathogenic free-living amoebas. Science 178:869–870.

Hecht, R.H., Cohen, A., Stoner, J. & Irwin, C., 1972. Primary amebic meningoencephalitis in California. Calif. Med. 117:69–73.

Maitra, S.C., Krishna Prasad, B.N., Das, S.R. & Agarwala, S.C., 1974. Study of *Naegleria aerobia* by electron microscopy. Trans. R. Soc. Trop. Med. Hyg. 68:56–60.

McDade, J.E., Shepard, C.C., Fraser, D.W., Tsai, T.R., Redus, M.A. & Dowdle, W.R., 1977. Legionnaires' disease. Isolation of a bacterium and demonstration of its role in other respiratory disease. N. Engl. J. Med. 297:1197–1203.

Miller, R.P., 1979. Cooling towers and evaporative condensers. Annals of Internal Med. 90:667–670.

Orrison, L.H., Cherry, W.B., Tyndall, R.L., Fliermans, C.B., Gough, S.B., Lambert, M.A., McDougal, L.K., Bibb, W.F. & Brenner, D.J., 1983. *Legionella oakridgensis*: An unusual new species isolated from cooling tower water. Appl. envir. Microbiol. 45:536–545.

Page, F.C., 1976. An illustrated key to freshwater and soil amoebae. Freshwater Biological Association, Scientific Publication 34:155–173.

Robinson, B.S. & Lake, J.A., 1981. The influence of temperature on growth and distribution of *Naegleria* species. Proc. Aust. Water and Wastewater Association, 9th Federal Convention, Perth, April 6 to 10.

Rowbotham, T.J., 1980. Preliminary report on the pathogenicity of *Legionella pneumophila* for freshwater and soil amoebae. J. Clin. Pathol. 33:1179–1183.

Scaglia, M., Strosselli, M., Grazioli, V., Gatti, S., Bernuzzi, A.M. & De Jonckheere, J.F., 1983. Isolation and identification of pathogenic *Naegleria australiensis* (Amoebida, Vahlkampfidae) from a spa in northern Italy. Appl. envir. Microbiol. 46(6):1282–1285.

Stevens, A.R., Tyndall, R.L., Coutant, C.C. & Willaert, E., 1977. Isolation of the etiological agent of primary amoebic meningoencephalitis from artificially heated waters. Appl. envir. Microbiol. 34:701–705.

Stout, J., Yu, V.L., Vickers, R.M., Zuravleff, J., Best, M., Brown, A., Yee, R.B. & Wadowsky, R., 1982. Ubiquitousness of *Legionella pneumophila* in water supply of a hospital with endemic Legionnaires' Disease. New Engl. J. Med. 306:466–468.

Tyndall, R.L., Willaert, E. & Stevens, A.R., 1979. Presence of pathogenic amoebae in power plant cooling waters. Final report for the period October 15, 1977 to September 30, 1979. NUREG/CR–1761. Oak Ridge National Laboratory, Oak Ridge, TN.

Tyndall, R.L., 1981. Presence of pathogenic micro-organisms in power plant cooling waters. Report for October 1, 1979 to September 30, 1981. NUREG/CR–2980. Oak Ridge National Laboratory, Oak Ridge, TN.

Tyndall, R.L. & Domingue, E.L., 1982. Co-cultivation of *Legionella pneumophila* and free-living amoebae. Appl. Envir. Microbiol. 44:954–959.

Tyndall, R.L. & Domingue, E.L., 1984. Analysis of selected sites of the Illinois Power Company for the presence of pathogenic free-living amoebae. Illinois Power Report. Feb., 1984, in press.

Tyndall, R.L., Gough, S.B., Fliermans, C.B., Domingue, E.L. & Duncan, C.B., 1983. Isolation of a new *Legionella* species from thermally altered waters. Curr. Microbiol. 9:77–80.

Tyndall, R.L., Kuhl, G. & Bechthold, J., 1983. Chlorination as an effective treatment for controlling pathogenic *Naegleria* in cooling waters of an electric power plant. In Water Chlorination, Vol. 4, Book 2, 1097–1103.

Tyndall, R.L., Willaert, E. & Stevens, A.R., 1981. Pathogenic amoebae in power plant cooling lakes. EPRI EA–1897. Electric Power Research Institute, Palo Alto, CA.

Wadowsky, R.M., Yee, R.B., Mezmar, L., Wing, E.J. & Dowling, J.N., 1982. Hot water systems as sources of *Legionella pneumophila* in hospital and nonhospital plumbing fixtures. Appl. Envir. Microbiol. 43(5):1104–1110.

Wadowsky, R.M. & Yee, R.B., 1983. Satellite growth of *Legionella pneumophila* with an environmental isolate of *Flavobacterium breve*. Appl. Envir. Microbiol. 46(6):1447–1449.

Walters, R.P., Robinson, B.S. & Lake, J.A., 1981. Experiences in the control of *Naegleria* in public water supplies in South Australia. Proc. Aust. Water and Wastewater Association, 9th Federal Convention, Perth, April 6 to 10.

Wellings, F.M., Amuso, P.T., Lewis, A.L., Farmelo, M.J., Moody, D.J. & Osikowicz, C.L., 1979. Pathogenic *Naegleria*: Distribution in nature. US Environmental Protection Agency–600/1–79–018.

Wilkinson, H.W., 1982. Center for Disease Control, Atlanta, GA. Personal communication.

Willaert, E., 1974. Etude immunotaxonomique du genre *Naegleria*. Ann. Soc. Belge Med. Trop. 54:395–402.

Willaert, E., 1974. Primary amoebic meningo-encephalitis: A selected bibliography and tabular survey of cases. Ann. Soc. Belge Med. Trop. LIV/4–5: 54:429–435.

Wong, M.M., Karr, S.L., Jr. & Balamuth, W., 1975. Experimental infections with pathogenic free-living amoebae in laboratory primate hosts. I. (A) A study of susceptibility to *N. fowleri*. J. Parasitol. 61:199–208.

Author's address:
R.L. Tyndall
Zoology Department
University of Tennessee
Knoxville, Tennessee 37916, USA.

CHAPTER 9

Modeling geomicrobial processes in reservoirs

D. GUNNISON, J.M. BRANNON and R.L. CHEN

Abstract. To permit interpretation and evaluation of geomicrobial processes occurring in reservoirs, a model was developed that incorporates the salient features of pathways for major chemical species of interest in reservoir. The model, REDROX, includes an anaerobic stage that evolves reduced chemical species in the stepwise manner characteristic of natural systems. This is followed by an aerobic stage that oxidizes the reduced chemical species produced by the first stage. Measures for obtaining rate coefficients suitable for use in the model are considered, and representative data for each of the stages are presented. Interpretation and evaluation of the information obtained using these rate coefficients are discussed.

Introduction

Impoundment of natural waters in reservoir projects has occasionally resulted in oxygen depletion and the subsequent occurrence of anaerobic processes within the hypolimnion. Release of water from projects with bottom withdrawal can result in the appearance of nutrients and metals released by anaerobic processes in downstream areas. Following release, many of these anaerobic materials are oxidized within a short distance from their point of release. Adverse effects of this situation include harmful impacts upon aquatic biota, high operation and maintenance costs for reservoir projects, and increased treatment costs to downstream water users.

Models describing major geomicrobial processes provide one possible means to assist reservoir project managers with the evaluation of various operational alternatives to minimize the impact of microbial mobilization of nutrients and metals. Present-day water quality models are primarily of the chemical equilibrium type. Principal examples of these

are the REDEQL2–MINEQL–GEOCHEM family of chemical equilibrium models.

REDEQL2 (McDuff & Morel, 1975) utilizes an approach based on the Newton-Raphson method for digital computation of chemical equilibrium (Conte & de Boor, 1972). Here, chemical compounds or complexes in certain aqueous systems are expressed as functions of free metal and free ligand concentrations. The program stores equilibrium constants based on 'critical stability constants' (Martell & Smith, 1976–1977) for each independent reaction; these are used to calculate rate-balance relationships. Capabilities of REDEQL2 include: computation of solubility and complexation reactions between 35 metals and 58 ligands; calculation of 24 standard couples of redox reactions; computation of mixed solid reactions, such as chlorite, illite, microcline, and dolomite; and tracking of adsorption and desorption of metal ions on metal oxides. REDEQL2 has been applied to situations involving the relationship between Mn^{2+} and Fe^{2+} in the hypolimnion of a stratified lake (Hoffmann & Eisenreich, 1976) and the concentra-

Gunnison, D. (ed.) Microbial Processes in Reservoirs.
© 1985, Dr W. Junk Publishers, Dordrecht, Boston, Lancaster. ISBN 90 6193 751 5.

tion of trace metals in a wastewater discharge (Morel *et al.*, 1975).

MINEQL applies much of the same information and definitions as those used in REDEQL2, but in a more compact form. MINEQL utilizes the equilibrium constant method for dealing with problems in aqueous systems (Westall *et al.*, 1976). MINEQL can be applied to situations involving speciation of metals in algal culture media, solubility constants for metal chelates in complex media, degradation of nitriloacetic acid in natural water, and chemical equilibria in aqueous solutions (Morel *et al.*, 1976; Westall *et al.*, 1976).

GEOCHEM, rather than being a model for purely aqueous solutions, is restricted to the soil solution system. GEOCHEM calculates equilibrium speciation of chemical elements in a soil solution based on chemical thermodynamics and is a modified version of REDEQL2, albiet with twice the thermodynamic data of REDEQL2. GEOCHEM utilizes thermodynamic data selected specifically for soil systems, employing a separate subroutine to apply corrections for the effect of nonzero ionic strength to the thermodynamic equilibrium constants. Applications of GEOCHEM include: prediction of concentrations of inorganic and organic complex of a metal cation in the soil solution; calculation of the concentration of a particular chemical form of a nutrient element in a solution bathing plant roots to correlate that form with nutrient uptake; prediction of the fate of a pollutant metal added to a soil solution of known characteristics; and estimation of the effect of changing pH, ionic strength, redox potential, water content, or the concentration of a given element on the solubility of some selected element in a soil solution (Sposito & Mattigod, 1979).

The practicality of using chemical equilibrium models for applications in natural water systems has been examined by several authors (see Morgan & Sibley, 1975; Morel *et al.*, 1975; Sibley & Morgan 1977; Westall, 1977; Jackson & Morgan, 1978; Giamatteo *et al.*, 1980; Zimmerman, 1980). Nevertheless, chemical equilibrium models, based on thermodynamics and used to calculate chemical equilibria in aquaous systems and soil solutions, have had only limited success in predicting seasonal changes and chemical speciation of a few elements. In natural water systems, equilibrium models have difficulty in adequately predicting biogeochemical processes. This is particularly true for the shift from anaerobic to aerobic conditions in reservoirs where reaction kinetics are very important. In this case, rate models that utilize kinetic data are a better choice for accurate prediction of water quality.

Gunnison and Brannon' (1980) produced a conceptual model describing anaerobic processes in reservoirs, while a conceptual model depicting the reoxidation of principal products of anaerobic processes has been described by Chen *et al.* (1983). Portions of these conceptual models are capable of further refinement into subroutines for use either on their own or as a part of a larger, more comprehensive water quality model. In the latter case, the subroutines will add to a model the capacity to simulate anaerobic transformations and releases and then to simulate the effects of reaeration, either directly through natural destratification or induced through mechanical measures.

We present here a combined aerobic-anaerobic model that focuses on major components of biogeochemical transformation. This model draws upon anaerobic and aerobic conceptual descriptions presented in detail elsewhere (Gunnison & Brannon, 1980; Chen *et al.*, 1983). A compilation of anaerobic release rates and aerobic oxidation rates are presented for nine reservoir sediments based on data obtained using large-scale laboratory reactor units (Gunnison *et al.*, 1980; Chapters 3 and 7, this volume). While detailed attention is not given to the roles of individual microorganisms in mediating these processes, it is necessary to emphasize that all the anaerobic processes and most of the aerobic processes are carried out by the microbial communities present in reservoir sediments and the overlying water column. Consideration is also given to recommended procedures for selection of appropriate rate coefficients.

Model organization and function

For simplicity, the combined conceptual model depicting anaerobic processes [previously termed DO

SAP (Gunnison & Brannon, 1980)] and reaeration [previously named RE–AERS (Chen *et al.*, 1983)] is referred to as REDROX. The model is divided into 2 stages. In Stage A, the reservoir becomes stratified, and the hypolimnion undergoes oxygen depletion followed by release of anaerobic products. In Stage B, destratification is initiated and reoxidation of anaerobic products is tracked through up to and including maintenance of dissolved oxygen at 5–6 mg l^{-1}, once all reduced constituents have been oxidized.

Stage A

Stage A contains all processes formerly contained in the DO SAP Model and was described in detail by Gunnison & Brannon (1980). The salient features of this model are recapitulated here for clarity. Stage A is divided into 7 steps or phases that gradually and continuously proceeds from aerobic (oxic) to anaerobic (anoxic) conditions and then proceeds through a limited succession of anaerobic product-generating loops, each of which is more intensely anaerobic (reducing) than its predecessors.

Phase I: DO depletion
This phase is initiated at the onset of thermal stratification in the reservoir. During stratification, the diffusive exchange of dissolved oxygen (DO) between the epilimnion and the hypolimnion and advective transport of oxygen-bearing water into the hypolimnion are greatly diminished. Decomposition of organic matter proceeds together with oxidation of reduced anaerobic constituents released from reservoir bottom sediments. Through these processes, water column DO may be depleted down to a level of 2 mg l^{-1}, the initiation point for Phase II.

Phase II: DO exhaustion
Depletion of DO continues in this phase until all of the DO has been consumed; this is the signal to initiate Phase III. As a consequence of the decreasing concentration of DO, nitrate begins to replace oxygen as an inorganic electron acceptor for microbial respiration processes. As a result, the concentration of nitrate begins to decrease during Phase II.

Phase III: Ammonium accumulation
Nitrate is rapidly depleted and lost to the system. Ammonium begins to accumulate in the water column. This is a consequence of the nonavailability of oxygen for use in the microbial oxidation of ammonium that is derived from decomposition of nitrogenous organic matter and from anaerobic reduction of nitrate to ammonium. Once all nitrate has been depleted, the system shifts to Phase IV.

Phase IV: Manganese accumulation
Soluble reduced manganese released from the sediments begins to accumulate in the hypolimnetic water column. The oxidation-reduction sequential process continues until reducing conditions become more intense, as indicated by maximum rates of manganese reduction and accumulation. A decrease in the rate of manganese reduction signals the start of Phase V.

Phase V: Iron accumulation
Reduction of ferric iron (Fe^{3+}) to ferrous iron (Fe^{2+}) begins and the subsequent accumulation of soluble ferrous iron in the water column continues until the rate of iron accumulation begins to approach zero. Phase VI is then initiated.

Phase VI: Sulfate reduction
Reduction of sulfate present in the water and/or any sulfate entering the hypolimnion with inflows is initiated, and sulfide begins to appear in the water solumn. Sulfide released from the sediment to the hypolimnion is stable due to the intense anaerobic conditions. The rotten egg odor characteristic of hydrogen sulfide will be evident in any hypolimnetic waters that reach the surface, either upon destratification and mixing of the reservoir, or in samples of the hypolimnion that are brought up for examination. In addition, sulfide and its attendant odors will also occur in any releases made from the hypolimnion. Finally, the sulfide will easily combine with any reduced iron present in the water to form insoluble iron sulfide, which is then precipitated. This phase is terminated when

the supply of sulfate has been exhausted, and Phase VII is initiated.

Phase VII: Methanogenesis
Methane formation is initiated and continues until the system is perturbed or the reservoir destratifies.

Stage A function in relation to the reservoir ecosystem
An example of the manner in which Stage A might function as a model depicting a reservoir ecosystem is shown in Figure 1. The hypolimnion of the reservoir may be modeled as a series of horizontal slices of variable thickness stacked from the bottom of the water column. Stage A is initiated following the onset of thermal stratification. According to this example, all slices of the hypolimnion would be in Phase I of the Stage A at this time (Figure 1a).

After a period of thermal stratification, the oxygen demand of the bottom sediments and bottom layers of water may become large enough to deplete these layers of DO, thereby pushing these sediments and layers into Phase II. With continued oxygen depletion, resulting from oxygen depletion exceeding oxygen gains, the upper layers of the hypolimnion could also become depleted and these layers would enter Phase II as shown in Fig. 1b.

Continued thermal stratification and development of anaerobic conditions would gradually result in each of the layers in the hypolimnion progressing through the seven Stage A phases with eventual development of an anaerobic hypolimnion (Fig. 1c). Hypolimnetic discharges would contain dissolved manganese, ammonium, and traces of iron. Because of the prolonged period of oxygen depletion and the large requirement for oxidizable carbon sources, it is doubtful that, in most reservoirs, the topmost layer of the hypolimnion would proceed much beyond Phase III or IV. In addition, it is doubtful also that Phase VII would ever be found above the bottommost layers.

Stage A can also be used to simulate the effects of a metalimnetic oxygen demand where high oxygen demand associated with interflows of decomposition of settling organic particulates causes depletion of oxygen in a layer of the metalimnion; therefore, this layer(s) enters Phase II.

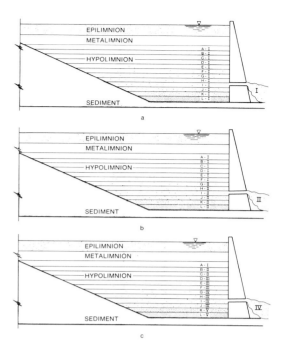

Fig. 1. Diagram depicting hypothetical operations of Stage A in simulating conditions in the hypolimnion of a reservoir being modeled as a series of horizontal layers. Stippled areas in layers E through L represent suspended sediments. Above-Early stratification; Middle-Several weeks after onset of stratification with dissolved oxygen becoming depleted; Below-Dissolved oxygen totally depleted with ammonium, manganese, and iron being accumulated.

Stage B

Stage B changes emphasis from a redox-centered orientation to a dissolved oxygen-centered orientation. There are only two phases in Stage B, although these can be applied to a number of different kinds of aeration – i.e., natural destratification, hypolimnetic aeration, or mechanical aeration.

Natural destratification
This process begins with the onset of loss of stratification in late summer or early fall when air temperatures begin to decrease. During natural destratification, convection currents and epilimnetic circulation expand the thickness of the oxygen-bearing epilimnion. The decomposition rate of organic matter increases, and reduced chemical components, originally in the top layer of the anaerobic

hypolimnion, are oxidized. The water column above the metalimnion becomes nearly saturated with dissolved oxygen.

Turnover begins when the epilimnion has been thickened to such an extent that stratification loses its stability with a resulting decrease in the reservoirs resistance to mixing. At this point, the wind can mix the water. The initiation of mixing (circulation) can occur within a few hours, days, or weeks, depending upon wind velocity, temperature gradient, reservoir depth, reservoir hydrodynamics, and other environmental factors. With the onset of circulation, dissolved oxygen is again present throughout the entire water column.

Hypolimnetic aeration

In hypolimnetic aeration, either air or gaseous oxygen is bubbled into the hypolimnion. If the reservoir is small enough, or the aeration devices are spread over a large enough area, the entire hypolimnion can become aerobic within a short time frame. Often, however, aeration is restricted to a particular area of interest, as for example near the dam when reaeration of water prior to release is desired. Once initiated, hypolimnetic aeration can cause partial mixing of hypolimnetic and metalimnetic waters, and a slight ($0.6–1.1°C$) increase in hypolimnetic water temperatures may result. On occasion, metalimnetic oxygen levels are unstable during aeration, and a zone of low dissolved oxygen or anaerobiosis may occur near the metalimnion. In either event, addition of dissolved oxygen to the hypolimnion results in conditions suitable for the oxidation processes depicted by Stage B.

Mechanical aeration

Mechanical aeration involves the use of a mechanical device to mix the entire water column within the locus of influence of the apparatus–i.e. mechanical destratification. Upon initiation, mechanical destratification lifts cold, anaerobic, hypolimnetic water to the surface, mixing it with the warmer, aerobic, epilimnetic water. The new equilibrium depth for oxygen and the rate of oxygenation depends upon the capacity of the mechanical circulator, oxygen supply rate, and rate of oxygen demand of the reservoir. Aerobic

decomposition of organic matter and the chemical and microbial oxidation of reduced constituents begin in the aerated water column. Having an isothermal reaerated water column means that equilibrium constants and activity coefficients do not require temperature correction for use in an equilibrium model (Ingle *et al.*, 1980).

Activities associated with the stage B aeration process

The phases in Stage B are based on (a) the pH range over which the phase is functional, (b) the dissolved oxygen concentration, and (c) the sequences of major events. To prevent confusion with Stage A phases, the phases in Stage B are designated as a continuation of Stage A.

Phase VIII: Pre-aeration

The assumption here is that, although conditions may be more intensely reducing at the bottom of the reservoir, conditions within most of the water column are best represented by a phase that averages the conditions throughout most of the hypolimnion. This places overall conditions in the hypolymnion in a situation somewhere between Phase IV and V of Stage A. The redox potential is at or below $50 mV$ within a pH range of 6.6 to 7.4. Ammonium, Mn^{2+}, and Fe^{2+} along with PO_4^{3-} are present along with suspended ferrous sulfide. Figure 2 depicts the major chemical changes and pathways associated with this phase. Table 1 gives a listing of the specific processes associated with each arrow in Fig. 2.

Phase IX: Reaeration

This phase allows an instantaneous shift in certain conditions in hypolimnion. The dissolved oxygen rises from 0 to $5–6 mg l^{-1}$, the redox potential goes above $+400 mV$, and pH range expands to 6.3 to 7.5. Oxidation reactions of reduced chemicals proceeds rapidly upon resumption of aerobic conditions. Sulfide is the most labile constituent and undergoes rapid chemical oxidation to elemental sulfur ($S°$) followed by microbial oxidation to sulfate; this is followed in short order by iron oxidation to yield ferric oxides, hydroxides, or iron phosphates that precipitate onto the sediment surface.

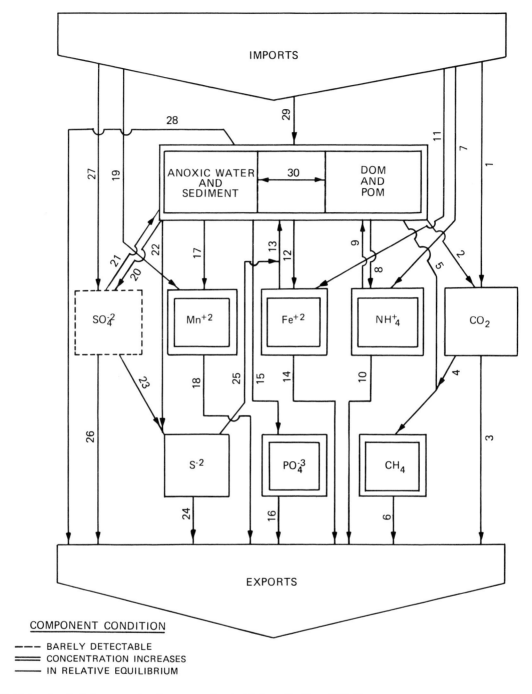

Fig. 2. Major chemical changes and pathways associated with Stage B, Phase VIII. For processes represented by arrows, see Table 1.

Reduced manganese undergoes an oxidation process similar to that of iron, but reaction rates are slower than those for iron in aerated water. However, reduced manganese may be chemically oxidized by other chemical species, such as nitrate, if the hypolimnion is oxygenated. Hydrolysis and/or organic complexation of oxidized manganese (IV) enhances colloidal stability as manganese settles onto and is incorporated into organic matrices. Microbial oxidation of ammonium to nitrite and

Table 1. Specific processes represented by arrows in figure 2.

Arrow No.	Processes represented by arrow
1	Import of dissolved carbon dioxide (as HCO_3^-)
2	Diffusion of dissolved CO_2 (HCO_3^-) from sediment
3	Export of dissolved CO_2 with outflows
4	Reduction of CO_2 to CH_4
5	Release of methane to the water layer from sediment underlying anoxic water layer
6	Export of methane from system, primarily as rising bubbles of gas
7	Import of ammonium with inflow
8	Diffusion of ammonium from sediment, interstitial water or anoxic water layer beneath the water layer of concern into overlying water
9	Immobilization of ammonium
10	Export of ammonium in outflows
11	Import of reduced iron with inflow
12	Diffusion of dissolved reduced (mostly chelated) iron
13	Precipitation of ferrous iron from water layer of concern to sediment
14	Export of dissolved reduced iron in outflows
15	Release of dissolved inorganic phosphate from sediment interstitial water into water layers of concern (two major mechanisms for producing dissolved inorganic PO_4^{-3} are desorption of occluded and nonoccluded inorganic PO_4^{-3} from sediment and mineralization of organic PO_4^{-3} to inorganic PO_4^{-3} in sediments)
16	Export of dissolved inorganic phosphate with outflows
17	Diffusion of reduced manganese from sediment into water layer of concern
18	Export of dissolved reduced manganese with outflows
19	Import of dissolved reduced manganese with outflows
20	Release of inorganic sulfate from anaerobically decomposing organic sulfate
21	Diffusion of dissolved inorganic sulfate into sediment or underlying anoxic water layers
22	Diffusion of dissolved inorganic sulfide from sediment or underlying anoxic water layers
23	Reduction of sulfate to sulfide
24	Export of dissolved sulfide with outflows and advection
25	Precipitation of dissolved sulfide to sediment or underlying anoxic water layer as a consequence of the formation of insoluble ferrous sulfide
26	Export of dissolved inorganic sulfate in outflows
27	Import of dissolved inorganic sulfate with inflows

Table 1. Continued.

Arrow No.	Processes represented by arrow
28	Export of dissolved and particulate organic matter and sediment (suspended) with outflows
29	Import of dissolved and particulate organic matter and settling of suspended solid into the sediment-water system
30	Interchange of dissolved organic matter between sediment and interstitial water

nitrate is somewhat slower and is counterbalanced to a small extent by diffusion of nitrate into the highly reduced sediment followed by denitrification. The major chemical changes and pathways associated with this phase are shown in Fig. 3, and the processes indicated by the arrows in Fig. 3 are shown in Table 2.

Stage B function in relation to the reservoir ecosystem

The manner in which Stage B functions is depicted in Figs. 4 and 5. Fig. 4 presents the scenario for natural destratification, while Fig. 5 displays hypolimnetic aeration. In Fig. 4a, the hypolimnion of the reservoir is again shown as a series of horizontal slices stacked from the bottom of the water column. As Stage B is initiated, the majority of the slices are in Phase VIII, while the bottommost layers are in Phase VI or VII of Stage A.

As the cooling of the epilimnion continues and the thermocline moves deeper into the reservoir, fewer layers in the hypolimnion are left in Phase VIII (Fig. 4b). Finally, overturn occurs and the entire reservoir is mixed (Fig. 4c). In the case of hypolimnetic aeration, reaeration occurs virtually instantaneously relative to natural destratification. While some loss of the thermocline occurs, the basic stratified nature of the reservoir is retained (Fig. 5.).

Selection of appropriate rate coefficients

Selection of anaerobic and aerobic rate coefficients appropriate for use in water quality models may be

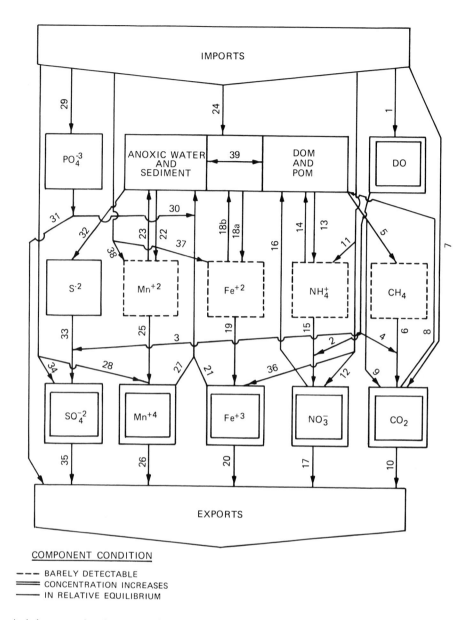

Fig. 3. Major chemical changes and pathways associated with Stage B, Phase IX. For processes represented by arrows, see Table 2.

approached in a number of ways, including (a) use of direct laboratory measurements or use of existing data from the reservoir concerned, (b) use of release rates obtained from reservoirs near the project of interest, (c) use of release rates from reservoirs with similar chemical and physical properties to the project of interest, or (d) for some parameters, measurement of sediment interstitial water concentrations and sediment porosity.

Direct measurements, such as those described by

Brannon *et al.* in Chapter 7, are one of the more accurate but difficult methods of determining release rate coefficients. Modelers may also have at their disposal actual field measurements of parameters measured over time during anaerobic conditions and under aerobic conditions as they are re-established during turnover. If such field data exist, information concerning the volume of hypolimnetic water, concentration changes over time, and the surface area of sediment exposed to the hypolim-

Table 2. Specific processes represented by arrows in figure 3.

Arrow No.	Processes represented by arrow
1	Import of DO with inflow, circulation, mechanical aeration
2	O_2 requirement for nitrification ($NH_4^+ \rightarrow NO_2^- \rightarrow NO_3^-$)
3	O_2 requirement for sulfide oxidation
4	O_2 requirement for methane oxidation
5	Release of methane from sediment to overlying water column
6	Methane oxidation
7	Import of dissolved carbon dioxide (predominantly as HCO_3^-) with inflows
8	Photosynthesis by algae in sediment
9	Diffusion of carbon dioxide from sediment into water layer of concern
10	Export of dissolved carbon dioxide (as HCO_3^-) with outflows
11	Import of ammonium with inflows
12	Import of nitrate and nitrite with inflows
13	Diffusion of mineralized ammonium from sediment to water layer of concern
14	N immobilization and fixation
15	Nitrification
16	Nitrate reduction
17	Export of nitrate and nitrite with outflows
18a	Release of reduced dissolved iron from anoxic sediment to overlying water column
18b	Precipitation of reduced iron to sediment as a consequence of organic and inorganic complexation
19	Homogenetic and heterogenetic oxidation of ferrous iron to ferric iron
20	Export of ferric iron with outflows
21	Colloidal growth of ferric iron followed by adsorption and/or settlement onto sediment as $Fe(OH)_3 \cdot nH_2O$
22	Release of dissolved reduced manganese from anoxic sediment to overlying water column
23	Precipitation of reduced manganese (Mn^{+2}) to sediment as a consequence of organic and inorganic complexation
24	Import of dissolved organic particulate matter and settling of the suspended solids into the sediment-water system
25	Oxidation of manganous manganese
26	Export of manganese (Mn^{+4}) with outflows
27	Colloidal growth of manganese followed by adsorption and/or settlement onto sediment as $MnOOH$
28	Import of manganese with inflows
29	Import of dissolved phosphate with inflows
30	Formation of iron (ferric) phosphate (ppt)
31	Export of dissolved phosphate with outflows

Table 2. Continued

Arrow No.	Processes represented by arrow
32	Diffusion of dissolved inorganic sulfide out of sediment
33	Oxidation of sulfide to sulfate
34	Import of dissolved inorganic sulfate with inflows
35	Export of dissolved inorganic sulfate with outflows
36	Import of suspended inorganic and organic ferric compounds with inflows
37	Import of soluble reduced ferrous iron with inflows
38	Import of soluble reduced Mn^{+2} with inflows
39	Interchange of dissolved organic matter between sediment and interstitial water

nion will allow calculations of fluxes ($mg\,m^{-2}$ day $^{-1}$) to be made.

If a reservoir is in the vicinity of a project for which rate coefficients exist, there is a strong possibility that rate coefficients will be similar. Prior to acting on such an assumption, modelers should ascertain that general water circulation and stratification in the reservoirs being compared are simi-

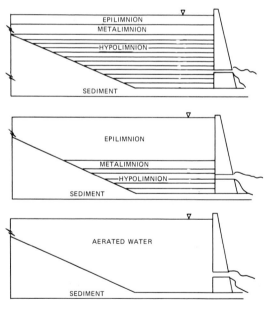

Fig. 4. Thermal destratification cycle: Above-Beginning of Stage B with no dissolved oxygen present and reduced forms of elements accumulated; Middle-After destratification has been initiated; Below-Completion of destratification with a well-mixed water column having adequate levels of dissolved oxygen.

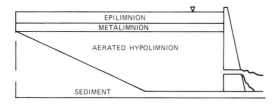

Fig. 5. After hypolimnetic aeration. Thermal stratification continues in the reservoir; dissolved oxygen in the hypolimnion is maintained at 5–6 mg l⁻¹.

lar and that sediment accumulation and general sediment characteristics are also comparable.

Anaerobic rate coefficients

Using the laboratory procedures described in Chapter 7, rate coefficients were obtained for sediment–water interactions under anaerobic conditions by performing linear regression analyses of mass release per unit area (mg m⁻²) versus time. Anaerobic rate coefficients are in the form of fluxes (mg m⁻² day⁻¹) and are summarized in Table 3. The rate coefficients presented in Table 3 were not related to any of the sediment properties (see Table 1, Chapter 7). This was not surprising in view of the complexity of the physical and chemical processes influencing release of materials from sediments (Berner, 1971; Lerman, 1979; Klump & Martens, 1981). However, if such data exist or can be ob-

tained for an existing project of interest, similarities in organic matter content and reducible substances such as iron and manganese should allow similar release rates to be used as a first approximation. Comparisons should be limited to existing projects since soil organic matter in preimpoundment areas differs substantially from that of sediments in existing projects (Gunnison *et al.*, 1983).

Results reported in Chapter 7 indicate that processes at the sediment–water interface in addition to simple molecular diffusion were affecting anaerobic releases. This complicates any approach to predicting anaerobic releases that relies strictly on interstitial water profile data and diffusion equations. Our results showed, however, that empirical relationships exist between predicted and observed fluxes of Fe and NH₄⁺-N. These relationships are presented in Figs. 5 and 6 of Chapter 7. Regression equations associated with each figure allow estimations of realistic Fe and NH₄⁺-N fluxes for use in water quality models. The relatively close correspondence between predicted and observed NH₄⁺-N fluxes is reflected by the proximity of the slope (1.32) to unity. Similar empirical relationships did not exist for Mn and orthophosphate-P. The average depth of the sediment concentration gradient was 8 cm for Fe and 11 cm for NH₄⁺-N.

Table 3. Anaerobic rate coefficients (mg m⁻² day⁻¹) derived from reaction chamber studies.

Sediment source	Parameter											
	Fe	Mn	OPO₄-P	NH₄-N	TKN	TP	NO₂ + NO₃-N	DO	Dissolved organic carbon	SO₄²⁻	S²
Beech Fork	111	148	NR	11.3	33.1	7.5	− 8.8	− 226	NR	− 1064	ND
Browns Lake	54.4	91.3	8.9	19.5	14.4	7.4	− 6.2	− 226	NR	− 165	ND
DeGray Lake	34.1	9.7	NR	15.2	20.0	7.2	− 2.3	− 373	17.1	NR	17.0
Eagle Lake	243	33.1	5.6	ND	ND	ND	ND	− 138	NR	− 170	ND
East Lynn	49.8	13.8	NR	7.3	11.1	NR	− 7.1	− 62.4	NR	NR	ND
Eau Galle	77.4	45.0	NR	NR	NR	NR	− 27.8	− 236	NR	NR	ND
Greers Ferry	72.0	36.1	NR	3.2	ND	ND	ND	− 95.0	ND	ND	ND
Lavon Lake	14.9	4.5	2.3	3.8	NR	NR	− 1.57	− 72.8	NR	NR	NR
Red Rock	66.3	39.1	2.6	6.7	46.9	7.7	− 10.5	− 125	NR	NR	289

NR = No release detected
ND = Not determined

Aerobic Rate Coefficients

Using the laboratory procedures described in Chapter 7, rate coefficients were obtained for re-aeration of the water column and reoxidation of reduced chemical constituents with the following additional methods. Following the 100 day incubation of the 250 l reactor units as described in Chapter 7, an initial water sample was taken to provide baseline information under anaerobic conditions. At this point, aeration was initiated by purging with filtered laboratory air at a flow rate of 60 ml min^{-1} into the system through an airstone diffuser approximately 10 cm above the sediment water interface. Dissolved oxygen concentrations were monitored and maintained at 7.0 ± 0.5 mg l^{-1} to simulate the estimated DO concentration of aerated anoxic bottom water (Chen *et al.*, 1979). Aeration studies were continued for approximately 30 days, with periodic water sampling for determination of various chemical parameters. Methods for quantitation of the chemical constituents were the same as those described in Chapter 7. Aerobic transformation rates (K) were obtained through application of the following equation:

$$-Kt = \ln \frac{Ct}{Co}$$

Where: t = time

Co = Baseline concentration of the chemical constituent at 0 time (anaerobic conditions)

Ct = Concentration of the chemical constituent at time t.

Aerobic rate coefficients obtained during the aeration studies presented in Table 4, are expressed in units of day^{-1} and are site specific. A negative K value indicates that during the transition from anaerobic to aerobic conditions, the oxidized constituent was produced in the water column. A positive K value indicates that during this period, the constituent of interest decreased in the overlying water.

Interpretation and evaluation of rate constants

Rate data presented in this chapter were obtained at 20° C. However, if hypolimnetic waters of a project are substantially lower than 20° C (i.e., 10° C), then the rate reported herein may have to be reduced to adequately represent release rates at the lower temperature. In another study, Gunnison *et al.* (1983) examined the effect of temperature on anaerobic release rates and oxygen depletion. It appeared that lowering temperatures approximately 10° C would result in halving of anaerobic release rates and oxygen depletion rates. This agreed well with the concept of Q_{10}, that for a 10° C rise in temperature, a doubling in reaction rates occurs (Hoar, 1966). Halving reaction rates obtained at 20° C for reservoir waters at 10° C, if release rates would otherwise be similar, therefore appears to be a valid means of obtaining a rate coefficient estimate.

Interpretation and evaluation of the information obtained using these rate constants must be con-

Table 4. Aerobic transformation rates (K day^{-1}) of selected nutrients and metals derived from reaction chamber studies.

Sediment source	Parameter							
	Soluble organic carbon	Total inorganic carbon	NH$_4$-N	NO$_3$-N	OPO$_4$-P	Fe^{2+}	Mn^{2+}	SO$_4^{2-}$
Beech Fork	0.106	0.007	0.206	− 0.235	0.150	0.281	0.085	ND
Browns Lake	0.073	0.071	0.131	− 0.251	− 0.035	0.461	0.405	− 0.061
Eagle Lake	0.052	0.034	0.480	− 0.445	0.329	0.840	0.777	− 0.111
Eau Galle	0.091	0.0120	0.430	− 0.538	0.600	1.88	0.323	− 0.014
Greers Ferry	− 0.120	0.015	0.201	− 0.247	− 0.037	0.444	0.045	− 0.120
Red Rock	ND	ND	ND	ND	0.235	0.111	0.123	− 0.025

ND = Not determined

ducted within the context of water quality models. Incorporation of appropriate rate constants into anaerobic as well as aerobic subroutines of a model will allow evaluation of the impact of anaerobic conditions and reaeration on a reservoir ecosystem.

Conclusion

Major pathways for microbial reduction and microbial and chemical oxidation of major chemical species of concern in reservoir ecosystems are well enough understood to permit formulation of conceptual models describing interaction under anaerobic and aerobic conditions. Rate coefficients for these processes can be obtained using reservoir sediments under controlled laboratory conditions. However, the rate data obtained will be site specific. Thus, the selection of rate coefficients must be done carefully, preferably based on data taken using sediments from the reservoir being examined or from reservoirs in the immediate vicinity. Interpretation and evaluation of the information obtained using these rate coefficients must be conducted within the context of the specific model being utilized.

Acknowledgements

This research was supported by the Environmental and Water Quality Operational Studies Program of the US Army Corps of Engineers. Permission was granted by the Chief of Engineers to publish this information.

References

Berner, R.A., 1971. Principles of chemical sedimentology. McGraw-Hill, NY, 240 pp.

Chen, R.L., Keeney, D.R. & Sikoray, L.J., 1979. Effects of hypolimnetic aeration on nitrogen transformation in simulated lake sediment-water systems. J. Environ. Qual. 8:429–433.

Chen, R.L., Gunnison D. & Brannon, J.M., 1983. Characterization of aerobic chemical processes in reservoirs: Problem description and model formulation. Technical Report E-83-16. US Army Engineer Waterways Experiment Station, Vicksburg, Mississippi, 79 pp.

Coute, S.D. & deBoor, C., 1972. Newton Raphson Method. McGraw-Hill, NY, pp. 22–84.

Giamatto, P.A., Schindler, J.E., Waldron, M.C, Freedman, M.L., Speziale, B.J. & Zimmerman, M.J., 1980. Use of equilibrium programs in predicting phosphorus availability. International Symposium on Environmental Biogeochemistry, Stockholm (In Press).

Gunnison, D. & Brannon, J.M., 1980. Conceptual model depicting anaerobic geomicrobial processes in reservoirs. In Symposium on Surface Water Impoundments. American Society of Civil Engineers, Minneapolis: 381–389.

Gunnison, D., Chen, R.L. & Brannon, J.M., 1983. Relationship of materials in flooded soils and sediments to the water quality of reservoirs –I. Oxygen consumption rates. Water Res. 17:1609–1617.

Hoar, W.S., 1966. General and comparative physiology Prentice-Hall, Inc., Englewood Cliffs, 975 pp.

Hoffman, M.R. & Eisenreich, E.J., 1976. Development of a computer-generated chemical equilibrium model for the variation of iron and manganese in the hypolimnion of Lake Mendota: Thermodynamic, kinetic, and extrathermodynamic considerations. Mineographed paper, Environmental Engineering Program, University of Minnesota, Minneapolis, 45 pp.

Ingle, S.E., Keniston, J.A. & Schults, D.W., 1980. REDEQL–EPAK, Aqueous chemical equilibrium computer program. EPA Report 600/3–80–049, Corvallis Environmental Research Laboratory, Environmental Protection Agency, Corvallis.

Jackson, G.A. & Morgan, J.J., 1978. Trace metal-chelator interactions and phytoplankton growth in seawater media: Theoretical analysis and comparison with reported observations. Limnol. Oceanogr. 23:268–282.

Klump, J.V. & Martens, C.S., 1981. Biochemical cycling in an organic rich coastal marine basin-II. Nutrient sediment-water exchange processes. Geochim. Cosmochim. Aeta, 45:101–121.

Lerman, A., 1979. Geochemical processes. Wiley, NY, 481 pp.

Martell, A.E. & Smith, R.M., 1976–1977. Critical stability constants, Volumes 1–4, Plenum Press, New York.

McDuff, R.E. & Morel, F.M., 1975. Description and use of the chemical equilibrium program REDEQL2. Technical report EQ–73–02. W.M. Keck Laboratory of Environmental Engineering Science, California Institute of Technology, Pasadena.

Morel, F.M., Westall, J.C., O'Melia, C.R. & Morgan, J.J., 1975. Fate of trace metals in Los Angeles County Wastewater discharge. Environ. Sci. Technol. 9: 756–761.

Morel, F.M., McDuff, R.E., & Morgan, J.J., 1976. Theory of interaction intensities, buffer capacities, and pH stability in aqueous systems with application to the pH of seawater and a heterogeneous model ocean system. Marine Chem. 4: 1–28.

Morgan, J.J. & Sibley, T.H., 1975. Chemical models for metals in coastal environments. American Society of Civil Engineers

Conference on Ocean Engineering.

Sibley, T.H. & Morgan, J.J., 1977. Equilibrium speciation of trace metals in freshwater – seawater mixtures. Proceedings of the International Conference on Heavy Metals in the Environment. 1: 319–338.

Sposito, G. & Mattigod, S.V., 1979. GEOCHEM: A computer program for the calculation of chemical equilbria in soil solutions and other natural water systems. Mimeographed paper, Department of Soil and Environmental Science, University of California, Riverside.

Westall, J.C., 1977. Chemical methods for the study of metal-liqand interactions in aquatic Environments. Ph. D. thesis, Massachusetts Institute of Technology, Cambridge.

Westall, J.C., Zachery, J.L., & Morel, F.M., 1976 MINEQL: A computer program for the calculation of chemical equilibrium composition of aqueous systems. Technical report 18. Department of Civil engineering, Massachusetts Institute of Technology, Cambridge.

Zimmerman, M.J., 1980. Aquatic chemistry: A chemical equilibrium approach. Ph. D. Thesis, University of Georgia, Athens.

Author's address:
D. Gunnison
J.M. Brannon
R.L. Chen
Environmental Laboratory
U.S. Army Engineer Waterways Experiment Station
P.O. Box 631,
Vicksburg, Mississippi 39180 U.S.A.

CHAPTER 10

Microbial ecology and acidic pollution of impoundments

AARON L. MILLS and ALAN T. HERLIHY

Abstract. Many impoundments are becoming acidified by acid precipitaion (AP) and acid mine drainage (AMD). Because of its more dilute constitution, AP is not expected to significantly affect microbial processes in mesotrophic or eutrophic impoundments, although some reduction in heterotrophic activity might be expected in oligotrophic situations. AMD can cause extensive alteration of the microbial community and associated functions. The input of sulfate in the pollution tends to enhance bacterial sulfate reduction in the anaerobic waters and sediments of the impoundment. Sulfate reduction generates carbonate alkalinity which can effectively neutralize the acid pollution if the alkalinity is not consumed by CO_2 fixation during reoxidation of the reduced sulfur species. Presence of reduced iron precipitates the sulfides preventing diffusion into the oxic zones. Bacterial sulfate reduction represents an important homeostatic mechanism in acidified impoundments, and should be explored as a possible management tool for impoundments acidified by AP or AMD.

Introduction

Acid pollution of surface waters comes largely from two sources, acid precipitation and acid mine drainage. While the qualitative constituencies of these contaminants is similar, quantitative differences and input differences dictate dissimilar effects on the receiving waters. Furthermore, major differences between lakes and reservoirs may generate dissimilarities in effect within each pollutant type. This chapter will examine problems of acidification of surface waters with special attention paid to reservoirs in light of their many differences from lakes. The effects of acidification on the microbial processes and the concomitant effect on overall water quality will be discussed, and details of the microbial effect on the acid pollutants will be presented. Because of a lack of direct information on acid pollutants and microorganisms in impound-

ments, it will be necessary to draw on knowledge of other systems, and to extrapolate to impoundments using knowledge of the similarities and differences between those systems.

Effects of any environmental perturbation on microbes may be expressed through an alteration of the structure of the community, a qualitative or quantitative alteration in the ecosystem functions carried out by the community, or a combination of structure/function changes. Community structure encompasses the identity of the organisms present and the distribution of individuals among the various guilds or taxonomic groups (Mills & Wassel, 1980). While changes in community structure may be of interest to the academic ecologist, such alterations are of importance to the ecosystem only if they are associated with changes in the functional status of the community. Thus, studies which merely list species present before and after acid-

Gunnison, D. (ed.) Microbial Processes in Reservoirs.

ification or in acidified vs non-acidified waters are irrelevant from a system viewpoint and will be largely ignored in this discussion. Instead, acid-ification effects will be approached from a func-tional viewpoint, i.e. how do acid precipitation and acid mine drainage affect the ability of the micro-bial community to carry out essential ecosystem functions such as primary productivity, organic matter decomposition, nutrient cycling, and pollu-tant diminution.

Magnitude of the problem

Acid rain

The most important acidic pollutants affecting streams, lakes, and impoundments are acid pre-cipitation and acid mine drainage. Acid rain (de-fied as having a pH less than 5.0) is generated by the dissolution in the atmosphere of emissions from the burning of fossil fuels. Of primary importance are the oxides of sulfur and nitrogen, which, when dissolved, form sulfuric and nitric acid, respec-tively. Acid precipitation was described in England in 1955 (Gorham, 1955), and the Scandinavians first recognized the problem as regional in the late 1960's (Haines, 1981). In the United States, the most seriously affected region is the norteast where typical pH values for precipitation have been below

5 for many years. Acidity of the rainfall is still increasing, and the area impacted by acid rain con-tinues to increase as well (Fig. 1). In addition to acidity, much of the precipitation that falls from air parcels contaminated with industrial emissions also may contain many metals (Table 1).

Acid precipitation must be considered as a non-point source pollutant, and the extent of alteration of the lake or impoundment by acid precipitation is dependent not only on the precipitation composi-tion, but also on the geologic and hydrologic characteristics of the entire watershed feeding the reservoir. Except for that portion of the rain that falls directly on the lake, the composition of the water actually entering the water body will depend on the ability of the watershed to neutralize the acidity, and remove the trace element pollutants. As seen in Table 1, stream water entering lakes is generally higher in pH than the incoming precipita-tion. However, the stream chemistry is altered by the precipitation, usually including a reduction in pH, so that some portion of the acidity of the precipitation is transferred to the reservoir. The watershed serves as a buffer, and in many cases, the buffer capacity is quite weak. In such areas, lakes and reservoirs may lose alkalinity and suffer declining pH. The most notable changes have been reported for lakes in Scandinavia, Nova Scotia, Ontario, Quebec, Maine, New Hampshire, New York, New Jersey, North Carolina, and Florida (Haines, 1981).

Table 1. Constituents in acid mine waters, acid precipitation, and precipitation affected streams from several geographical locations and mineral sources. Units of concentration are $\mu M/l$.

	Balaklala CA Metal	Weil CA Metal	Mammoth CA Metal	Argo CO Metal	Contrary VA Sulfur	Karvatn[a] Denmark		Birkenes[b] Denmark Precip
						Precip	Stream	
pH	2.75	1.61	2.83	2.60	2.85	4.90	5.95	–
Sulfate	7450	120000	10400	23700	3000–20000	34	24	5.1
Fe2+	68	13000	32	100	35	–	–	–
Fe3+	1840	35000	2970	3090	2010	–	–	–
Cu	255	4890	330	109	46	–	–	–
Al	626	5000	447	663	1073	9.0	–	–
Cd	–	15.0	2.7	1.0	–	0.002	–	0.0004
Zn	274	4800	449	675	172	0.274	–	0.046
Pb	–	3.4	–	–	–	0.052	–	0.007

[a] Norwegian State Pollution Control Authority, 1983, and Hannsen et al., 1980

[b] Hannsen et al., 1980

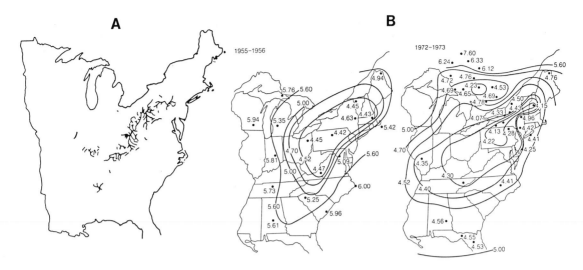

Fig. 1. Major areas of the United States affected by acid mine drainage (A) and acid precipitation (B). Data taken from Council on Environmental Quality (1982).

The degree which bodies of water are affected by acidic inputs is greatly dependent on the amount of bicarbonate present. Lakes with high alkalinities are not greatly influenced by acid rain. Goto *et al.* (1978) found that the pH of Lake Toya did not decrease appreciably despite the addition of acid mine drainage until the lake's alkalinity was depleted. The major factors effecting the vulnerability of a lake to acid rain are the watershed bedrock geology, soil cation exchange capacity, and permeability. Weathering reactions consume protons so that a watershed composed of an easily weatherable bedrock can buffer acidic inputs. Watersheds made up of resistant rocks such as granite or shale do not buffer acidity well and are susceptible to acid rain (Hendrey *et al.*, 1980; Root *et al.*, 1980; Galloway & Cowling, 1978). Watersheds that have soils with a high cation exchange capacity are also more resistant to acid rain since the protons can be trapped in the soil layer (McFee *et al.*, 1977). Galloway *et al.* (1983) found that two lakes with similar bedrock geology had different pHs due to different soil permeabilities. The watershed with the low permeability effectively funneled the acid precipitation into the low pH lake while the lake with the higher permeability allowed more rain water to infiltrate so that cation exchange and weathering reactions could buffer the acidity before it reached the lake.

Acid mine drainage

Acid mine drainage (AMD) is the effluent formed by the oxidation of reduced iron and sulfur to generate ferric iron, SO_4^{2-} and acidity.

$$2FeS_2 + 2H_2O + 7O_2 \rightarrow 2Fe^{2+} + 4SO_4^{2-} + 4H^+ \quad (1)$$
$$2Fe^{2+} + \tfrac{1}{2}O_2 + 2H_2O \rightarrow 2Fe(OH)_3 + H^+ \quad (2)$$

Recent reports (Council on Environmental Quality, 1981) estimate that over 17,000 km of major streams in the United States are contaminated with AMD (Fig. 1). Many of those streams have been dammed for one reason or another, and the AMD contamination then becomes important to the impoundment as well as the stream. As seen in Fig. 1, the most serious AMD problem (on a regional basis) occur in the eastern portion of the United States, in an area that is also receiving acid precipitation. Thus, reservoirs in the eastern US are receiving acid pollutants both from the atmosphere and from mining activities. Acid mine drainage is also a serious problem on a local scale in other parts of the country where mining activity is intense. Examples include the iron-rich region near Redding, California and the metal-rich region west of Denver, Colorado. For example, drainage from mines in the former area (See Table 1) enters Lake Shasta via Squaw Creek.

Acid mine streams are very different from those

polluted by acid rain. Stream pH values are often below 2, and the concentration of dissolved solids, including metals, is very high (Table 1). In addition to the obvious chemical and biological effects, the acid mine streams may also have altered physical properties. The density contributed to the stream-water by the dissolved solids causes the water to behave like much colder water or like weakly saline water. The result is an immediate plunge of the inflow stream to a level of neutral bouyancy. In some cases, that level may be near the thermocline (Koryak *et al.*, 1979), while in shallow, unstratified impounds, the flow may be along the sediment surface (Rastetter *et al.*, 1984, Herlihy & Mills, 1985). This behavior is important in terms of the microbial response to the acid inflow and will be discussed in detail later.

Microbiological effects

As pointed out by Wright *et al.* (1976), the decline in fish populations is well documented, but the effects of acidification on other aquatic life is poorly studied. Surveys suggest that reducing the ambient pH will tend to inhibit the decomposition of allochthonous organic matter, and decrease the number of species of phytoplankton, zooplankton and benthic invertibrate while encouraging the growth of benthic mosses.

The recovery of a community or ecosystem from any perturbation may be depicted as in Fig. 2. The horizontal axis may represent units of time if the environment is static or if the perturbation ceases. A lake-watershed complex acidified by acid precipitation might recover from the insult if the pollution is halted. In dynamic systems, such as streams or impounds, the horizontal axis might represent distance from the pollution source. Herein lies a set of differences for both lakes vs. reservoirs and for acid precipitation vs. AMD. Reservoirs frequently function more like streams than lakes. Furthermore, point source pollutants (e.g. AMD) behave differently than non-point source contamination (e.g. acid precipitation).

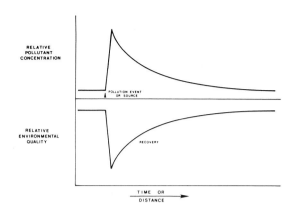

Fig. 2. General model of the abatement of pollution and the recovery of an ecosystem along a gradient of either time or space. Lakes would be expected to recover from a non-point source pollutant with time, while a reservoir receiving a point source discharge would demonstrate the response with distance from the insult.

Primary productivity

The effects of acidification on primary productivity by phytoplankton are not well studied. The most abundant evidence for an effect is in terms of a change in the number and type of species present. In fact, this may be the major effect. The number of species of phytoplankton in acidified lakes decreases as the pH declines (Almer *et al.*, 1974, 1978; Hendrey *et al.*, 1976; Kwiatkowski & Roff, 1976; Leivestad *et al.*, 1976; Yan & Stokes, 1978; Yan 1979), and the number of acidophilic species increases as the acidification proceeds (Leivestad *et al.*, 1976). Diatom diversity and the number of desmid species present in acidified Dutch moorland pools decreased over time as compared with unacidified pools (Van Dam *et al.*, 1980; Coesel *et al.*, 1978). Almer *et al.* (1974, 1978) reported that in Swedish lakes with pH above 6, the algal groups studied were about equally represented. As the pH declined, a reduction was observed for the Chrysophyta, Chlorophyta, and Cyanobacteria, and at pH 4 the Pyrrophyta became dominant. A similar pattern was obtained in several Ontario lakes (Yan & Stokes, 1976, 1978; Yan, 1979), however in other Ontario lakes, the Cyanobacteria became dominant as the pH decreased (Kwiatkowski & Roff, 1976). In the experimental acidification of Lake 223 in the Experimental Lakes Area, no change was observed in the community structure of the phytoplankton (Schindler et al., 1980). The acidification of that lake only changed the pH from 6.6. to 5.6.

A great many species of acid tolerant algae exist, and Rodhe (1981) suggested that species such as those found in some naturally acidic volcanic lakes in Japan be inoculated into culturally acidified lakes elsewhere to serve as a biological buffer against the acidification process. On the basis of observations of several Japanese lakes of different pH, Watanabe & Yasuda (1982) suggested that differences in the dominant diatoms in lakes of different pH values could be used for an acidity index.

The assumption is made that it is important to maintain the level of primary productivity in order to provide energy to higher trophic levels. There is little evidence to suggest that total phytoplankton production decreases with acidification of the lake or reservoir. Even in cases of extreme acidification, as by acid mine drainage, the acidophilic species take over and frequently generate extreme blooms. Chlorophytes such as *Ulothrix* are commonly found attached to rocks in acid mine streams with pH values from 2 to 3.

In cases where acidified lakes have been experimentally neutralized, the treatment initially decreased the standing crop of phytoplankton, and also zooplankton and benthic invertebrates (Scheider & Dillon, 1976). Within a year, the biomass of the phytoplankton was at the level observed prior to treatment, and species composition had begun to resemble that of unacidified lakes of the area. Similar effects have been observed in laboratory simulations (Yan & Stokes, 1978).

Measurements of primary productivity and production in acidified lakes are sparse, but those few studies suggest that acidification seems to have little effect on the biomass and productivity of phytoplankton as long as an adequate supply of phosphorus is maintained (Hendrey et al., 1976; Almer et al., 1978; Yan & Stokes, 1978; Dillon et al., 1979; Yan, 1979; Raddum et al., 1980; Wilcox & DeCosta, 1982). Experimental acidification of Lake 223 did not chang the biomass or productivity of the phytoplankton (Schindler et al., 1980). In several studies, growth of phytoplankton in lakes acidified by acid precipitation or acid mine drainage was stimulated by the addition of phosphorus (Dillon et al. 1979; Wilcox & DeCosta, 1982). In experimentally neutralized lakes, the biomass of the phytoplankton is increased above that of pre-neutralization only by the addition of phosphate (Dillon et al., 1979).

In summary, if acidification is to significantly impact the phytoplankton productivity of a reservoir, the mechanism is most likely to be alteration of the nutrient status as opposed to a direct pH effect on the algae.

Microbial heterotrophy

Similar to the studies on phytoplankton, much of the literature on bacterial and fungal effects of acidification consists of lists of organisms recovered from the polluted environment by various culture techniques (e.g. Weaver & Nash, 1968; Guthrie et al., 1978). For the purposes of the present discussion, emphasis will be placed on those studies which provide information of relevance to ecosystem function.

Many heterotrophic microorganisms are inhibited at pH values below 5 or above 9 (Alexander, 1977). Concomitantly, any associated activities of those organisms also should be reduced in magnitude. In addition to the simple pH effects, other constituents of the pollutants may affect microbial activity. Such effects are most likely to be seen in the case of pollution by acid mine drainage where the concentrations of the compounds are much higher (See Table 1). Metals such as Cu, Cd, Pb, and Zn inhibit the functions of many microbes in polluted environments (Bhuiya & Cornfield, 1974; Mikkelsen, 1974; Babich & Stotzky, 1977; Jensen, 1977; Lawrey, 1977 a, b; Mills & Colwell, 1977; Bitton & Freihofer, 1978; Gadd & Griffiths, 1978; Bollag & Barabasz, 1979; Sunda & Gillespie, 1979; Zevenhuisen et al., 1979; Sterritt & Lester, 1980).

In a comprehensive study of the combined effects of iron (the dominant metal in acid mine waters), H^+, and SO_4^{2-} on bacterial growth, acidity was determined to be the most significant factor, with some concern for the amount of SO_4^{2-} present (McCoy & Dugan, 1968). Using test organisms (strains of *Pseudomonas, Bacillus, Flavobacterium*) isolated from an unpolluted stream, it was concluded that the bacteria should grow if the pH were above 5.3, the iron concentration between 1 and 100 mg/l, and the SO_4^{2-} concentration between 5 and 500 mg/l. Growth of the organisms increased with increasing iron concentrations up to 100 mg/l at pH values greater than 5. The authors inferred that iron would not inhibit heterotrophic growth and activity unless the concentrations greatly exceeded 100 mg/l. The iron concentration in acid mine water may range as high as 10000 mg/l (see Table 1), so that the parameters applied by McCoy and Dugan would not be appropriate until the pH had increased to above 5 and the iron level had dropped to around 100 mg/l. In that state, the system might still be considered polluted, but certainly not so severely so as the majority of AMD contaminated waters.

In lakes acidified by acid precipitation, the action of the precipitation on the minerals in the watershed often leads to elevated levels of aluminum in the lake itself. The presence of the aluminum at levels in excess of 200 μg/l (7.4 μM) in combination with low pH is toxic to fish (Grahn, 1980). The effect of elevated aluminum in acidified waters on the heterotrophic microbes has not been studied. Elevated aluminum in waters acidified by AMD is of little concern because of the comparatively greater effect of other dissolved metals and acidity (refer to table 1). Physiological data collected in laboratory experiments demonstrates that aluminum has variable effects on microbes; in some cases it is stimulatory, in others inhibitory, and in still others, no effect can be observed. For example, *Aspergillus niger* was stimulated by the addition of aluminum at concentrations of 0.19 to 3.7 μM (Bertrand, 1963), but aluminum was inhibitory to bacterial growth at all concentrations between 1850 and 14800 μM (Sulochana, 952). Obviously, the great difference in concentration of the added aluminum precludes any conclusions drawn on the basis of comparisons of the two studies. The form in which the aluminum is added to the culture also influences the effect. Levels of aluminum above 370 μM added as an EDTA chelate were toxic to *Torula* and *Acetobacter* and similar additions at concentrations of 926 – 1850 μM aluminum were toxic to *Saccharomyces* and *Pseudomonas* (Matsuda & Nagata, 1958). When aluminum oxide (Al_2O_3) was the amendment, *Staphylococcus aureus* was not inhibited at levels of 0.4% but was inhibited at 6 and 10% (Bunemann *et al.*, 1963). *Escherichia coli* was unaffected at all amendment levels. The variable results with aluminum additions suggest that in an acidified lake, the low levels of aluminum (as compared with those used in most of the cultural studies) might have an inhibitory effect on some specific member of the community, but that a more resistant strain might be expected to replace the functions vacated by the susceptible organism.

Although heterotrophic bacteria exist that are capable of growth in acid mine streams (Wichlach & Unz, 1981), the bacteria appear to be specialized organisms that have adapted to the extreme environment. Almost nothing is known of their activity, of their role in the ecosystem, or of their contribution, if any, to the rehabilitation of the polluted water. Some heterotrophs are capable of recovering and regrowing after exposure to concentrated AMD (Hackney & Bissonnette, 1978). In cases where heterotrophs are cultured from AMD, they are frequently considered to be transients that have entered the system from other aquatic or terrestrial sources (Wassel & Mills, 1983); Tuttle *et al.*, 1968).

The small amount of information available on the recovery of aquatic microbial communities from the effects of AMD and similar pollutants suggests that a successional pattern of increased heterotrophic biomass is observed along with a concomitant increase in the total activity of that community. Heterotrophs are known to be able to survive immersion in AMD. Although Carpenter & Herndon (1933) reported that wastewater was sterilized by exposure to AMD, all later work has demonstrated that viable heterotrophs can be recovered from mixtures of wastewater and AMD (Joseph & Shay, 1952; Hackney & Bissonnette, 1978). While many of the populations survive the exposure, most are injured or debilitated by the contact, such that attempts at isolation on selective media are often unsuccessful (Hackney & Bissonnette, 1978). Use of a less selective medium (Roth & Keenan, 1971; Hackney & Bissonnette, 1978) allows the organisms to recover from the effects of the AMD and to regain their former activity.

Natural situations in which heterotrophs encounter AMD undoubtedly work in the same way. Assuming the contact with the AMD is sufficiently brief, dilution of the pollutant by additional waters or removal of the toxicants by other means, allows the populations to gradually resume their normal role in that environment. Hackney & Bissonnette (1978) were able to recover significant numbers of a variety of indicator bacteria from membrane diffusion chambers submerged in an acid mine stream for up to 24 h. Survival of cells at levels below the limit of detectability of that study could provide a sufficient inoculum for recovery of the population upon lessening of the pollution stress. Such survival and gradual recovery is consistent with the suggestion that the low levels of heterotrophic activity observed in slightly diluted acid mine streams is due to the presence of transient populations that may be injured but survive the exposure (Tuttle *et al.*, 1969a; Wassel & Mills, 1983).

Given a source of inoculum, viz. injured or transient heterotrophs, the general homeostatic model (Fig. 2) will apply for microbial communities, either in time, as in small lakes with no additional input of AMD, or with distance from the pollution source, as in streams or impoundments with a reasonable current. The work of Tuttle *et al.* (1968) demonstrated that fact for streams using cultural counts as the quantitative method, but the only comprehensive study of the behavior of aquatic microbial communities affected by acid mine drainage is that of the Lake Anna-Contrary Creek system, which will be used as the primary example of acidification effects in an impoundment.

Lake Anna is an impoundment in east-central Virginia (Fig. 3). It was flooded in 1971 to provide cooling water for the North Anna Power Station. Contrary Creek, which flows into the arm used as the study site, drains an area that was extensively mined for pyrite for the 70 years prior to 1920. Although the mining operations have ceased, the flow of acid mine drainage has continued to the present. The primary objectives of the study, which

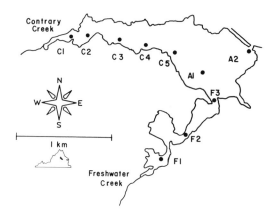

Fig. 3. Contrary Creek and the study sites in Lake Anna.

is still ongoing, were to determine the extent and types of alteration of the microbial communities of the lake due to inputs of acid mine drainage from Contrary Creek, and to determine the fate of the pollutant constituents once in the reservoir.

Another stream, Freshwater Creek, also flows into the arm, but contains no acid mine drainage. This stream and the area into which it flows served as control sites for the study. The mean discharge for Contrary Creek for the 1979 water year was 7.02 ft^3 sec^{-1}, with a maximum flow rate of 259 ft^3 sec^{-1} and a minimum of 0.45 ft^3 sec^{-1} (USGS/Va. State Water Control Board, 1979). Discharge measurements have not been made in Freshwater Creek, but the watershed areas drained by the two streams provide a crude basis for comparing the relative contribution of water from the two streams to the lake. The area drained by the unpolluted steam comprises 2288 ha as opposed to 1820 ha for the Contrary Creek drainage basin. These values represent 44 percent and 35 percent of the total watershed areas which empty out through the causeway at station A2 (See fig. 3) for Freshwater and Contrary Creek, respectively.

A series of stations was established in the Contrary and Creek Freshwater Creek areas of the arm. The major physical parameters associated with some of the stations are presented in Table 2. Data are reported as the annual averages for the three year period from June 1978 to May 1981. An obvious gradient of pH and conductivity exists in the Contrary Creek arm with the pH increasing and the conductivity decreasing with increasing distance from the mouth of the stream. In the Freshwater Creek area, the trend is reversed, although the gradient there is not as steep as in the Contrary Creek arm.

When heterotrophs are recovered from contaminated waters, the abundance and diversity are observed to decrease with decreasing pH. Guthrie *et al.* (1978), observed a 44 percent reduction in total culturable bacteria as the pH was lowered from 6.8 to 5.4 by fly ash contamination of a stream, and a concomitant redution in diversity (number of colony types) of 30 percent. Furthermore, a shift in the prevalent organisms inhabiting the system was observed. In order of quantitative importance, *Bacillus, Sarcina, Achromobacter, Flavobacterium,* and *Pseudomonas* were dominant at the higher pH, and *Pseudomonas, Flavobac-*

terium, Chromobacterium, Bacillus, and *Brevibacterium* were dominant at the lower pH. A noticeable increase in the number of pigmented forms accompanied the reduction in pH. Similar data were obtained in Lake Anna by Wassel and Mills (1983). The planktonic bacterial community was smaller and less diverse in the mouth of Contrary Creek (pH 3.2) than at the mouth of Freshwater Creek (pH 6.2).

The densities of bacterial communities were determined using both plate count techniques and acridine orange direct counts (AODC). In the 12 month period from June 1978 to May 1979 the sizes of the communities at C1, A2, and F1 (Fig. 3) as measured by AODC were not significantly different. Results using the plate count method revealed significant differences among the stations,

Table 2. Values of pH, conductivity, and temperature for selected stations in Contrary Creek and Freshwater Creek. Data are presented as annual averages and standard deviations for each of three years. Stations designated as T and B refer to surface and bottom water conditions respectively. Reproduced by permission from Mills (1985).

Station depth (m)		C1	C4T	C4B	A2T	A2B	F1
	0.5		3.0		7.0	1.5	
pH							
78–79	$\bar{x}$	3.5	5.0	4.1	5.9	5.3	6.4
	s.d.	.29	1.1	.29	.99	1.0	.60
79–80	$\bar{x}$	3.4	5.1	5.4	5.9	5.5	6.3
	s.d.	.43	.74	1.0	.97	.78	.41
80–81	$\bar{x}$	3.3	5.5	5.2	6.5*	6.1	6.3*
	s.d.	.26	.93	1.9	1.0	.83	.78
Conductivity							
78–79	$\bar{x}$	337.4	117.8	124.0	74.0	83.7	55.4
	s.d.	87.9	19.9	45.8	10.9	36.2	20.8
79–80	$\bar{x}$	259.3	116.7	77.6	59.4	71.5	52.2
	s.d.	100.8	130.6	22.8	10.9	22.4	23.9
80–81	$\bar{x}$	252.2	98.5	90.5	81.3	91.0	78.9
	s.d.	76.3	28.9	25.3	9.0	24.2	28.9
Temperature							
78–79	$\bar{x}$	17.7	17.8	17.4	18.2	17.6	18.3
	s.d.	9.9	8.2	8.9	9.6	8.3	9.8
79–80	$\bar{x}$	17.4	19.3	18.9	19.6	17.6	19.3
	s.d.	10.3	9.2	8.9	8.2	7.9	9.1
80–81	$\bar{x}$	15.0	17.8	16.9	18.5	17.4	15.7
	s.d.	8.4	8.4	9.0	8.2	8.0	9.2

* During this year, a pH value of 8.6 was reached in March. That was the highest value ever obtained, and deletion of that reading would lower the average pH for 1980–81 to 6.3. Similarly, deletion of an extremely low reading of 4.3 at F1 in May 1981 would raise the average pH at that station to 6.5.

with F1 >A2 >C1 (Wassel & Mills, 1983). Similar results were obtained for the same time period in 1979–1980 and again in 1980–81.

If all the stations included in the AODC determinations are examined, a more refined relationship is seen in the results. While statistical significance could not be demonstrated for differences in AODC values among the various stations, an interesting trend was observed (Fig. 4). Bacterial numbers in the unpolluted situation (Freshwater Creek) deceased from a maximum in the shallow productive areas near the stream mouth to a minimum in the deeper, open waters of the lake. However, the presence of acid mine drainage in the Contrary Creek arm has altered the situation there to one in which the bacterial densities increase in a downstream direction as the intensity of the pollution damage becomes less and less. The shape of the curve closely resembles that of the general recovery model presented in Fig. 2. The convergence of the AODC curves for the two arms of the lake prior to the point where the waters mix is an important observation indicating essentially complete recovery of the water column microbes within the Contrary Creek arm itself.

The usual criticism of cultural count methods notwithstanding, the lack of significant differences among stations for the AODC results is likely due to an abundance of chemoautotrophic bateria such as *Thiobacillus* sp. in the acid impacted areas. Using the FAINT procedure (fluorescent antibody for identification combined with INT reduction to indicate an active cytochrome system), the contribution of *T. ferrooxydans* to the AODC was determined to be about ten percent of the AODC obtained at C1, but only one percent at station A2 (Baker & Mills, 1982). The numbers of *T. ferrooxidans* and *T. thiooxidans* in Contrary Creek are approximately equal (Scala *et al.*, 1982), and if it is assumed that the relative proportions of organisms hold for the entire year, the points shown at station C1 in Fig. 4 would be lower, yielding an even more significant recovery of the heterotrophic fraction in the Contrary Creek arm.

The diversity of the heterotrophic communities was examined on a quarterly basis by isolating organisms from the enumeration plates, subjecting

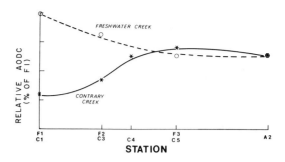

Fig. 4. Annual average of the AODC for several stations in the Contrary Creek and Freshwater Creek arms of Lake Anna. Reproduced with permission from Mills (1985).

the cultures to a battery of physiological tests, clustering the isolates into groups using numerical taxonomic techniques, and using these resultant clusters to perform diversity analysis using the rarefaction method (Mills & Wassel, 1980; Wassel & Mills, 1983). Organisms from both sediment and water showed similar diversity patterns, viz., the community at F1 was more diverse than at either A2 or C1, which, in turn, were not significantly different. The low diversity in the areas receiving the mine wastes was explained by the absence of strains found at station F1 but not at the affected sites. That is, a community exists for the entire area (both polluted and unaffected sites) which has, as a large portion of its makeup, a series of common organisms.

At station F1 however, a large number of individuals present had no counterparts at the other stations sampled. Furthermore, on the basis of the patterns of response of the bacterial isolates to the battery of physiological tests, the strains unique to F1 (not present elsewhere, resulting in a reduced diversity at A2 and C1) were different from the general community (common strains) and from the unique strains found at the sites receiving the mine drainage. The F1 unique strains had a lower ability to withstand the metals Pb, Cu, Zn, and Cd in the culture medium. All of those metals are abundant in the Contrary Creek AMD. There was no difference, however, in the ability of the organisms from the various stations to withstand acidity as determined by culturing each isolate at pH 4.0.

In contrast, the sessile bacterial communities at sites C1 and F1 were highly dissimilar in composition (L.A. Mallory & A.L. Mills, unpublished data), and displayed a great difference in their ability to tolerate the conditions at sites other than their native habitat. These findings imply that, in fact, the planktonic bacteria are transients, whereas the attached forms are able to grow because of an enhanced resistance to the extreme conditions in the acid mine drainage.

The effects of the acid mine drainage on the rate of leaf litter decomposition were investigated at two sites in the Contrary Creek arm (C1 and C5) in comparison to a control site (F1) in the Freshwater Creek arm (Carpenter *et al.*, 1983). Litterbags containing leaves of yellow birch (*Betula lutea*), flowering dogwood (*Cornus florida*), white oak (*Quercus alba*), and soft rush (*Juncus effusus*) were incubated for up to 149 days in the littoral zone at each site. The decomposition rate was measured by loss of ash free dry weight of the leaf litter, and was expressed as both a linear and an exponential decay rate. Interspecific differences in decomposition rates were found at all sites, with dogwood decaying the most rapidly, followed by birch, oak, and rush (Fig. 5). The decomposition rates also differed significantly among the study sites. For all leaf species, the decay rate coefficients at F1 were at least twice those obtained at C1. At C5, values were generally intermediate between those obtained for the other two sites. Bacterial counts (AODC) in the decomposing material did not differ sigificantly among the sites at any time. The decomposition process was primarily microbially mediated, as very few direct detritus processors (e.g. shredding and scraping invertebrates) were ever found in the litter bags. Thus, although microbial activity in the Contrary Creek arm was inhibited by the presence of the acid mine drainage, some activity was seen even in the most severely affected area.

Glucose assimilation measurements made in various parts of the lake (Carpenter *et al.*, 1983) supported the observations of the litter bag incubations. A pattern of assimilation was observed that resembled that of the distribution of AODC in the two arms, and the pattern was valid for the 10 month period from spring of 1980 to late winter of 1981 (Table 3). Assimilation of ^{14}C–glucose over the sampling period was lowest at C1 and increased

in a downstream direction moving away from the acidic environment. Conversely, the activity was highest at station F1 and decreased in a downstream direction in that arm; the general recovery pattern (Fig. 2) was seen in a more integrated examination. The heterotrophic activity results are entirely consistent with those of both the direct and cultural count analyses and show a recovery of the heterotrophic community that is essentially complete before the water from the two arms mix.

Actual rates of heterotrophic activity were consistent with all of the other results (Table 4). Values of V_{max} and turnover time (T_t) showed a depression of activity at the acid impacted area as compared with the other sites examined (Carpenter *et al.*, 1983). The time for glucose turnover at C1 was an order of magnitude longer than at F1, supporting the observation of a reduced rate of decomposition of organic matter there.

The microbiological effects of acid mine drainage observed in Lake Anna are quite similar to those obtained in ponds and streams artificially acidified to study the effects of acid precipitation, and others which have become acidified by the precipitation itself. Early observations on acidified Swedish lakes reported that particulate organic matter (litter) accumulated faster in those lakes than in unacidified ones (Grahn *et al.*, 1974), and

Table 3. Comparative levels of ^{14}C-glucose assimilation in the Contrary Creek area of Lake Anna. To clarify the relationships, all values are expressed as the percentage of the activity observed at station F1. Values presented are the averages of 10 months from November, 1979 to August, 1980. Comparisons are made on equal sized water samples. Reproduced by permission from Mills (1985).

Station	Heterotrophic activity	
	(% of F1)	S.D.
C1	8.3	11.9
C3	15.0	14.1
C4	29.0	19.2
C5	31.0	22.0
A2	36.0	22.6
F3	46.1	30.0
F2	49.0	38.2
F1	100	–

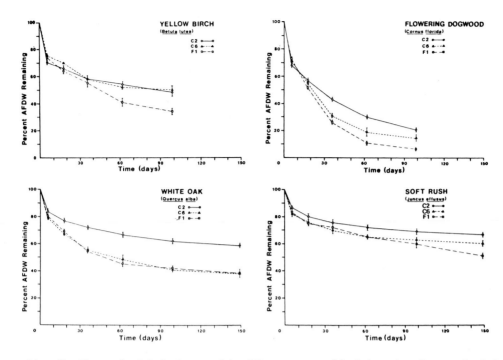

Fig. 5. Decomposition of leaf litter at sites in Lake Anna receiving different amounts of the drainage from Contrary Creek. Reproduced with permission from Carpenter *et al.*, 1983.

that mats of fungal hyphae also were found in the more acidic lakes. Given that acidic environments are thought to select for fungi over bacteria, Grahn *et al.* (1974) hypothesized that decomposition in the acid lakes was accomplished by fungi and that overall decomposition in the acidified lakes was reduced.

Because of the low pH of waters contaminated with AMD, it has been suggested that fungi may play an exceptionally important role. In strip mine soils, filamentous fungi are prolific and diverse (Wilson, 1965). Uptake of organic pollutants from AMD contaminated waters by fungi may result in the removal of iron by the production of organ-

ically stabilized colloids, but the observed reduction in acidity in such streams cannot be attributed to those microbes. (Weaver & Nash, 1968). Representatives of the yeasts *Candida, Rhodotorula,* and *Trichosporon* have often been isolated from streams carrying acid mine wastes (Rogers & Wilson, 1966; Weaver & Nash, 1968).

Transformations of other nutrients

Concomitant with decomposition of organic matter is transformation of nutrient elements such as sulfur and nitrogen. The biogeochemistry of sulfur in waters acidified by acid precipitation or acid mine drainage is currently the focus of a great deal of research. The transformations are intimately related to the ecosystem response of a reservoir to acidification, and the reactions will be treated in detail in a later section.

Literature on the effects of acidification on nitrogen transformations in impoundments is non-existent. In fact, only soil microbes have been examined with regards to the impact of acid pollutants

Table 4. Kinetic values for heterotrophic uptake of ^{14}C-glucose at stations in the Contrary Creek area (April, 1981). Reproduced by permission from Carpenter *et al.* (1983).

Station	V_{max-1} (nMh)	T_t (h)
C1	10.2	133
A2	78.2	52.6
F1	124.3	9.38

on the nitrogen cycle. For the heterotrophic reactions, the effects are linked to general effects on heterotrophs. That is, if heterotrophic activity is reduced, the nitrogen cycling reactions associated with the community will likewise be reduced. For autotrophic reactions such as nitrification, evidence from soil studies suggests that small amounts of acidity may be detrimental to the process (Alexander, 1980). In the presence of simulated acid rain, the amount of nitrate formed from added ammonium was much lower than in unamended controls. In a set of samples amended with nitrapyrin, the amount of nitrate formed during the incubation period was no different for amended or unamended samples. Alexander (1980) concluded that the autotrophic nitrifiers in the soil samples were more acid sensitive than the heterotrophic nitrifiers. The relationship of results such as these to lakes and/or impoundments remains to be determined.

As pointed out later, it is conceivable that acid precipitation might induce an increase in the rate of nitrate reduction (denitrification or dissimilatory nitrate reduction) in sediments or anoxic hypolimnia. To this date; however, no data have been published on such a stimulation, nor has anyone speculated as to the potential contribution of denitrification in buffering acid pollutants.

Microbiological effects on acid pollutants

There are a number of internal lake processes that can buffer or otherwise affect acidic inputs. Photosynthesis is a net consumer of H^+ but most of the fixed carbon is subsequently decomposed and the reaction is probably of little importance in buffering acidification. The production of organic acids during decomposition may contribute to alkalinity if their pK_a is greater than 1.92, the pK_1 of H_2SO_4. However, the concentration of these organics is usually small so that they are of minor quantitative importance in acid neutralization. The release of Fe^{2+} or manganese from the sediment under anoxic conditions results in an increase in alkalinity (Mortimer, 1971) as bicarbonate forms from dissolved inorganic carbon to balance the input of the positive cations. However upon the introduction of oxygen into the hypolimnion during turnover, these metals are oxidized and the alkalinity is lost when bicarbonate is turned into CO_2 (Schindler et al., 1980).

Due to the increased concentrations of sulfate and nitrate in acidic pollution, the most important reactions involved in acid neutralization in lakes are bacterial sulfate reduction and nitrate reduction (denitrification). The reason these reactions are effective is that they generate alkalinity and yield end products that can be stable in the presence of oxygen so that the alkalinity is conserved after turnover. The nitrogen gas formed by denitrification can easily escape the system and the sulfide from sulfate reduction can escape the immediate system as H_2S or can react with ferrous iron and precipitate into the sediment as iron sulfide.

A generalized reaction for sulfate reduction (SR) is give in equation 3 (Richards, 1965).

$$1/_{53} (CH_2O_{106}(NH_3)_{16}H_3PO_4) + SO_4^{2-} \rightarrow CO_2 + HCO_3^- + HS^- + H_2O + 16/_{53} NH_3 + 1/_{53} H_3PO_4 \quad (3)$$

Equation (3) shows that SR will generate carbonate alkalinity (bicarbonate) which can raise the pH of acidic waters. Berner et al. (1970) concluded that SR was the principal factor affecting carbonate alkalinity in the pore waters of coastal marine sediments. Berner formulated the generation of alkalinity from SR as a two step process.

$$2CH_2O + SO_4^{2-} \rightarrow S_2^{2-} + 2CO_2 + 2 H_2O \quad (4)$$
$$S_2^{2-} + 2CO_2 + 2H_2O \rightarrow H_2S + 2HCO_3^- \quad (5)$$

Ben-Yaakov (1973) used a chemical model of anoxic pore waters, to show that SR is the most important process controlling pH. In equation (4), there is a net transfer of charge from a non-protolytic species (sulfate) to a protolytic species (weak acid or base, i.e. HS^-) which can remove protons from solution. Ben-Yaakov's model predicted a pH of 6.9 in an anoxic sulfate reducing sediment in which all the produced sulfide remained in solution. If all the sulfide precipitated out of solution, the pH was predicted to be 8.3 because a weak acid would have been removed from the system.

Kelly et al. (1982) derived the following equations to predict the change in hypolimnetic alkalinity based on the changes in concentrations of the major cations and anions. The major protolytic species contributing to alkalinity in freshwater systems can be written as:

$$alk = 2CO_3^{2-} + HCO_3^- - H^+ + HS^- + 2S^{2-} + NH_3 \tag{6}$$

A neutral or lower pH values CO_3^{2-}, S^{2-}, and NH_3 are insignificant and equation 6 reduces to:

$$alk = HCO_3^- - H^+ + HS^-. \tag{7}$$

A charge balance of the major hypolimnetic ions can be expressed as:

$$O = 2\Delta Fe^{2+} + 2\Delta Mn^{2+} + \Delta NH_4^+ + \Delta H^+ - \Delta HCO_3^- - \Delta NO_3^- - 2\Delta SO_4^{2-} - \Delta HS^- \tag{8}$$

Rearranging equation (8) and substituting in equation (7) yields:

$$\Delta alk = 2\Delta Fe^{2+} + 2\Delta Mn^{2+} + \Delta NH_4^+ - \Delta NO_3^- - 2\Delta SO_4^{2-} \tag{9}$$

Equation (9) demonstrates that the removal of sulfate or nitrate or the release of ferrous iron or manganese will increase the alkalinity. However the alkalinity generated by iron, manganese, and NH_4^+ is not permanent and is lost upon oxidation of these cations. Also, any sulfide that has not precipitated as FeS may also be oxidized so the equation for persistent alkalinity formation is best represented by:

$$\Delta alk = \Delta NO_3^- + 2\Delta(SO_4^{2-} - H_2S) \tag{10}$$

where $(SO_4^{2-} - H_2S)$ represents the fraction of the sulfate reduced that is resistant to oxidation. Using equations (6–10), Cook (1981) observed good agreement between calculated and measured alkalinities in the hypolimnion of Lake 223.

Several examples from the literature demonstrate the ability of sulfate and nitrate reduction to neutralize acidic pollution. Tuttle et al. (1969a) described an AMD stream blocked by a porous wood dust dam. The wood dust provided organic matter for the sulfate reducers. The pH increased from 2.84 to 3.38, and sulfate level decreased from 8.8 mM to 6.1 mM from the pond upstream of the dam to the downstream pond. No sulfate reducing bacteria were observed in the upstream pond but were numerous in the wood dust and the downstream pond. Black iron sulfide precipitates were also observed in the lower pond. In laboratory incubations with AMD waters and wood dust, the pH increased and sulfate concentrations decreased over time. Tuttle et al. (1969b) further investigated this system as a potential AMD pollution abatement procedure, and concluded that once anaerobic conditions are established, and a carbon source supplied, it is theoretically possible for SR to alleviate AMD pollution.

During the experimental acidification of Lake 223 with sulfuric acid, Schindler et al. (1980) observed that the pH did not decrease as much as predicted. Weatherable bedrock or weak organic acids could have accounted for the observed buffering, but they were not found in sufficient concentrations in this region. An input-output mass balance for sulfate suggested significant sulfate reduction in the anoxic hypolimnion, enough to cause the observed neutralization.

In a follow-up study, Schindler & Turner (1982) reported that after five years of acidification, one fourth to one third of the added sulfate has precipitated into the sediment as FeS, reducing the efficiency of acidification. In modeling the same region, Kelly et al. (1982) demonstrated that as sulfate and nitrate inputs from acid rain increased, the dominant anaerobic metabolic pathway shifted from methanogenesis (no alkalinity generation) to sulfate and nitrate reduction (alkalinity generation). The alkalinity generated from this shift was enough to neutralize 'typical' acid deposition in eutrophic lakes.

Kilham (1982), in studying records of Weber Lake in Michigan over the last 25 years, reported an increase in alkalinity despite a twenty fold increase in proton loading from acid precipitation. His budget showed that alkalinity from SR and nitrate uptake by plants was sufficient to neutralize the incoming acidity.

Recently, Wieder & Lang (1982) observed a freshwater wetland which receives AMD. Although Tub Run Bog in West Virginia receives AMD along its eastern edge, the stream draining the bog showed little or no influence from the

AMD. The interstitial water in the bog showed a general decline in acidity and sulfate concentration with increasing distance from the mines, and hydrogen sulfide was present in the pore water. In this environment, SR may also be the important process in modifying the AMD.

Sulfate reduction and denitrification require both organic substrates and anoxic conditions. Thus the major factors controlling the capacity of a lake to neutralize incoming acidity are the amounts of organic matter input and lake hydrology. Campbell *et al.* (1965) observed a series of acid strip mine lakes in Missouri and found that there was a natural succession in these lakes to an alkaline state with a corresponding increase in the abundance and diversity of the biota. King *et al.* (1974) attributed the recovery to biological sulfate reduction and in a series of microcosm studies showed that the addition of raw wastewater sludge to acid strip mine lake water caused a decrease in sulfate, iron and acidity. They concluded that the amount of mine spoils remaining in the watershed and the rate of organic matter input controlled the recovery process in these lakes.

Schindler & Turner (1982) concluded that the alkalinity generated by sulfate reduction in Lake 223 was a function of hypolimnetic volume, light penetration, and the concentrations of sulfate and oxygen. Larger anoxic hypolimnia will entrap more sulfate for reduction so that more alkalinity will be produced. The amount of mixing that occurs during turnover will determine the oxygen and sulfate concentration in the hypolimnion. While less mixing means less oxygen in the hypolimnion and a quicker start to sulfate reduction, it also means that sulfate movement to the reducing zone will similarly be impeded. Schindler & Turner (1982) found that with more complete mixing more sulfate was reduced. It is also interesting to note that in experimentally acidified Lake 114 there was no anoxic hypolimnion; however, sulfate reduction still occurred as the sulfate diffused into the anoxic sediments beneath the aerobic water. Although the calculated rates of sulfate reduction were lower in Lake 114 than in Lake 223, the percentages of added sulfate precipitated as iron sulfides were almost the same (Schindler & Turner, 1982).

Acidic pollution has different attributes in impoundments than it does in natural lakes primarily due to the differing hydrological characteristics of the two water bodies. Impoundments usually have more shoreline than lakes and a higher drainage area to surface area ratio indicating the potential for higher organic matter inputs (Gunnison, 1981). Also, there is often a supply of organic matter on the bottom of the reservoir resulting from the flooding of a terrestrial environment. Water flow into impoundments is usually dominated by one or two major tributaries and the residence times are shorter than natural lakes (Thornton *et al.*, 1980; Gunnison, 1981). Thus pollutants entering the reservoir will be carried to the outflow faster.

Acid mine drainage is different from acid rain in terms of input source and chemistry. Acid rain is a diffuse, dilute source of acidity while acid mine drainage is a concentrated, point source of acidity that enters from streams. Acid mine drainage will often enter an impoundment as a coherent mass either overflowing, interflowing, or underflowing depending on the thermal density stratification of the lake and the temperature of the inflow. Koryak *et al.* (1979) observed all three types of flow in the Tygart River Impoundment. In spring the lake was homothermal and the more easily warmed acid inflow overflowed on the surface of the lake. During the summer a thermal stratification developed and interflow was observed. In the fall the lake was again homothermal but the inflow water was cooler than the warm lake water and caused an underflow. Underflows can be a problem in impoundments with hypolimnetic withdrawl since acid mine waters will not mix in the lake and be diluted but will plunge directly to the outflow resulting in a release of acidic waters that can cause fish kills downstream (Koryak *et al.*, 1979). The same problem can occur with surface withdrawls and overflows. Since neutralization of acidity occurs in the anoxic hypolimnion and sediments, the flow pattern of acid mine drainage inflows is also important in determining how fast neutralization can occur. Underflows would concentrate the acidity in the region where neutralization occurs. On the other hand, overflows and acidity from acid rain enter the epilimnion and lake mixing processes are needed to bring the acidity down to the hypolimnion. These mixing processes can be quite slow especially if a thermal stratification is present.

The chemical composition of the two types of acidic pollution are different. Acid rain has enhanced concentrations of nitrate and sulfate so both denitrification and sulfate reduction are important. On the other hand, the dominant anion in acid mine drainage is sulfate so sulfate reduction is the most important alkalinity producing reaction. Also the elevated metal concentration in acid mine drainage is important in buffering the acidity. Most heavy metals form insoluble precipitates with sulfides (Hutchinson, 1975). Thus in lakes receiving acid mine drainage most of the sulfide produced by sulfate reduction will precipitate in the sediment conserving the alkalinity generated by sulfate reduction. Since acid rain generally has low concentrations of heavy metals the effectiveness of sulfate reduction in removing acidity is dependent on the iron content of the lake and the soil of the surrounding watershed. In iron poor regions sulfate reduction will not be as important a mechanism in reducing acidity since a significant fraction of the produced sulfides will be oxidized with the concomitant loss of alkalinity.

In many impoundments acid rain will probably not be a major problem in the near future due to the high alkalinities and organic matter present in most of these systems. Due to the increases in coal mining and reservoir construction, acid mine drainage is the more serious acid problem in impoundments. Bacterial sulfate reduction can reduce the effect of the AMD given sufficient organic matter and the right kind of flow pattern. However, if a large part of the inflow into an impoundment is made up of strong AMD or if the reservoir has a short residence time, sulfate reduction may be incapable of buffering the acid inflow.

In the Lake Anna study site, the concentration of the pollutants from the acid mine drainage decreases in a down lake direction at a rate faster than can be explained by dilution alone (Mills, 1985; Herlihy & Mills, 1985, Rastetter *et al.*, 1984). A conceptual model was formulated that would explain the observed removal of the AMD (Fig. 6). Sulfate diffuses into the sediment where it is re-

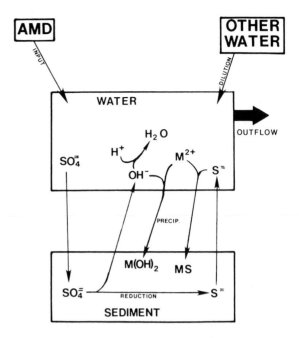

Fig. 6. Conceptual model of the relationship of SO_4^{2-} reduction in sediment and the removal of AMD from the overlying water column. Although SO_4^{2-} reduction generates alkalinity as carbonate, OH^- has been used here for simplicity in depicting the subsequent reactions. Reproduced by permission from Mills (1985).

duced. Sulfate reduction in the lake sediment generates alkalinity to neutralize the pH, and the produced sulfide causes metals to precipitate as metal sulfides. Sulfate reduction is modeled as occurring in the sediments because most of the Contrary Creek arm is shallow and the water is oxygenated to the bottom all year. Only at station A2, and then only for brief periods in the summer, is the bottom water anaerobic. Thus, little or no sulfate reduction occurs in the water column.

In order to see if the addition of acid mine drainage to freshwater lake sediments enhanced sulfate reduction rates, sediment sulfate reduction rates were measured seasonally over a one year period at four stations in the Contrary Creek arm of Lake Anna (Herlihy & Mills, 1985). One of the stations (F1 in the Freshwater Creek section, Fig. 3) receives 'clean' water ($SO_4^{2-} = 10$–$80 \mu M$, pH = 5.6–7.1) from Freshwater Creek and was used as a control for comparative purposes.

Sulfate reduction rates in the Contrary Creek section of the lake were significantly higher than in the Freshwater Creek section in the summer (Table 5). Rates in the two sections were similar in the fall, winter, and spring, presumably due to low sediment temperatures limiting the reaction. Acid volatile sulfide concentrations in the Contrary Creek section were always an order of magnitude higher than those in the Freshwater Creek section (Fig. 7).

In order to verify the model, an input-output budget for sulfate was calculated for the Contrary Creek arm of Lake Anna (Herlihy & Mills, in review). The sulfate inputs included Contrary Creek, Freshwater Creek, and a direct input component. The direct input component included precipitation and all surface water that did not enter the lake via Contrary or Freshwater Creeks. The sulfate output was based on the water leaving the arm under the Route 652 bridge near station A2. Sulfate input-output loads were determined on a daily basis from October 1, 1982 to September 30, 1983 by multiplying the appropriate discharge by its sulfate concentration. The daily discharges for all components of the lake were based on the daily Contrary Creek discharge values. The discharges for Freshwater Creek, the direct input component, and the outflow were calculated from the Contrary Creek discharge on the basis of watershed area ratios. The sulfate concentration of the outflow was based on the depth averaged monthly sulfate concentration at station A2. The average sulfate

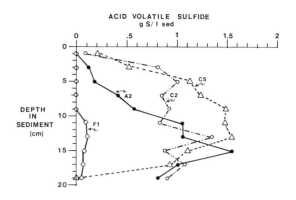

Fig. 7. Acid volatile sulfide concentration with depth in the sediment in the winter of 1983. Data points are the mean of triplicate samples. Coefficients of variance were usually between 10 and 60%. Fall, summer, and spring acid volatile sulfide concentration profiles were similar in shape and magnitude to the winter profile. Reproduced with permission from Herlihy & Mills (1985).

concentration of Freshwater Creek was $46.3\,\mu M$ (SD = 18.6) based on 9 observations taken during the last half of 1983. The sulfate concentration of the direct input component was assumed to be equal to the Freshwater Creek concentration. The sulfate concentration in Contrary Creek was found to be related to creek discharge. The sulfate concentration in the creek for any given day was interpolated from a graph of the sulfate-discharge relationship. The daily sulfate input loads were summed to determine the amount of sulfate added over the year. The outflow sulfate load for the year was subtracted from the input load to determine the amount of sulfate that remained within the system.

The integrated SR rates from Herlihy and Mills (1985) were used to calculate the amount of sulfate removed by SR in the Contrary Creek arm of Lake Anna. The lake was divided into four regions (C2, C5, A2 and F1) and the surface area of that region (m^2) was multiplied by the seasonal SR rate ($mmol\,m^{-2}\,day^{-1}$). Each season was taken to be 91.25 (365/4) days long. The amount of sulfate reduced in each region and each season was summed to yield the total sulfate removal via SR for the year. These calculations assumed that the SR activity was uniform within the four lake regions. Sulfide oxidation was assumed to be equal to the

Table 5. Integrated SO_4^{2-} reduction rates in Lake Anna sediment. Rates are mmol SO_4^{2-} reduced $m^{-2}\,day^{-1}$. Data from Herlihy and Mills (1985).

	Station			
	C2	C5	A2	F1
Fall 1982	.720	1.81	.227	2.10
	(.045)	(1.02)	(.041)	(.119)
Winter 1983	2.10	39.8	.409	4.78
	(.486)	(60.6)	(.095)	(1.96)
Spring 1983	5.46	.551	1.27	.730[b]
	(2.48)	(.648)	(.452)	(.051)
Summer 1983	225	26.8	31.9	8.39
	(192)	(3.43)	(3.58)	(2.45)

[a] Values in parentheses are one SD of triplicate sediment cores
[b] Data obtained from duplicate, not triplicate cores

difference between the measured SR rates and the sulfate input-output budget.

The sulfate fluxes (Table 6) shows that 57 percent of the sulfate that enters the lake does not leave. Of the sulfate that remains in the lake, 125 percent ($\pm$55 percent) can be accounted for by measured rates of SR (Herlihy & Mills, in review). Despite the large variances in the budget, the good agreement between observed rates of SR and sulfate disappearance from the lake indicates that sulfate reduction is the major mechanism removing AMD derived sulfate.

The depth distribution of sulfate concentration in Lake Anna demonstrates the spatial and temporal inflow pattern of the AMD (Fig. 8). The observed summer minimum and winter maximum in sulfate reported by Herlihy and Mills (in review) can be explained by the observed SR rates and Contrary Creek discharge. Contrary Creek has a winter-spring maximum and a late summer minimum in discharge (USGS/Va. State Water Control Board, 1979). Even though the AMD pollutants were more concentrated in the creek in the summer, the low flow rate reduced the pollutant load entering the lake. Also, in the summer, SR rates were highest, so more sulfate was being removed from the water. The combination of these two factors caused the observed minimum and maximum in sulfate, acidity, and conductivity.

Throughout much of the year there was a strong stratification of sulfate and pH. In July, the top water at C2 had a pH of 5.5 and a sulfate concentration of 304 μM. The bottom water had a pH of 3.9 and a sulfate concentration of 638 μM. The stratification was even more striking in August. C2 water pumped from the bottom had a sulfate concentration of 198 μM. Pumped bottom water samples were usually taken about 20–30 cm above the sediment to avoid sucking up the loose iron floc on the sediment surface. C2 bottom water from an equilibrator right at the sediment-water interface had a sulfate concentration of 1190 μM, a five–fold increas in 20–30 cm. The plunging of AMD increases rates of sulfate removal from the lake by concentrating sulfate in the bottom water where it would more rapidly diffuse into the sedement and be reduced.

In an attempt to better understand the causes of the observed stratification a modified version of DYRESM-6 was applied to the Contrary Creek arm of Lake Anna. DYRESM-6 is a one dimensional computer simulation model for predicting temperature and salt structure in medium sized lakes (Imberger & Patterson, 1981). Ferrous sulfate was used as the dominant salt to compute salt density and the model was modified to simulate the outflow under the bridge near station A2. The results of the simulation indicate that thermal stratification in the lake is much stronger than salt stratification, however, a strong thermal stratification effectively traps the sulfate, acid plume in the hypolimnion. A weak salt stratification did develop when lake temperatures were homogeneous. When a strong wind was added to the model it inhibited the formation of the thermocline and caused the lake to mix (Rastetter *et al.*, 1984). The results of the computer simulation in conjunction with the observed increase in sulfate reduction rates in the summer indicate that the sediments are most effective at sulfate removal during calm summer days when thermal stratification is strong.

Conclusions

It should be clear that present knowledge of the behavior of acidic pollutants in impoundments is not well understood. For small reservoirs that are overwhelmed by concentrated pollutants like acid

Table 6. Sulfate budget for the Contrary Creek arm of Lake Anna.

	10^6 moles of SO_4^{2-}
INPUTS	
Contrary Creek	6.3
Freshwater Creek	0.23
Direct input	0.10
	———
Input sum	6.6
OUTPUT	
Outflow under bridge	3.0
Input – Output	3.6
Sulfate reduction	4.5 (125%)

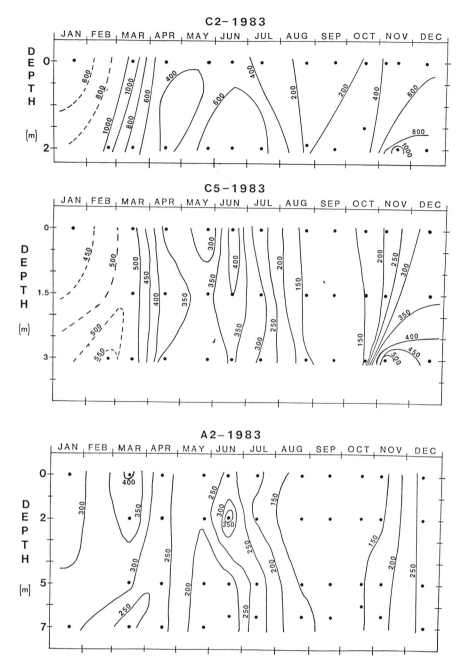

Fig. 8. Sulfate concentration (μM) isopleths in the Lake Anna water column at stations C2, C5, and A2 during 1983. Isopleths are given in 200 μM intervals at C2 and 50 μM intervals at C5 and A2. Dashed lines indicate approximate isopleths due to the scarcity of the surrounding data points. Reproduced with permission from Herlihy & Mills (1985).

mine drainage, the damage may be complete and the water converted to that similar to an acid mine stream. The more prevalent cases are those in which a small proportion of the inflow water is acidified, or the precipitation is acidic. In such instances, many factors, physical, chemical, and microbiological, combine to buffer the acidity. The current lack of information prevents definitive statements abbout the combined effects at present. Continued research in the area should be aimed at those interactions, and attempt to quantify the processes involved so that accurate predictions of the behavior of the polluted impoundments may be made.

References

Alexander, M., 1977. Introduction to Soil Microbiology, 2nd ed., J. Wiley & Sons, NY, 467 pp.

Alexander, M., 1980. Effects of acid precipitation on biochemical activities in soil. In Drablos, D. & Tollan, A., (eds.), Proc. Internat. Conf. Ecological Impact of Acid Precipitation. Sandefjord, Norway. SNSF Project: 47–52.

Almer, B., Dickson, W., Ekstrom, C., Hornstrom, E. & Miller, U., 1974. Effects of acidification on Swedish lakes. Ambio 3:30–36.

Almer, B., Dickson, W., Ekstrom, C., Hornstrom, E., & Miller, U., 1978. Sulfur pollution and the aquatic ecosystem. In Nriagu, J., (ed.), Sulfur in the environment. Part II: Ecological impacts. John Wiley & Sons, NY: 273–311.

Babich, H. & Stotzky, G., 1977. Effect of cadmium on fungi and on interactions between fungi and bacteria in soil: Influence of clay minerals and pH. Appl. environ. Microbiol. 33:1059–1066.

Baker, K.H. & Mills, A.L., 1982. Determination of the number of respiring *Thiobacillus ferrooxidans* cells in water samples by using combined fluorescent antibody-2-(p-iodophenyl)-3-(p-nitrophenyl)-5-phenyltetrazolium chloride staining. Appl. environ. Microbiol. 43:338–344.

Ben-Yaakov, S., 1973. pH Buffering of pore water of recent anoxic marine sediments. Limnol. Oceanogr. 18:86–94.

Berner, R.A., Scott, M.R., Thomlinson, C., 1970. Carbonate alkalinity in the pore waters of anoxic marine sediments. Limnol. Oceanogr. 15:544–549.

Bertrand, D., 1963. Aluminum – a dynamic trace element for *Aspergillus niger*. Compt. Rend. 257:3057–3059.

Bhuiya, M.R.H. & Cornfield, A.H., 1974. Incubation study of effect of pH on nitrogen mineralization and nitrification in soils treated with 1000 ppm lead and zinc oxides. Environ. Pollut. 7:161–164.

Bitton, G. & Freihofer, V., 1978. Influence of extracellular polsaccharides on the toxicity of copper and cadmium toward Klebsiella aerogenes. Microb. Ecol. 4:119–125.

Bollag, J.-M. & Barabasz, W., 1979. Effect of heavy metals on the denitrification process in soil. J. environ. Qual. 8:196–201.

Bunemann, von G., Klosterkotter, W. & Ritzerfeld, W., 1963. Effects of inorganic dusts on microorganisms. Arch. Hyg. Bakteriol. 147:58–65.

Campbell, R.S., Lind, O.T., Harp, G.L., Geiling, W.T. & Letter, J.E., Jr., 1965. Water pollution studies in acid strip-mine lakes: Changes in water quality and community structure associated with aging. Proc. Symp. Acid Mine Drainage Res., Mellon Inst., Pittsburgh, Pa.: 188–198.

Carpenter, L.V. & Herndon, L.K., 1933. Chemical analysis of acid mine drainage. West Virginia Univ. Eng. Exper. Sta. Res. Bull. No. 10. Morgantown.

Carpenter, J.M., Odum, W.E. & Mills, A.L., 1983. Leaf litter decomposition in a reservoir affected by acid mine drainage. Oikos 41:165–172.

Coesel, P., Kwakkestein, R., & Verschoor, A., 1978. Oligotrophication and eutrophication tendencies in some Dutch moorland pools, as reflected in their desmid flora. Hydrobiologia 61:21–31.

Cook, R.B., 1981. The biogeochemistry of sulfur in two small lakes. Ph.D. thesis, Columbia Univ. 246 pp.

Council on Environmental Quality, 1981. Environmental Trends. US Government Printing Office. 346 pp.

Dillon, P., Yan, N.D., Scheider, W., & Conroy, N., 1979. Acidic lakes in Ontario: Characterization, extent, and responses to base and nutrient addition. Arch. Hydrobiol. 13:317–336.

Gadd, G.M. & Griffiths, A.J., 1978. Microorganisms and heavy metal toxicity, Microb. Ecol. 4:279–387.

Galloway, J.N. & Cowling, E.B., 1978. The effects of precipitation on aquatic and terrestrial ecosystems: A proposed precipitation chemistry network. APCA Jour. 28:229–235.

Galloway, J.N., Schofield, C.L., Peters, N.E., Hendrey, G.R. & Altwicker, E.R., 1983. Effect of atmospheric sulfur on the composition of three Adirondack lakes. Can. J. Fish. Aquat.Sci. 40:799–806.

Goto, K., Tanemura, T. & Kawamura, S., 1978. Effect of acid mine drainage on the pH of Lake Toya, Japan. Water Res. 12:735–740.

Gorham, E., 1955. On the acidity and salinity of rain. Geochim. Cosmochim. Acta 7:231–239.

Grahn, O., 1980. Fishkills in two moderately acid lakes due to high aluminum concentrations. In D. Drablos & A. Tollan (eds.), Proc. Internat. Conf. Ecological Impact of Acid Precipitation. Sandefjord, Norway. SNSF Project: 310–312.

Grahn, O., Hultberg, H. & Landner, L., 1974. Oligotrophication – a self-accelerating process in lakes subject to excessive supply of acid substances. Ambio 3:93–94.

Gunnison, D., 1981. Microbial Processes in recently impounded reservoirs. ASM News 47:527–531.

Guthrie, R.K., Cherry, D.S. & Singleton, F.L., 1978. Responses of heterotrophic bacterial populations to pH changes in coal ash effluent. Water Resour. Bull. 4:803–808.

Hackney, C.R. & Bissonette, G.K., 1978. Recovery of indicator bacteria in acid mine streams. J. Water Pollut. Contr. Fed. 50:775–780.

Hanssen, J.E., Rambaek, J.P., Semb, A. & Steinnes, E., 1980. Atmospheric deposition of trace elements in Norway. In D. Drablos & A. Tollan (eds.), Proc. Internat. Conf. Ecological Impact of Acid Precipitation. Sandefjord, Norway. SNSF Project: 116–118.

Haines, T.A., 1981. Acidic precipitation and its consequences for aquatic ecosystems: A review. Trans. Am. Fish. Soc. 110:669–707.

Hendry, G.R., Baalsrud, K, Traen, T., Laake, M. & Raddum, G., 1976. Acid precipitation: Some hydrobiological changes. Ambio 5:224–227.

Hendrey, G.R., Galloway, J.N., Norton, S.A., Schofield, C.L., Shaffer, P.W. & Burns, D.A., 1980. Geochemical and hydrochemical sensitivity of the eastern United States to acid precipitation. EPA-600/3-80-024.

Herlihy, A.T. & Mills, A.L., 1985. Sulfate reduction in freshwater sediments receiving acid mine drainage. Appl. environ. Microbiol. 49:719–186.

Hutchinson, G.E., 1975. A treatise on limnology, Vol. 1, Part 2. Chemistry of lakes. J. Wiley & Sons, NY 1015 pp.

Imberger, J. & Patterson, J.C., 1981. A dynamic reservoir simulation model – DYRESM5 In Fischer, H.B. (ed.), Transport models for inland and coastal waters. Academic Press, NY: 310–361.

Jensen, V., 1977. Effects of lead on biodegradation of hydrocarbons in soil. Oikos 28:220–224.

Joseph, J.M. & Shay, D.E., 1952. Viability of Escherichia coli in acid mine water. Am. J. Pub. Health 42:795.

Kelly, C.A., Rudd, J., Cook, R.B. & Schindler, D.W., 1982. The potential importance of bacterial processes in regulating rate of lake acidification. Limnol. Oceanogr. 27:868–882.

Koryak, M., Stafford, L.J. & Montgomery, W.H., 1979. The limnolgical response of a West Virginia multipurpose impoundment to acid inflows. Water Resour. Res. 15:929–934.

Kilham, P., 1982. Acid precipitation: Its role in the alkalization of a lake in Michigan. Limnol. Oceanogr. 27:856–867.

King, D.L., Simler, J.J., Decker, C.S. & Ogg, C.W., 1974. Acid strip mine lake recovery. WPCF Journal 46:2301–2315.

Kwiatkowski, R. & Roff, J, 1976. Effects of acidity in the phytoplankton and primary production of selected northern Ontario lakes. Can. J. Bot. 54:2546–2561.

Lawrey, J.D., 1977a. The relative decomposition potential of habitats variously affected by surface coal mining. Can. J. Bot. 55:1544–1552.

Lawrey, J.D., 1977b. Soil fungal populations and soil respiration in habitats variously influenced by coal strip-mining. Environ. Pollut. 14:195–205.

Leivestad, H., Hendrey, G.R., Muniz, I. & Snekvik, E., 1976. Effects of acid precipitation on freshwater organisms. In F. Braekke (ed.), Impact of acid precipitation on forest and freshwater ecosystems in Norway. Acid Precipitation – Effects on Forest and Fish Project, Research Report 6, Aas, Norway: 86–111.

Matsuda, K. & Nagata, T., 1958. Effects of aluminum concentration on growth of microorganisms. Nippon Dojo-Hiryogaku Zasshi 28:405–408.

McCoy, B. & Dugan, P.R., 1968. The activity of microorganisms in acid mine water. II. The relative influence of iron, sulfate, and hydrogen ions on the microflora of a non-acid stream. Proc. Symp. on Coal Mine Drainage Res., 2nd. Mellon Inst., Pittsburgh: 64–79.

McFee, W.M., Kelley, J.M. & Beck, R.H., 1977. Acid precipitation effects on soil pH and base saturation of exchange sites. Water Air Soil Pollut. 7:401–408.

Mikkelsen, J.P., 1974. Effect of lead on the microbiological activity in soils. Tidsskr. Planteavl. 78:509–516.

Mills, A.L., 1985. Acid mine waste drainage: Microbial impact on the recovery of soil and water ecosystems. In R.L. Tate and D. Klein (eds.). The Microbiology of Reclamation Processes. Marcel Dekker, Inc. pp. 35–81.

Mills, A.L. & Colwell, R.R., 1977. Microbiological effects of metal ions in Chesapeake Bay water and sediment. Bull. environ. Contam. Toxicol. 18:99–103.

Mills, A.L. & Wassel, R.A., 1980. Aspects of diversity measurement for microbial communities. Appl. environ. Microbiol. 40:578–586.

Mortimer, C.H., 1971. Chemical exchanges between sediments and water in the Great Lakes – speculations on probable regulatory mechanisms. Limnol. Oceanogr. 16:387–404.

Norwegian State Pollution Control Authority. 1983. Environmental Monitoring Report 108/83.

Raddum, G., Holbaek, A., Lomsland, E. & Johnsen, T., 1980. Phytoplankton and zooplankton in acidified lakes in south Norway. In Drablos D. & Tollan A., (eds.), Proc. Internat. Conf. Ecological Impact of Acid Precipitation. Sandefjord, Norway. SNSF Project: 332–333.

Rastetter, E.B., Hornberger, G.M., Mills, A.L. & Herlihy, A.T., 1984. Mathematical model of vertical stratification in a lake receiving acid mine drainage. Annual Meeting of the American Geophysical Union.

Richards, F.A., 1965. Anoxic basins and fjord. In Riley, J.P. & Skirrow, G. (eds.), A treatise on chemical oceanography, Vol. 1. Academic Press, NY: 611–645

Rodhe, W., 1981. Reviving acidified lakes. Ambio 10:195–196.

Rogers, T.O. & Wilson, H.A., 1966. pH as a selecting mechanism of the microbial flora in wastewater: Polluted acid mine drainage. J. Water Pollut. Contr. Fed. 38:990.

Root, J., McColl, J. & Niemann, B., 1980. Map of areas potentially sensitive to wet and dry acid deposition in United States. In Drablos, D. & Tollan, A.,. (eds.), Proc. Internat. Conf. Ecological Impact of Acid Precipitation. Sandefjord, Norway. SNSF Project: 128–129.

Roth, L.A. & Keegan, D., 1971. Acid injury of Escherichia coli. Can. J. Microbiol. 17:1005–1008.

Scala, G.S., Mills, A.L., Moses, C.O. & Nordstrom, D.K., 1982. Distribution of autotrophic Fe and sulfur oxidizing bacteia in mine drainage from several sulfide deposits measured with the FAINT assay. Abstr. Meet. Am. Soc. Microbiol. N68.

Scheider, W. & Dillon, P., 1976. Neutralization of acidified lakes near Sudbury, Ontario. Water Pollut. Res. Canada 11:93–100.

Schindler, D.W. & Turner, M.A., 1982. Biological, chemical and physical responses of lakes to experimental acidification. Water Air Soil Pollut. 18:259–271.

Schindler, D.W., Wagemann, R., Cook, R.B., Ruszczynski, T. & Prokopowich, J., 1980. Experimental acidification of Lake 223, Experimental Lakes Area: Background data and the first three years of acidification. Can. J. Fish. Aquat. Sci. 37:342–354.

Simmons, G.M. & Reed, J.R., 1973. Mussels as indicators of biological recovery zone. J. Water Pollut. Contr. Fed. 45:2480–2492.

Sterritt, R.M. & Lester, J.N., 1980. Interactions of heavy metals with bacteria. Sci. Tot. environ. 14:5–7.

Sulochana, C.B., 1952. The effect of microelements on the occurrence of bacteria, actinomycetes and fungi in soil. Proc. Indian. Acad. Sci. B36:19–33.

Sunda, W. & Gillespie, P., 1979. The response of a marine bacterium to cupric ion and its use to estimate cupric ion activity in seawater. J. Mar. Res. 37:761–777.

Thornton, K.W., Kennedy, R.H., Carroll, J.H., Walker, W.W., Gunkel, R.C. & Ashby, S., 1980. Reservoir sedimentation and water quality – an heuristic model. Am. Soc. of Civil Engineers symposium on surface water impoundments, Minneapolis, Minn., 2–5 June 1980.

Tuttle, J.H., Dugan, P.R., MacMillan, C.B. & Randles, C.I., 1969a. Microbial dissimilatory sulfur cycle in acid mine water. J. Bacteriol. 97:594–602.

Tuttle, J.H., Dugan, P.R. & Randles, C.I., 1969b. Microbial sulfate reduction and its potential utility as an acid mine water pollution abatement procedure. Appl. Microbiol. 17:297–302.

Tuttle, J.H., Randles, C.I. & Dugan, P.R., 1968. Activity of microorganisms on acid mine water. I. Influence of acid mine water on aerobic heterotrophs of a normal stream. J. Bacteriol. 95:1495–1503.

United States Geological Survey, 1979. Water Resources Data for Virginia. USGS Water Data Report, Va-79-1. Nat. Tech. Inform. Serv., Springfield, VA.

Van Dam, H., Suurmond, G. & ter Braak, C., 1980. The impact of acid precipitation on diatom assemblages and chemistry of Dutch moorland pools. In Drablos D., & Tollan, A. (eds.), Proc. Internat. Conf. Ecological Impact of Acid Precipitation. Sandefjord, Norway. SNSF Project: 298–299.

Watanabe, T. & Yasuda, I., 1982. Diatom assemblages in lacustrine sediments of Lake Shibu-ike, Lake Misumi-ike, Lake Naga-ike, Lake Kido-ike in Shiga Highland (Japan) and a new biotic index based on the diatom assemblage for the acidity of lake water. Jpn. J. Limnol. 43:237–245.

Wassel, R.A. & Mills, A.L., 1983. Changes in water and sediment bacterial community structure in a lake receiving acid mine drainage. Microb. Ecol. 9:155–169.

Weaver, R.H. & Nash, H.D., 1968. The effects of strip mining on the microbiology of a stream free from domestic pollution. 2nd Symp. Coal Mine Drainage Res. Mellon Inst., Pittsburgh, PA pp. 80–97.

Wichlacz, L. & Unz, R.F., 1981. Acidophilic, heterotrophic bacteria of acid mine waters. Appl. environ. Microbiol. 41:1254–1261.

Wieder, R.K. & Lang, G.E., 1982. Modification of acid mine drainage in a freshwater wetland. In McDonald, B.R. (eds.), Proceedings of the symposium on wetlands of the unglaciated Appalachian region. West Virginia University, Morgantown.

Wilcox, G. & DeCosta, J., 1982. The effect of phosphorous and nitrogen addition on the algal biomass and species composition of an acidic lake. Arch. Hydrobiol. 94:393–424.

Wilson, H.A., 1965. The microbiology of strip-mine spoil. West Virginia Univ. Agric. Exp. Sta. Bull. no. 506T.

Wright, R.F., Dale, T., Gjessing, E.T., Hendrey, G.R., Henriksen, A., Johannessen, M. & Muniz, I.P., 1976. Impact of acid precipitation on freshwater ecosystems in Norway. Water Air Soil Pollut. 6:483–499.

Yan, N.D., 1979. Phytoplankton community of an acidified, heavy metal-contaminated lake near Sudbury, Ontario: 1973–1977. Water Air Soil Pollut. 11:43–55.

Yan, N.D. & Stokes, P., 1976. The effects of pH on lake water chemistry and phytoplankton in a LaCloche Mountain lake. Water Pollut. Res. Canada 11:127–137.

Yan, N.D. & Stokes, P., 1978. Phytoplankton of an acidic lake, and its responses to experimental alterations of pH. Environ. Cons. 5:93–100.

Zevenhuizen, L.P.T.M., Dolfing, J., Eshuis, E.J. & Scholten-Korselman, I.J., 1979. Inhibitory effects of copper on bacteria related to the free ion concentration. Microb. Ecol. 5:139–146.

Author's address:
Aaron L. Mills
Alan T. Herlihy
Department of Environmental Sciences
University of Virgina
Charlottesville, Virginia 22903

Index